CHEMISTRY RESEARCH AND APPLICATIONS

OXIDE ELECTRONICS AND FUNCTIONAL PROPERTIES OF TRANSITION METAL OXIDES

CHEMISTRY RESEARCH AND APPLICATIONS

Additional books in this series can be found on Nova's website
under the Series tab.

Additional e-books in this series can be found on Nova's website
under the e-book tab.

CHEMISTRY RESEARCH AND APPLICATIONS

OXIDE ELECTRONICS AND FUNCTIONAL PROPERTIES OF TRANSITION METAL OXIDES

ALEXANDER PERGAMENT
EDITOR

nova publishers
New York

For permission to use material from this book please contact us:
Telephone 631-231-7269; Fax 631-231-8175
Web Site: http://www.novapublishers.com

Library of Congress Cataloging-in-Publication Data

Oxide electronics and functional properties of transition metal oxides / editor, Alexander Pergament (Petrozavodsk State University, Russia).
pages cm. -- (Chemistry research and applications)
Includes index.
ISBN: 978-1-63321-499-6 (hardcover)
1. Transition metal oxides. 2. Oxides--Electric properties. I. Pergament, Alexander, editor.
QD172.T6O95 2014
621.381--dc23

2014026239

Published by Nova Science Publishers, Inc. † New York

CONTENTS

OXIDE ELECTRONICS: AN INTRODUCTION

Alexander Pergament[1]
Petrozavodsk State University, Petrozavodsk, Russia

ABSTRACT

MOSFETs (Metal-Oxide-Semiconductor Field-Effect Transistors) have for a long time been the workhorse of modern electronics industry. For the purpose of a permanent integration enhancement, the size of a MOSFET has been decreasing exponentially for over decades in compliance with the Moore's Law, but nowadays, owing to the intrinsic restrictions, the further scaling of MOSFET devices either encounters fundamental (e.g. quantum-mechanical) limits or demands for more and more sophisticated and expensive engineering solutions. Alternative approaches and device concepts are currently designed both in order to sustain an increase of the integration degree, and to improve the functionality and performance of electronic devices. Oxide electronics is one of such promising approaches which could enable and accelerate the development of information and computing technology. The behavior of *d*-electrons in transition metal oxides (TMOs) is responsible for the unique properties of these materials, causing strong electron-electron correlations, which play an important role in the mechanism of metal-insulator transition. The Mott transition in vanadium dioxide is specifically the effect that researchers consider as one of the most promising phenomena for oxide electronics, particularly, in its special direction known as a Mott-transition field-effect transistor (MTFET). Therefore, VO_2-based MTFET is one of the fields of oxide electronics. Also, oxide ReRAM is another rapidly growing field of oxide electronics. Finally, many other functional properties of TMOs, including, for example, optical and electrical switching of thermochromic VO_2 smart coatings, optical properties (especially Raman spectra) of single crystalline lithium niobate and tantalate ($LiNbO_3$ and $LiTaO_3$), as well as optical properties (near band gap optical absorption) of TMO-based nanolaminates, like e.g. ZrO_2-Al_2O_3, HfO_2-Al_2O_3, TiO_2-Al_2O_3, ZrO_2-TiO_2, and HfO_2-TiO_2, are extremely important to understand and estimate potential ability of different TMOs and TMO-based structures in diverse fields of oxide electronics.

Keywords: Oxide electronics, Transition metal oxides, Oxide ReRAM, Lithium Niobate and Tantalate, Vanadium dioxide, Oxide nanolaminates

---

[1] E-mail: aperg@psu.karelia.ru.

The term "oxide electronics" have emerged not so long ago in the everyday-life of scientific literature, but already firmly taken its place. The point is that the modern IT revolution is based on technological progress which enables an exponentially growing enhancement of the performance of electronic devices. During all the history of the development of electronic components, from a vacuum diode to modern highly integrated ICs with nanometer scale of individual elements, the question of the physical limitations on the further progress in this area arose repeatedly. After the invention of an IC by J. Kilby and R. Noyce in 1958 [1], the number of transistors on a chip roughly doubles every two years, and afterwards the processing speed and storage capacity increase correspondingly (Moore's Law). Such a dynamics is typical of all other key parameters of the ICs, the most important of which is a characteristic size of the active region d_m [2], for example, the FET effective channel length. In recent years, the issue of constraints for standard Si-based electronics has been widely discussed in the scientific literature, which is primarily associated with the possibility of further scaling toward nano-size. In this regard, in the 2007 edition of The International Technology Roadmap for Semiconductors (ITRS, http://www.itrs.net), a new section has appeared, namely "Emergent Research Device Materials", which indicates the need to develop a new generation of devices based on new physical principles [3].

Dimensional constraints of the conventional CMOS technology will not allow, apparently, overcoming the limit of d_m far beyond 10 nm, and this can be called as a "Moore's Law violation" [4] (or, so to say, "More than Moore", – the pun which seems to originate from the ITRS authors). Note, however, that the ITRS program still optimistically claims that a theoretical limit of scaling for Si is not seen, and by 2026 it is planned to achieve the level of $d_m = 6$ nm (and according to the Intel's road map – 10 nm by 2015, the so called "P1274 process" [5]). Recently, a laboratory prototype of a SOI-based FET with a 3 nm channel length has been reported [6]. Last years, technologies with characteristic topological dimensions of 45, 22 and 10 nm are being actively developed, and the main directions here are as follows: high-*k* gate dielectrics, multigate structures, the use of such materials as Ge, A3B5 and graphene, Si-Ge alloys in the source and drain regions and strained silicon, and finally, «tri-gate» FET configuration [5] (some of these directions have also been presented in the recent review «Technology Evolution for Silicon Nanoelectronics: Postscaling Technology» [7]). Simultaneously, new technical solutions for architecture optimization (such as, e.g., multi-core processors and the Blue Gene project), system integration and innovative design are developed (see, e.g., a corresponding discussion in the review [4]).

Alternative approaches are based on another mechanism (as compared to the field effect in Si CMOS FETs) or even on a drastic change in computational paradigm or architecture (quantum computers, neuroprocessors). Amongst the approaches utilizing new physical mechanisms, one can list, for example, spintronics, superconducting electronics, single-electronics, molecular electronics, as well as one more quite recent direction, so-called "soletronics" (single atom electronics) [8]. One of such novel directions, oxide electronics, is based on the idea of application of unique properties and physical phenomena in strongly correlated transition metal oxides (TMO). Metal-insulator transition (MIT) [9] belongs to the class of the aforementioned phenomena, and many TMOs, e.g. vanadium dioxide, undergo MITs as functions of temperature or electric field [4, 9, 10].

Complex strongly correlated TMOs, such as HTSC cuprates, CMR manganites or some interfaces (such as, for instance, $LaAlO_3/SrTiO_3$), had first been considered as candidate materials for oxide electronics [3], and the list of devices proposed had included, for example,

FETs with electron transport in complex oxide heterostructures [12, 14] (a "Sketch-FET" [13]), sensors, signal converters, memory elements, etc [3].

Afterwards, three main areas of research have emerged in the field of new functional devices of oxide electronics, namely:

- Elements of non-volatile memory – oxide ReRAM.
- Devices, mainly oxide-based transistors and diodes, for transparent electronics.
- FETs based on materials with MIT ("Mott-FET").

One cannot but admit that the above classification is rather relative. Particularly, the basic materials for transparent and flexible electronics are apparently not oxides: they are, for example, organic compounds and low-dimensional carbon materials (nanotubes, graphene) [17-19]. On the other hand, oxide heterostructure-based p-n junctions as access elements (selective diodes) for ReRAM might be considered as an independent branch of oxide electronics. Also, complex perovskite oxide ferroelectrics and multiferroics, garnets for magneto-optical and acoustoelectronic devices, photonic crystals, various TMO-based nanolaminates, thermochromic coatings for smart windows etc. can be utilized in various discrete oxide electronic devices [4, 20].

Transparent electronics and oxide ReRAMs are widely discussed in the literature and described in detail in several reviews. Note, however, that the memory effect, although manifested mainly by TMOs [21-29], is obviously not directly associated with the electron correlation phenomena. The most discussed models in the literature for the ReRAM mechanism in oxide structures are those based either on the growth and rupture of a metal filament inside the oxide matrix under the action of electric current, or on the redox processes responsible for the formation of some high-conductivity or low-conductivity local inclusions corresponding to a particular oxygen stoichiometry. The MIT ideology is also sometimes involved to explain the properties of the structures and the memory switching mechanism therein [27]. In any case, the memory switching phenomenon seems to be associated with the ion transport [23, 24, 26, 28]. It is also appropriate to mention here the works discussing the memory effects in a material with MIT (vanadium dioxide) associated with the presence of hysteresis in the temperature dependence of conductivity [30, 31].

Typical oxides for transparent electronics (ZnO, ITO, In-Ga-Zn oxide, Cu_xO, etc.) [32-38] do not belong to the class of TMOs, except for copper oxide, and, correspondingly, the phenomena therein are not connected specifically with the correlation effects. (Apropos, the work [38] is one of the most cited articles where the term "oxide electronics" has apparently first appeared.) Due to a sufficiently wide band gap and a large density of defect states, these oxides belong to the class of transparent conductors [39], i.e. they exhibit both a relatively high conductivity and a satisfactory transparency in the visible spectrum region. On the other hand, the developed low-temperature synthesis methods for the thin oxide films preparation allow deposition of these films onto flexible substrates which ensures their competitive ability as compared with conventional materials of stretchable transparent electronics [40], such as organic polymers and carbon nanotubes [17-19].

The third of the above listed three areas of oxide electronics, i.e. that connected with transistor structures based on materials with a MIT, dates back to 1997 when the work [41] has been published in which the idea of a FET on the basis of a hypothetical molecular layer, undergoing a Mott transition, has been proposed, and in 1996 the authors of [41] had patented their idea [42]. Such a device has been called as a "Mott-FET", or MTFET – Mott Transition Field Effect Transistor.

Vanadium dioxide is currently considered as the most suitable material for the MTFET implementation. It should be noted that a simpler material exhibiting the Mott MIT, such as e.g. heavily doped silicon, where this transition occurs at a free charge carrier density of $n_c \sim 3.5\times10^{18}$ cm^{-3} [9], would seem to be a more promising material for this purpose. However, the Mott transition in doped Si is the second order phase transition and hence it is not accompanied by a conductivity jump. On the other hand, in vanadium dioxide, the change of conductivity at the transition temperature (T_t = 340 K [9]) reaches 4-5 orders of magnitude.

In the work [41], a Mott transition field effect transistor, based on hypothetical molecular (Mott insulator) layers, in particular, such exotic materials as K^+TCNQ^- (the quasi-monomer organic conductor) or KC_{60} (the doped fulleren) have been proposed. The version of a MTFET based on VO_2 [44] seems to be more attractive. It demonstrates high speed, low dimensions, and (what is more important) it works on the basis of the well-studied, reliable material, which has already been tested as a laboratory prototype. In addition, the important merits of vanadium dioxide are that its transition temperature is very close to room temperature and that this material is thermodynamically stable [45] as compared to other oxides in the vanadium-oxygen system (in which, by the way, there are more than ten oxides exhibiting MITs at different temperatures).

In this edited collection entitled "Oxide Electronics and Functional Properties of Transition Metal Oxides", four papers concerning the above outlined issues are presented. The chapters presented herein were solicited from a selected group of researchers who are experts in the fields of TMOs, theirs properties, and oxide electronics. Rather brief, albeit very important in context of oxide electronics, Chapter I is devoted to unipolar resistive switching in TMO-based MOM structures. It is written by Doctor Tatiana V. Kundozerova and Professor Genrikh B. Stefanovich who were with the Department of Condensed Matter Physics, Royal Institute of Technology – KTH (Stockholm, Sweden). They are now with the Department of Information Measuring Systems and Physical Electronics, Faculty of Physical Engineering of Petrozavodsk State University, 185910 Petrozavodsk, Russia.

Chapter II "Some fundamental points of technology of lithium niobate and lithium tantalate single crystals" is written by Doctors Nikolay V. Sidorov and Mikhail N. Palatnikov who are with the Labs of Vibrational Spectroscopy and Electronics Materials, resp., of I.V. Tananaev Institute of Chemistry and Technology of Rare Elements and Mineral Raw Materials of Kola Science Centre of RAS, 184209 Apatity, Russia.

Chapter III is devoted to the sputter deposited nanolaminate containing of some group IVB (Ti, Zr, Hf) oxides, as well as to their phase compositions, crystal structures and near band gap optical properties. This Chapter is written by Professor Carolyn R. Aita who is with Department of Chemistry and Biochemistry of University of Wisconsin-Milwaukee P. O. Box 413 Milwaukee, Wisconsin 53201, USA.

And finally, Chapter IV "Optical and electrical switching of thermochromic VO_2 smart coatings" is written by Doctor Mohammed Soltani who was with INRS Energy Materials Telecommunications Research Centre, Qc, Canada, and now he is with RSL-Tech 9114 Descartes, Montreal, Qc, H1R 3P5 Canada.

Thus, in this edited collection we have tried to bring together the most important materials, properties and phenomena which are at the cutting edge of oxide electronics and related fields of condensed matter physics.

ACKNOWLEDGEMENTS

This my work as an editor was partly supported by the Strategic Development Program of Petrozavodsk State University (2012 – 2016) and by the RF Ministry of Education and Science state contract no. 2014/154 through the project no. 1704. I would also like to express my heartfelt gratitude to all the authors who contributed to this book for their support and assistance.

REFERENCES

[1] J. S. Kilby. Turning Potential into Reality: The Invention of the Integrated Circuit. Nobel Lecture, 2000 [Online]. Available: http://www.nobelprize.org/nobel_prizes/ physics.

[2] Yu. V. Gulyaev, V. B. Sandomirskiĭ, A. A. Sukhanov, and Yu. Ya. Tkach, "Physical limitations on miniaturization in microelectronics," *Sov. Phys. Usp.*, vol. 27, pp.868-880, 1984.

[3] H. Takagi and H. Y. Hwang, "An Emergent Change of Phase for Electronics," *Science*, vol. 327, pp. 1601-1602, 2010.

[4] A. L. Pergament, G. B. Stefanovich, and A. A. Velichko, "Oxide Electronics and Vanadium Dioxide Perspective: A Review" Journal on Selected Topics in Nano Electronics and Computing, Vol. 1, no. 1, pp. 24-43, Dec. 2013. [Online]. Available: http://jstnec.petrsu.ru/journal/article_en.php?id=3002&seq=

[5] M. Bohr and K. Mistry. INTEL's Revolutionary 22-nm Transistor Technology, 2011 [Online]. Available: http://download.intel.com/newsroom/kits/22nm/pdfs/22nm-Details_Presentation.pdf.

[6] S. Migita, Y. Morita, M. Masahara, and H. Ota, "Fabrication and demonstration of 3-nm-channel-length junctionless field-effect transistors on silicon-on-insulator substrates using anisotropic wet etching and lateral diffusion of dopants," *Jpn. J. Appl. Phys.*, vol. 52, pp. 04CA01-5, 2013.

[7] S. Zaima, "Technology evolution for silicon nanoelectronics: postscaling technology," *Jpn. J. Appl. Phys.*, vol. 52, p.030001, 2013.

[8] J. F. Rossier, "Single-atom devices: Quantum engineering," *Nature Materials*, vol. 12, pp. 480–481, 2013; C. Schirm, M. Matt, F. Pauly, J. C. Cuevas, P. Nielaba, and E. Scheer, "A current-driven single-atom memory," *Nature Nanotechnology*, vol. 8, pp. 645–648, 2013.

[9] N. F. Mott, *Metal-Insulator Transition*, 2nd ed. London: Taylor and Francis, 1990.

[10] A. L. Pergament, G. B. Stefanovich, A. A. Velichko, and S. D. Khanin , "Electronic Switching and Metal-Insulator Transitions in Compounds of Transition Metals," in *Condensed Matter at the Leading Edge*. Nova Science Publishers, 2006, pp.1-67.

[11] A. P. Ramirez, "Oxide Electronics Emerge," *Science*, vol. 315, pp. 1377-1378, 2007.
[12] C. Cen, S. Thiel, J. Mannhart, and J. Levy, "Oxide Nanoelectronics on Demand," *Science*, vol. 323, pp. 1026-1030, 2009.
[13] P. Irvin, M. Huang, F. J. Wong, T. D. Sanders, Y. Suzuki, and J. Levy, "Gigahertz-frequency operation of a LaAlO3/SrTiO3-based nanotransistor," *Appl. Phys. Lett.*, vol. 102, p. 103113, 2013.
[14] P. Zubko, S. Gariglio, M. Gabay, P. Ghosez, and J.-M. Triscone, "Interface Physics in Complex Oxide Heterostructures," *Annu. Rev. Condens.Matter Phys.*, vol. 2, pp. 141–165, 2011.
[15] H. Y. Hwang, Y. Iwasa, M. Kawasaki, B. Keimer, N. Nagaosa, and Y. Tokura "Emergent phenomena at oxide interfaces," *Nature Materials*, vol. 11, pp. 103–113, 2012.
[16] G. Herranz, M. Basletić, M. Bibes, C. Carretero, E. Tafra, E. Jacquet, K. Bouzehouane, C. Deranlot, A. Hamzić, J.-M. Broto, A. Barthelemy, and A. Fert," High Mobility in LaAlO3/SrTiO3 Heterostructures: Origin, Dimensionality, and Perspectives," *Phys. Rev. Lett.*, vol. 98, no. 21, pp. 216803-216807, 2007.
[17] S. Pang, Y. Hernandez, X. Feng, and K. Mullen, "Graphene as Transparent Electrode Material for Organic Electronics," *Adv. Mater.*, vol. 23, pp. 2779–2795, 2011.
[18] J. Lewis, "Materials challenge for flexible organic devices," *Materials Today*, vol. 9, pp.38-45, Apr. 2006.
[19] S. Kumar, B. A. Cola, R. Jackson, and S. Graham, "A Review of Carbon Nanotube Ensembles as Flexible Electronics and Advanced Packaging Materials," *Journal of Electronic Packaging*, vol. 133, no. 2, p. 020906, 2011.
[20] D.H. Blank, D.S. Ginley, M.E. Hawley, S.K. Streiffer, and D.C. Paine (ed.), *Transport and Microstructural Phenomena in Oxide Electronics*, Mat. Res. Soc. Proc., vol. 666, 2001.
[21] Y. Fujisaki, "Review of Emerging New Solid-State Non-Volatile Memories," *Jpn. J. Appl. Phys*, vol. 52, p. 040001, 2013.
[22] R. Waser, "Resistive non-volatile memory devices", *Microelectronic Engineering*, vol. 86, pp. 1925-1928, 2009.
[23] H. Akinaga, "Recent Advances and Future Prospects in Functional-Oxide Nanoelectronics: The Emerging Materials and Novel Functionalities that are Accelerating Semiconductor Device Research and Development," *Jpn. J. Appl. Phys*, vol. 52, p. 100001, 2013.
[24] D. B. Strukov and R. S. Williams, "An ionic bottle for high-speed, long-retention memristive devices," *Appl. Phys. A*, vol. 102, pp. 1033–1036, 2011.
[25] L. Liu, B. Chen, B. Gao, F. Zhang, Y. Chen, X. Liu, Y. Wang , R. Han, and J. Kang, "Engineering oxide resistive switching materials for memristive device application," *Appl. Phys. A*, vol. 102, pp. 991–996, 2011.
[26] D. Ielmini, R. Bruchhaus, and R. Waser, "Thermochemical resistive switching: materials, mechanisms, and scaling projections," *Phase Transitions*, vol. 84, no. 7, pp. 570–602, Jul. 2011.
[27] S. Balatti, S. Larentis, D. C. Gilmer, and D. Ielmini, "Multiple Memory States in Resistive Switching Devices Through Controlled Size and Orientation of the Conductive Filament," *Adv. Materials*, vol. 25, no. 10, pp. 1474–1478, 2013.

[28] A. Pergament, G. Stefanovich, A. Velichko, V. Putrolainen, T. Kundozerova, and T. Stefanovich, "Novel Hypostasis of Old Materials in Oxide Electronics: Metal Oxides for Resistive Random Access Memory Applications," *Journal of Characterization and Development of Novel Materials*, vol. 4, no. 2, pp. 83-110, 2011.

[29] T. V. Kundozerova, A. M. Grishin, G. B. Stefanovich, and A. A. Velichko, "Anodic Nb2O5 Nonvolatile RRAM," *IEEE Trans. Electron Devices*, vol. 59, no. 4, pp. 1144-1148, Apr. 2012.

[30] T. Driscoll, H.-T. Kim, B.-G. Chae, M. Di Ventra, and D. N. Basov, "Phase-transition driven memristive system," *Appl. Phys. Lett.*, vol. 95, p. 043503, 2009.

[31] R. Xie , C. T. Bui , B. Varghese , Q. Zhang , C. H. Sow, B. Li , and J. T. L. Thong, "An Electrically Tuned Solid-State Thermal Memory Based on Metal–Insulator Transition of Single-Crystalline VO_2 Nanobeams," *Adv. Funct. Mater.*, vol. 21, pp. 1602-1607, 2011.

[32] S. J. Pearton, W. T. Lim, E. Douglas, H. Cho, and F. Ren, "Flexible Electronics Based on InGaZnO Transparent Thin Film Transistors," *Key Engineering Materials*, vol. 521, pp. 141-151, 2012.

[33] J. F. Wager, B. Yeh, R. L. Hoffman, and D. A. Keszler, "An amorphous oxide semiconductor thin-film transistor route to oxide electronics," *Curr. Opin. Solid State Mater. Sci.*, in press, 2013. http://dx.doi.org/10.1016/j.cossms.2013.07.002

[34] J. S. Park, W.-J. Maeng, H.-S. Kim, and J.-S. Park, "Review of recent developments in amorphous oxide semiconductor thin-film transistor devices," *Thin Solid Films*, vol. 520, pp. 1679-1693, 2012.

[35] D. Keszler, "Oxide electronics: Transistors pick up steam," *Nature Materials,* vol. 10, pp. 9-10, 2011.

[36] E. Fortunato and R. Martins, "Where science fiction meets reality? With oxide semiconductors!" *Phys. Stat. Solid.*, vol. 5, pp. 336 -339, 2011.

[37] R. F. P. Martins, A. Ahnood, N. Correia, L. M. N. P. Pereira, R. Barros, P. M. C. B. Barquinha, R. Costa, I. M. M. Ferreira, A. Nathan, and E. E. M. C. Fortunato, "Recyclable, Flexible, Low-Power Oxide Electronics." *Adv. Funct. Mater.,* vol. 23, pp. 2153–2161, 2013.

[38] M. Suzuki and T. Ami, "A proposal of epitaxial oxide thin film structures for future oxide electronics," *Materials Sci. Engineering*, vol. B41, pp. 163-174, 1996.

[39] C. G. Granqvist, "Transparent conductors as solar energy materials: A panoramic review," *Solar Energy Materials & Solar Cells*, vol. 91, pp. 1529-1598, 2007.

[40] J. A. Rogers, T. Someya, and Y. Huang, "Materials and Mechanics for Stretchable Electronics," *Science*, vol. 327, pp. 1603-1607, 2010.

[41] C. Zhou, D. M. Newns, J. A. Misewich, and P. C. Pattnaik, "A field effect transistor based on the Mott transition in a molecular layer," *Appl. Phys. Lett.*, vol. 70, no. 5, pp. 598-600, 1997.

[42] D.M. Newns, J.A. Misewich, and C. Zhou, "Nanoscale Mott-transition Molecular Field Effect Transistor," U.S. Patent YO996-06, 1996.

[43] A. Pergament, G. Stefanovich, N. Kuldin, and A. Velichko, "On the Problem of Metal–Insulator Transitions in Vanadium Oxides," *ISRN Condensed Matter Physics*, vol. 2013, 2013. Available: http://dx.doi.org/10.1155/2013/960627

[44] G. Stefanovich, A. Pergament, and D. Stefanovich, "Electrical switching and Mott transition in VO_2," *Journal of Physics: Condensed Matter*, vol. 12, no. 41, pp. 8837-8845, 2000.

[45] A. L. Pergament and G. B. Stefanovich, "Phase composition of anodic oxide films on transition metals: a thermodynamic approach," *Thin Solid Films*, vol. 322, no.1-2, pp. 33-36, 1998.

In: Oxide Electronics and Functional Properties …
Editor: Alexander Pergament
ISBN: 978-1-63321-499-6

Chapter 1

UNIPOLAR RESISTIVE SWITCHING EFFECT

Tatiana V. Kundozerova [*] ***and Genrickh B. Stefanovich*** [†]
Faculty of Physical Engineering,
Petrozavodsk State University, Petrozavodsk, Russia

ABSTRACT

Emerging memory technologies based on resistive random access memory (ReRAM) devices are considered as promising candidates to replace Flash in the next generation of a high density and high volume non-volatile memory. In this chapter we present an overview of unipolar nonvolatile resistive switching in a metal-oxide-metal thin-film memory cell. This phenomenon has been studied extensively for its functional properties, ON-OFF switching mechanism and its potential applications in computer memory matrixes and, particularly, in flexible electronic devices.

1. INTRODUCTION: RESISTIVE SWITCHING IN OXIDE RERAM

The effect of resistive switching is a sharp and reversible transition of materials between two states with a different resistance. Switching is observed in a large class of compounds: complex perovskite oxides, organic compounds, binary metal oxides such us NiO [1], CuO[2], ZnO [3], TiO_2[4], Nb_2O_5[5], Ta_2O_5[6], ZrO_2[6], HfO_x[7] etc. [8].

Resistive Random Access memory (ReRAM) it is an electronic memory which is based on resistive switching effect. The ReRAM memory cell has a capacitor-like structure (Metal–Insulator–Metal) in which an oxide layer is located between two metal electrodes (Figure 1).

[*] Email: tacan@mail.ru.
[†] Email: gstef@psu.karelia.ru.

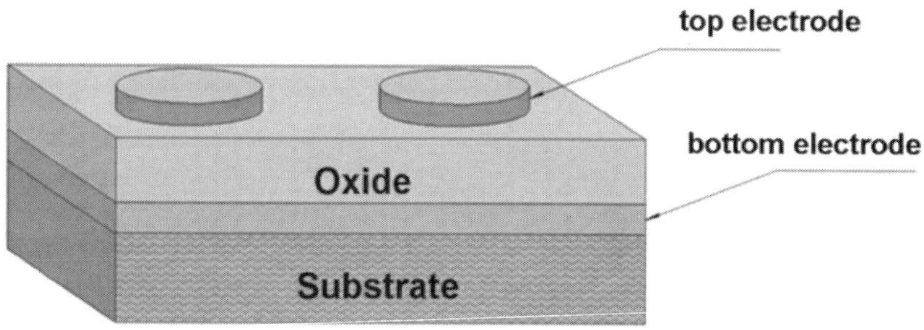

Figure 1. Scheme of ReRAM cell.

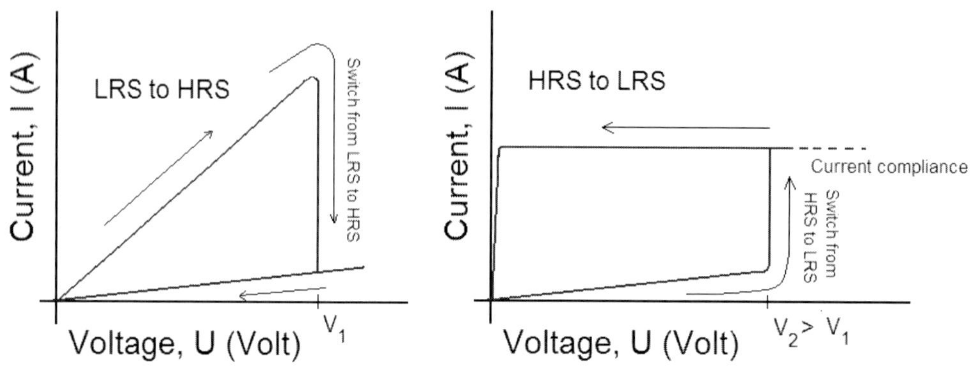

Figure 2. Typical *I-V* characteristics of unipolar resistive switching effect.

Under the voltage pulses ReRAM cells switch between high resistance state (HRS) and low resistance state (LRS). HRS and LRS represent a logical "1" and "0", it is stable in time nonvolatile states.

For the first time, an opportunity of application of the resistive switching effect in memory devices had been proposed in 1967 [9], though first experimental achievement of this idea has been made only in 2002 [10]. Since that time ReRAM starts to thrive, and nowadays, as compared, e.g., with flash-memory, ReRAM devices have a higher speed and endurance, while coupled with a smaller cell area and power consumption [8].

In appearance of current–voltage characteristics switching behavior (ReRAM operations) can be divided into two broad classes: unipolar and bipolar (Figures 2 and 3). Switching is called unipolar (or symmetric) when the switching procedure does not depend on the polarity of the voltage and current signal, it depends only on amplitude. Bipolar switching requires an alternating polarity of the applied signal. This type of switching is described in numerous papers [11-12]. The same material can show both bipolar and unipolar switching. Type of switching depends on material of electrodes, property of oxide layer, interface between oxide and electrode, condition of electroforming process. In this chapter only the unipolar resistive switching behavior is presented.

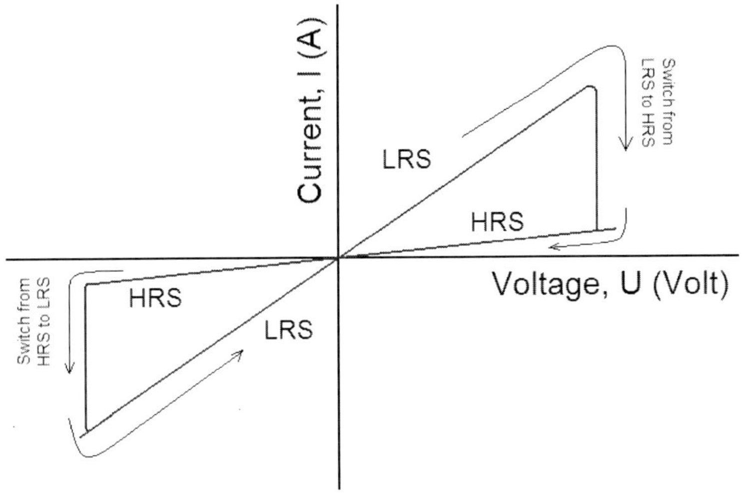

Figure 3. A typical *I-V* characteristic of bipolar resistive switching effect.

2. Unipolar Resistance Switching: Main Characteristics

The working process of ReRAM cell based on unipolar switching includes 3 stages: 1) electroforming (forming); 2) switching from low resistance state (LRS) to high resistance state (HRS); 3) switching from HRS to LRS. The forming process should be done only once at the beginning. Numerous switches between LRS and HRS are working cycle off memory cells.

The materials of oxide layers which used in resistance random access memory can be obtain by various method of thin film deposition: PLD (pulsed laser deposition), magnetron sputtering, sol-gel method, e-beam evaporation, thermal oxidation, anodic oxidation (anodizing). The reported oxides were obtained by anodic oxidation. It is simple, low cost, high deposition efficiency process to deposit thin dielectric films. This method now is a well known technique, it has various industrial applications and allows fabrication of various oxides such as Ta_2O_5, ZrO_2, Nb_2O_5, CuO, Al_2O_3, V_2O_5 etc.[13]. An oxide is formed on a metal surface by applying an electrical current or potential through an electrochemical cell with suitable electrolyte. Anodic oxidation is usually carried out isothermally (T = const) in galvanostatic (j = const) or voltstatic (V = const) regime. Formation of an oxide occurs on internal and external oxide's boundaries due to electric-field-induced migration of metal and oxygen atoms through the growing oxide layer. The thickness of the oxide layer can be controlled by anodization voltage V_a (the potential which drops through the oxide), according the anodization constant (Figure 4) [5].

As prepared ReRAM structures do not demonstrate a resistance switching effect. As it is shown in Figure 5, the current-voltage characteristic of the structure is nonlinear and asymmetric, which is typical for an anodic oxide layer.

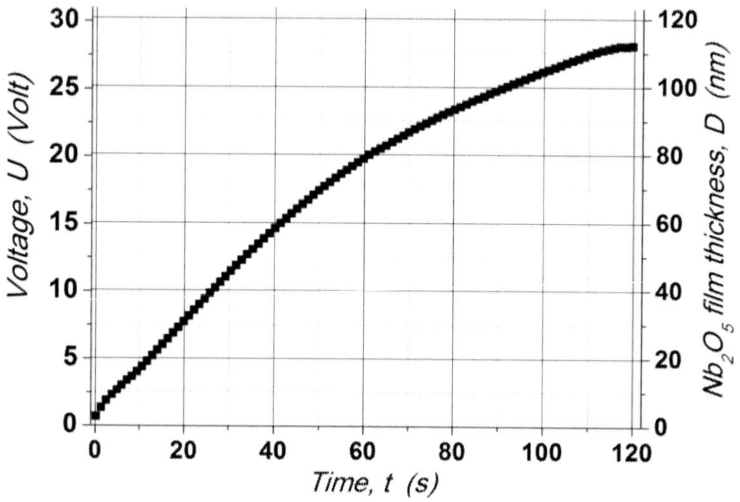

Figure 4. Thickness vs. time calibration curve for the anodization process of Nb film sputtered on Si wafer in aqueouse solution of H_3PO_4 acid. [5].

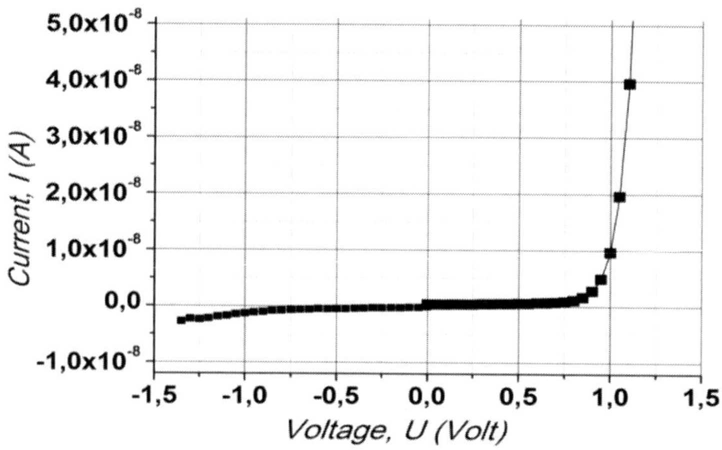

Figure 5. Current-voltage characteristic of $Si/Nb/Nb_2O_5/Au$ structure before forming.

The electroforming process is a dielectric breakdown (an abrupt increasing of the oxide layer's conductivity) of a metal-oxide-metal (MOM) structure with a current compliance. The electroforming is carried by the following way: on a top electrode of the structure a linearly increasing voltage is applied, the bottom electrode is grounded. After the threshold voltage is reached, the resistance of the structure abruptly (nanoseconds range) falls by several orders of magnitude. The current flowing through the structure during the electroforming process is limited at I_c = 5 mA. (Figure 6). Changes in a current value can be traced more carefully if electroforming is performed by the linearly increasing current (Figure 7). As is seen, after a negative differential resistance (NDR) region, the current-voltage characteristic becomes linear, i.e. it corresponds to the Ohm's law.

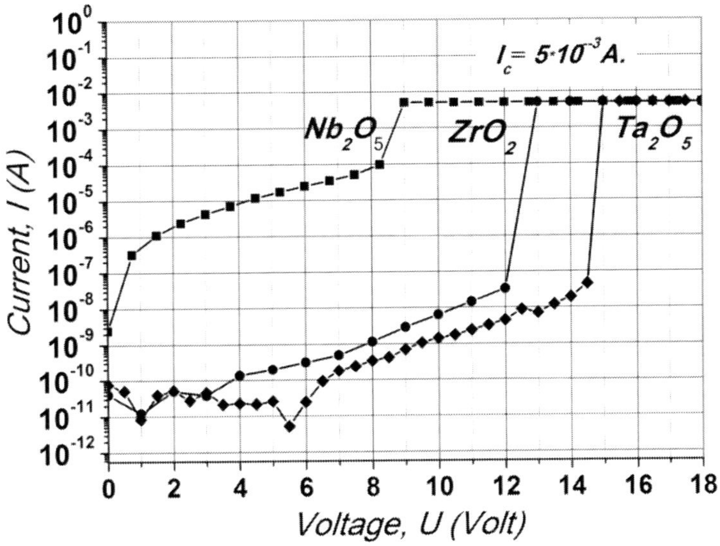

Figure 6. Current-voltage characteristics of forming process, voltage generation mode. Au/Nb_2O_5(80 nm)/Nb, Au/Ta_2O_5(55 nm)/Ta, Au/ZrO_2(40 nm)/Zr structures.

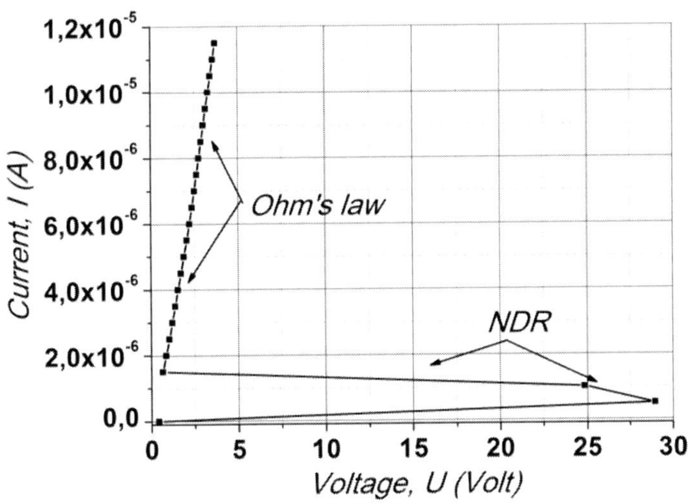

Figure 7. Current-voltage characteristics of forming process, current generation mode. Au/Nb_2O_5(98 nm)/Nb structure.

In case of anodic oxide, a positive polarity of a top electrode during the forming processes is required. Under negative voltages electroforming occurs at higher voltages, as a result an energy which released in this process increase and probability of irreversible breakdown increases.

As a result of electroforming a constant conductive filaments are generated in oxide (Figure 8). It is confirmed by the study of planar and sandwich structures [2, 11, 14] (Figure 8). A chemical composition of the filament is different for different structures and depends on the material of oxide [15]. As it will be shown later, a filament plays a main role in the processes of switching.

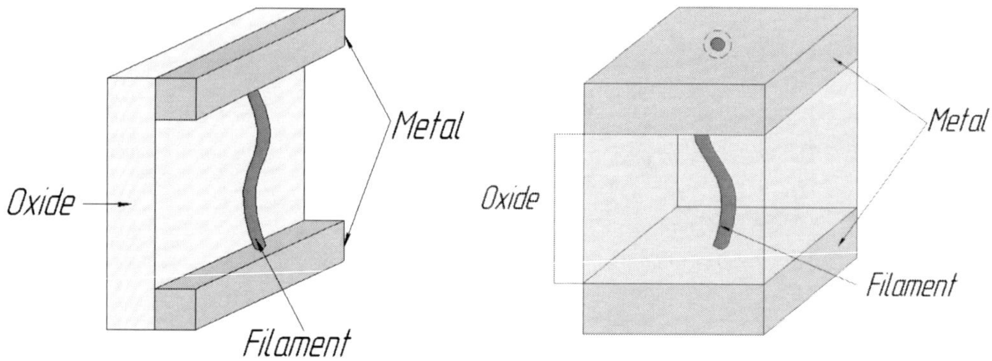

Figure 8. Illustration of a filamentary conducting path in a planar and sandwich configuration of structure.

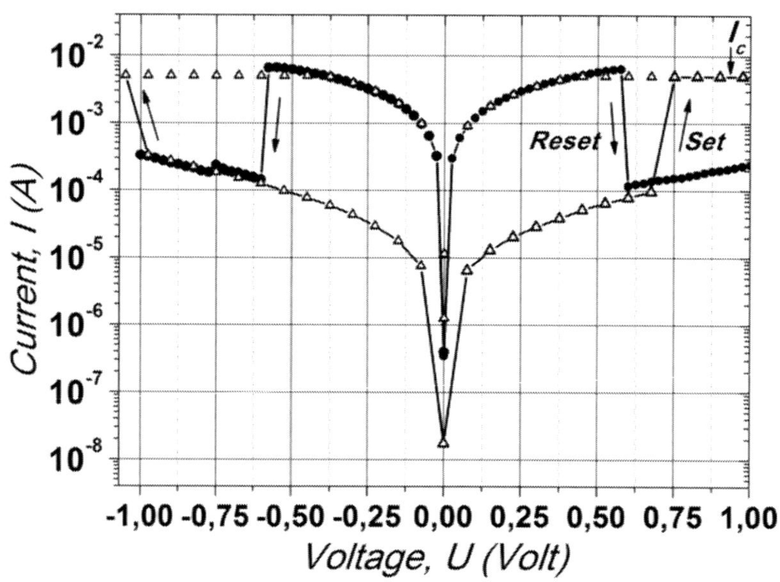

Figure 9. Resistance switching *I*-*V* characteristics of Au/Nb_2O_5(90nm)/Nb/Si memory cell.

Note that a setting of adequate current compliance is very important during the electroforming process. Without compliance, a structure switches to irreversible low resistance state. The lower range of current compliance value is determined by a threshold value. The following regularities were determined: 1) Electroforming of the structures with a high initial resistances lead to irreversible breakdown, no matter what is the value of current compliance. 2) In case of more conductive samples electroforming lead to reversible resistance switching between LRS and HRS. 3) If current compliance was fixed in a value corresponding to the threshold voltage, a structure demonstrates resistance switching without memory.

After the forming process, a system in low resistance state is switched to a high-resistance state by applying a threshold voltage ('reset process'). Switching from HRS to LRS ('set process') is achieved by applying a threshold voltage greater than the reset voltage (Figure 9). Note that, similarly to electroforming, the set process requires a current compliance. Without current compliance the structure is switched into a permanent low-resistance state.

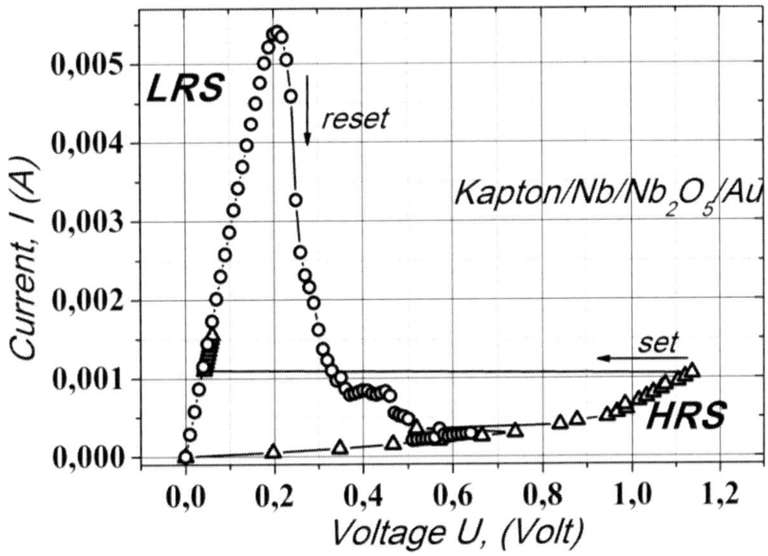

Figure 10. Resistance switching *I-V* characteristic of Au/Nb_2O_5/Nb/Kapton memory cell, current generation mode.

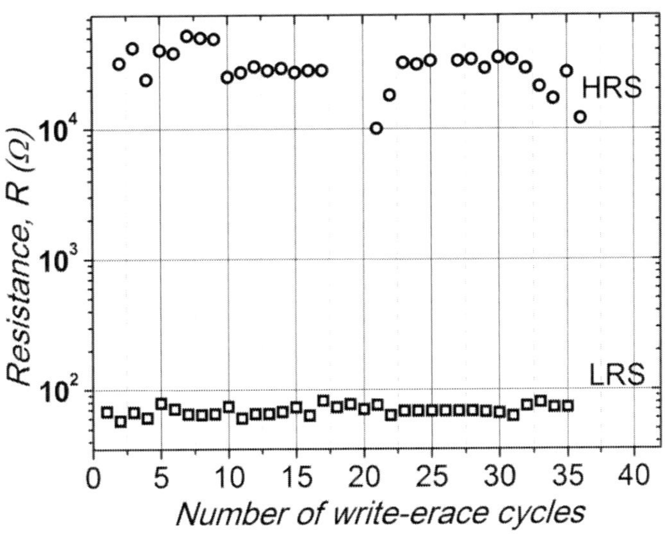

Figure 11. Switching cycling characteristics for LRS and HRS in Au//Nb_2O_5(90nm)/Nb/Si film memristor measured at 100 mV.

To trace changes in the current value during the set and reset processes, another mode of measurement can be used. A linearly increasing current is applied on the top electrode. As shown in Figure 10, the switching from HRS to LRS has an S-type of current-voltage characteristics with an NDR region. Reverse switching from LRS to HRS has an N-type current-voltage characteristics, transition is longer and occurs under a higher current. Resistance switching is also realized by means of short voltage pulses. Highly reproducible set and reset operation is observed with the pulse duration of 50 ns and several microseconds range, respectively.

Note that the resistance values of the LRS in relation to the number of cycles are less scattered than those of the HRS (Figure 11). We also investigated long term stability of the resistance in the reset and set states. Figure 12 shows that both resistance states are stable at a read out voltage of 0.1 V.

Both LRS and HRS are slightly increasing during the storage time. Retention of such behavior indicates a stable and reliable storage of information. All working parameters has a wide operating window, it eliminates errors during write/erase and reading of information.

As readily seen in Figures 9 and 10, reset (LRS-to-HRS switching) operation is performed by increasing bias voltage. Typical reset voltage is in the range V_{reset} = 0.4-0.9 V (Au/Nb_2O_5(90nm)/Nb/Si memory cell). Limitation of feeding current is not required for reset process (self-compliance regime). Reset voltages V_{reset} are less scattered than the voltages V_{set} required to set a LRS (see Figure 13). A set operation (reverse HRS-to-LRS switching) is produced by applying higher voltages V_{set}= 1.2-2.8 V with a current compliance I_c=5·10^{-3} A. Reset V_{reset} and set V_{set} voltages are slightly dependent on the magnitude of a set current compliance though they are usually within the TTL (transistor-transistor logic) range (0.5-2 V).

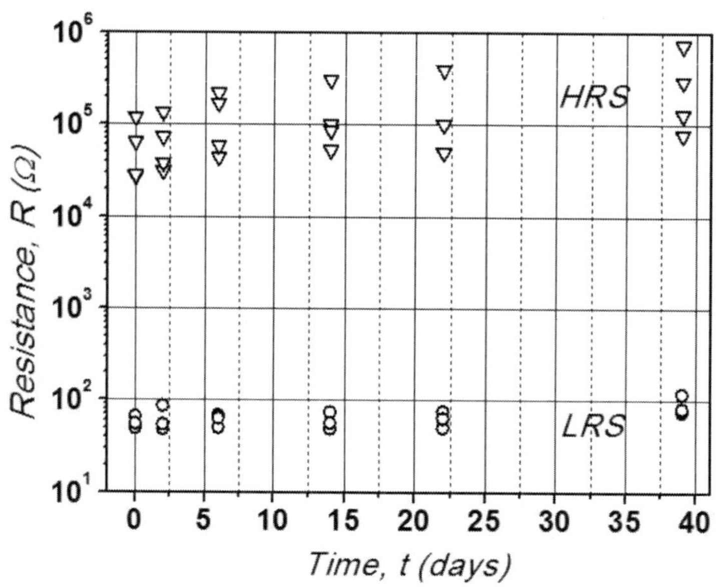

Figure 12. Room temperature retention characteristics for LRS and HRS in Au//Nb_2O_5(90nm)/Nb/Si film memristor measured at 100 mV.

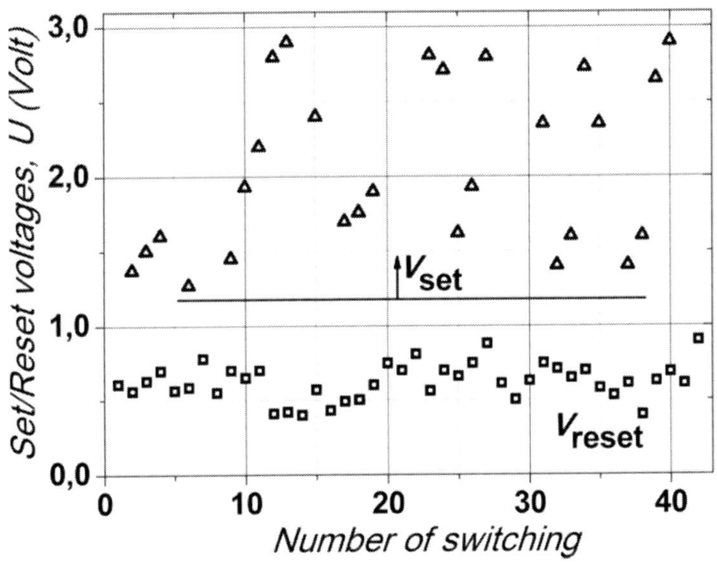

Figure 13. Scattering of operation voltages required to set (V_{set}) and reset (V_{reset}) low resistance state in Al//Nb_2O_5(130nm)/Nb(foil) memory cell.

Despite the final convention on the switching mechanisms is not achieved, it is nevertheless widely accepted that nonvolatile resistance switching occurs through the formation and rupture of nanoscale conducting filament [2, 16]. The presence of the filament in a memristor cell leads to metallic-type conductivity. During the reset process, the conductive filament is disrupted and semiconducting properties are restored in the memory cell. HRS can be developed by various metal-dielectric phase configurations whereas a high reproducibility of LRS attributes to unique conducting percolation path. This model is confirmed by a series of experiments: temperature dependence of resistance, scaling behavior of resistance states (oxide thicknesses, area of electrodes), FIB-SEM and XPS investigation of the structure and phase composition, frequency dependence of impedance etc. [8, 17-20].

3. The Properties of Unipolar Resistive Switching

3.1. Scaling Behavior

The switching voltages are only slightly dependent on thicknesses of oxide layers (Figure 14), unlike the forming voltages which are directly proportional to the oxide thicknesses. In the range of 90 – 450 nm, the forming voltage rises by 7 times from 5 to 35 V. Forming process is a dielectric breakdown which leads to growth of a filament through all the film thickness from bottom to top electrode; that is why it is rather obvious that the voltage is proportional to the thicknesses. During the next switching operation only a part of the metallic filament is changed. The filament region which is close to the electrode interface is destroyed and recovered under set and reset operations. The size of this region does not depend strongly on the entire oxide layer thicknesses.

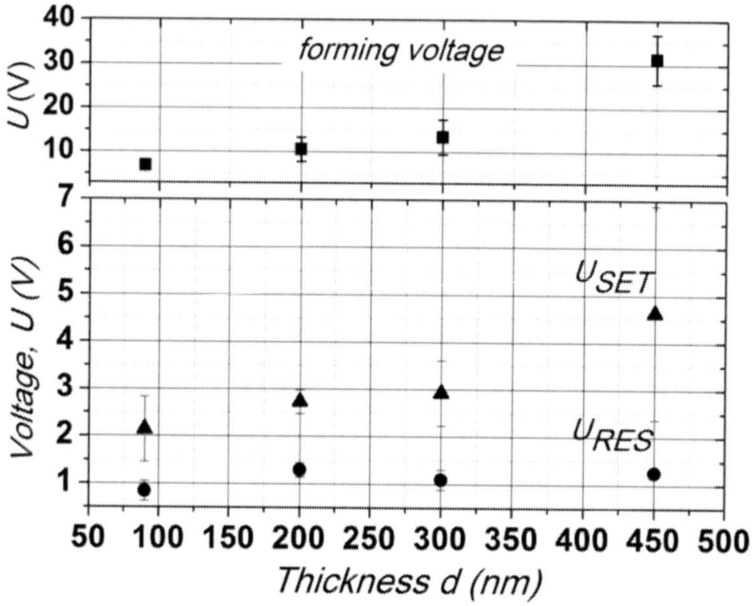

Figure 14. Voltage vs. thicknesses of oxide layer dependence. Memory cells based on Nb_2O_5 oxide with different thicknesses: 450, 300, 200, 90 nanometers.

A scaling of an electrode area effects slightly on resistance in LRS, whereas resistance of HRS decreases with increasing of electrodes area [8]. Total resistance ratio R_{HRS}/R_{LRS} increases and it is one more advantage of scaling of ReRAM cells. In terms of the filament model, resistance of the LRS is determined by resistance of the filaments and it does not depend on the electrode area. In case of HRS, all the volume of dielectric layer affects on resistance which rises with decreasing the contact area.

Physically, the size of ReRAM cells is limited by the size of conductive filament which appears in oxide volume during a forming process. The reached minimum of oxide thicknesses for NiO based structure is 10 nm [18], and the minimum of electrodes area is 10x10 nm. [19].

3.2. The Temperature Dependence of Resistance

It is known [5, 11, 20] that at switching, the temperature dependence of the resistance changes from semiconducting (initial state, HRS) to metallic (LRS). Temperature dependence of the resistance *R*(*T*) is shown in Figure 15. The resistance of LRS increases with temperature (metallic type). Calculated temperature coefficient $\alpha = 3.3 \cdot 10^{-3}$ 1/K corresponds to published data for Nb ($\alpha = 3.9 \cdot 10^{-3}$ 1/K). The resistance of HRS decreases with increasing temperature and fits to the exponential function. Such a behavior is typical for dielectrics and semiconductors.

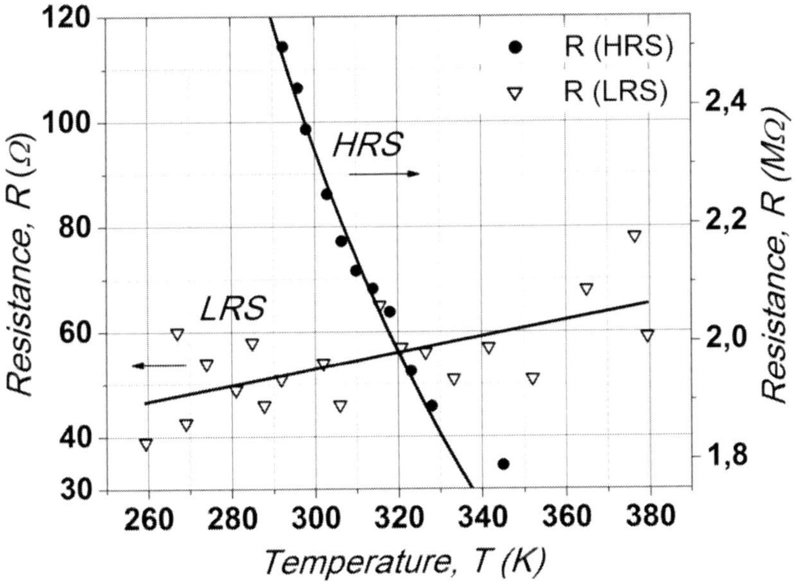

Figure 15. Temperature dependence of the resistance $R(T)$ in Au/Nb_2O_5(90nm)/Nb/Si memory cell.

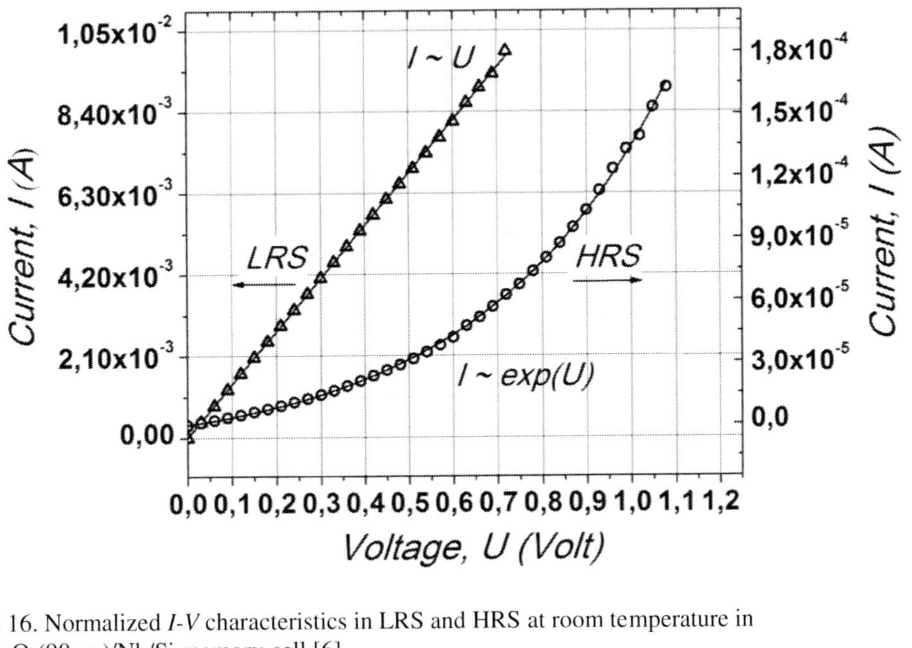

Figure 16. Normalized *I-V* characteristics in LRS and HRS at room temperature in Au/Nb_2O_5(90nm)/Nb/Si memory cell [6].

The current-voltage characteristics of LRS exhibit a pure ohmic conductivity with a linear *I* vs. *V* dependence. Being also linear beyond 300 mV, the *I-V* characteristic of HRS at high voltages follows the linear dependence in the ln(I/V) vs. $V^{1/2}$ coordinates which is typical for the effect of high electric field on the conductivity, for example the Poole-Frenkel type of conductivity (Figure 16) [5].

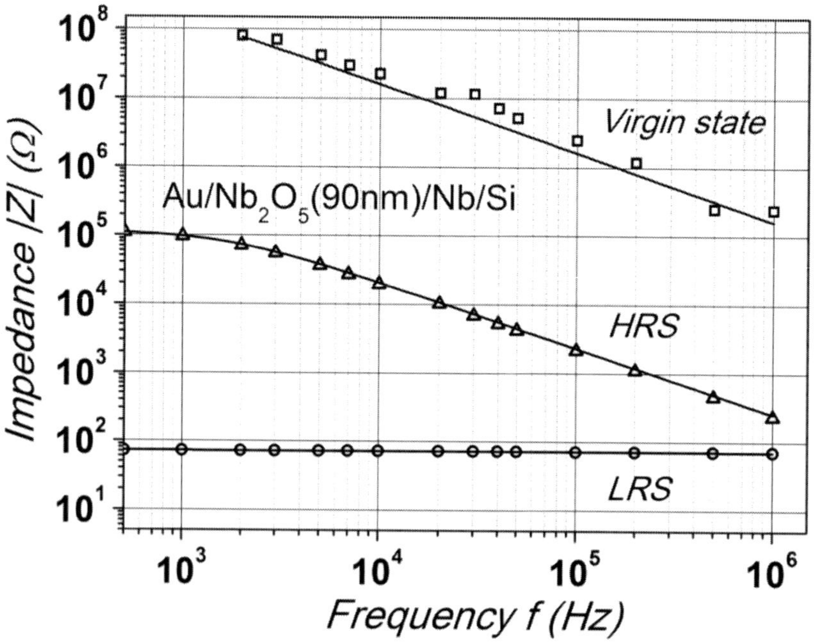

Figure 17. Frequency dependence of impedance in the Au/Nb_2O_5(90 nm)/Nb/Si memory cell in the virgin (as prepared nonformed) state, the HRS, and the LRS. (symbols) Experimental data. (solid lines) Impedance of the equivalent circuit.

3.3. The Frequency Dependence of Resistance

Figure 17 shows the frequency dispersion of the impedance in the virgin state, the LRS, and the HRS of the Au/Nb_2O_5(90 nm)/Nb cell made by anodization of the 300-nm thick Nb film on the Si wafer. Frequency-independent impedance indicates metallic-type conductivity in LRS meantime frequency dispersion in the virgin state and the HRS can be modeled with equivalent circuits.

An equivalent circuit of memory cell is commonly described in terms of constant phase element (CPE). Hereinafter CPEs having impedance $Z_{CPE} = Z_o(i2\pi f)^n$ with $n \geq 0.96$ and $n \leq 0.09$ we consider, respectively, as a capacitive and resistive elements [21]. The virgin state (see Figure 18) can be presented as a parallel connection of capacitor $C = 0.98$ pF and resistor $R = 810$ MΩ. Very high resistance R indicates a low leaky capacitive character of the as-grown Nb_2O_5 cell in a virgin nonformed state.

The corresponding dielectric permittivity ε and the loss tan(δ) at 50 kHz were found to be 31 and 0.02, respectively. These parameters are characteristic of the Nb_2O_5 phase, which is the most stable valence state of Nb ion referred to as the *n*-type semiconductor with a band gap of 3 to 4 eV. As an example, frequency dependence values of ε and tan(δ) for the Al/Nb_2O_5(130 nm)/Nb(foil) cell are presented in Figure 19.

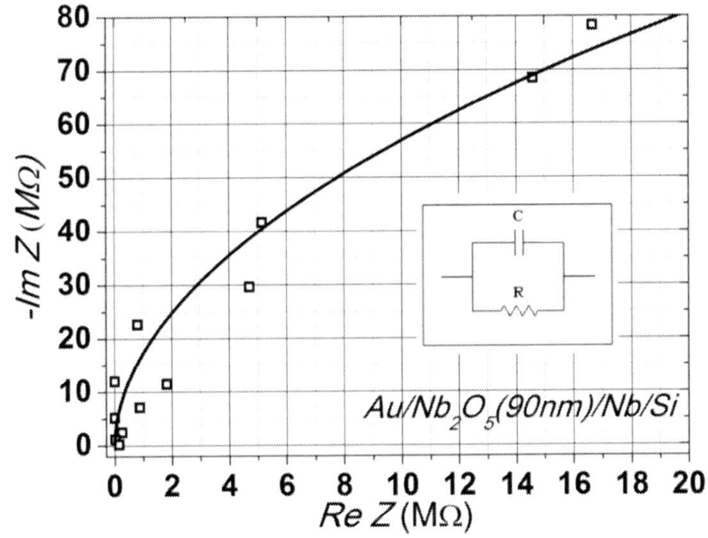

Figure 18. ImZ - ReZ plot for the virgin state of Au/Nb_2O_5(90nm)/Nb/Si memory cell (symbols) and Cole-Cole fit (solid line) with a parallel connected capacitor $C = 0.98$ pF and a resistor $R = 810$ MΩ.

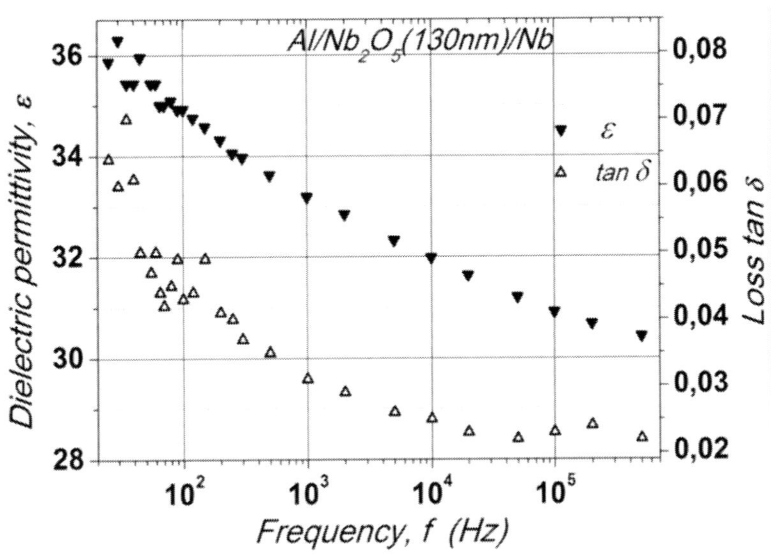

Figure 19. Frequency dependence of dielectric permittivity and loss tangent in the virgin state of Al/Nb_2O_5(130nm)/Nb(foil) film structure.

After electroforming, the memory cell was switched to the LRS. The equivalent circuit for the LRS (see Figure 20) comprises those connected in parallel: capacitor C = 525 pF series-connected to resistor $R_1 = 7$ Ω and inductor $L = 1.1$ μH series-connected to resistor R_2 = 73 Ω. ImZ – ReZ (Cole–Cole) diagrams for the virgin state and the LRS are presented in Figures 18 and 20, respectively. The inset in Figure 20 shows the phase shift $\theta = \tan^{-1}$ (ImZ/ReZ) versus frequency f.

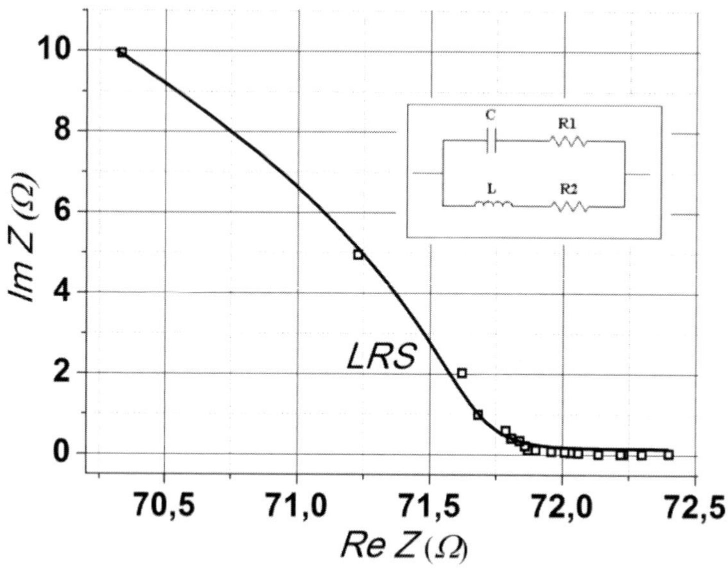

Figure 20. Cole-Cole (ImZ – ReZ) plots for the LRS of Au/Nb_2O_5(90nm)/Nb/Si cell: symbols – experiment, solid line – theoretical fit for the equivalent circuit that comprises a capacitor C = 525 pF, inductor with L = 1.1 μH, and two resistors R_1 = 7 Ω and R_2 = 73 Ω.

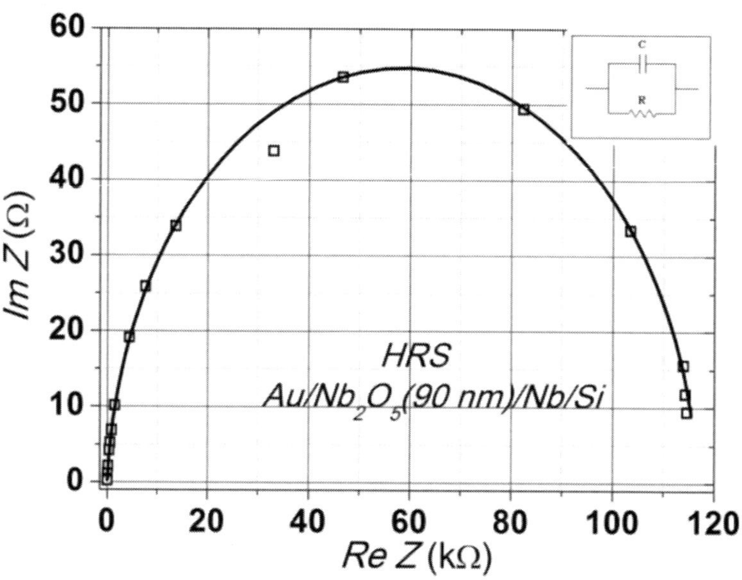

Figure 21. ImZ - ReZ plots for the HRS of Au/Nb_2O_5(90nm)/Nb/Si cell: symbols – experiment, solid line – theory for the model with a parallel connected capacitor C = 1200 pF and a resistor R = 120 kΩ.

The equivalent circuit for the HRS looks similar to the model for the virgin state: a parallel connection of capacitor C = 1200 pF and resistor R = 120 kΩ. The Cole–Cole plot for HRS is shown in Figure 21.

Comparing equivalent circuits of different resistance states, the following conclusions can be made:

1) High resistance states before and after electroforming can be explained by the same equivalent scheme (parallel connected capacitor and resistor). The difference is only in the values of *C* and *R*.
2) After switching of the structure into low resistance state, an inductor as a new element appears in an equivalent scheme. This inductance corresponds to metallic filament which appears during switching.

Thus, the experimentally obtained results which are presented in this section confirm one of the most predominant model of switching mechanism, based on the formation and rupture of conducting filaments.

4. A Model of Unipolar Switching

It has been shown previously [1, 11] that the model of the electrically actuated formation of the nanosize metal filament inside an oxide matrix is a most prevalent model for unipolar switching explanation. Nevertheless a composition of the filament, processes of its formation and rupture for a wide number of oxides are not determined completely. The most investigated oxide, at least experimentally, is NiO [1, 22-25]. The model metallic Ni filament formation during the switching of NiO ReRAM structure was first proposed by J.F. Gibbons and W.E. Beadle [1]. More detailed examination of this model is represented in our recent work [26].

Briefly, the switching process includes the following stages:

I. The forming process

1) A dielectric breakdown of the oxide layer with a required current compliance. 2) A discharge of a capacitor. Release of energy which ReRAM as a capacitor structure stored before forming process.	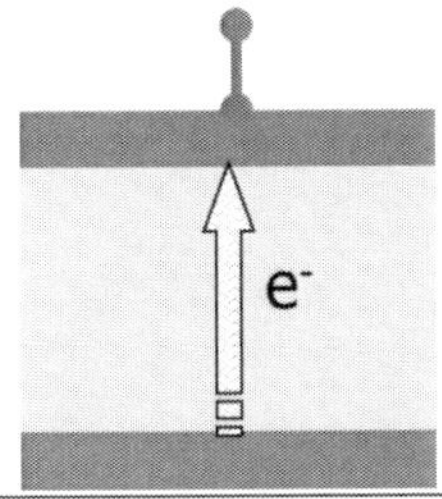
3) A sharp increase in temperature and, as a result, fast local redox reactions.	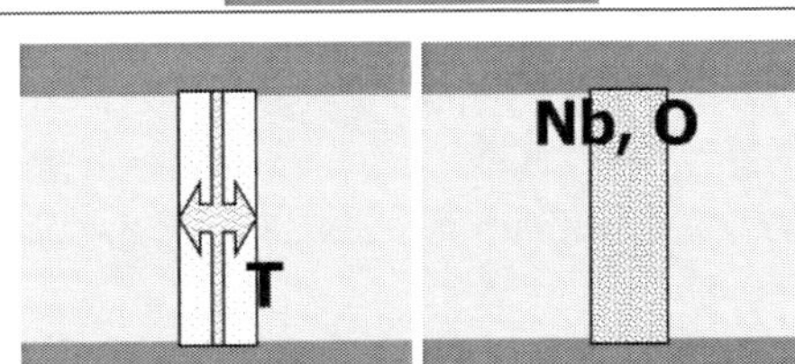

I. The forming process (Continued)

4) Under a gradient of temperature and diffusion process a Soret state is established. Segregation of metal in a center of high temperature region occurs.	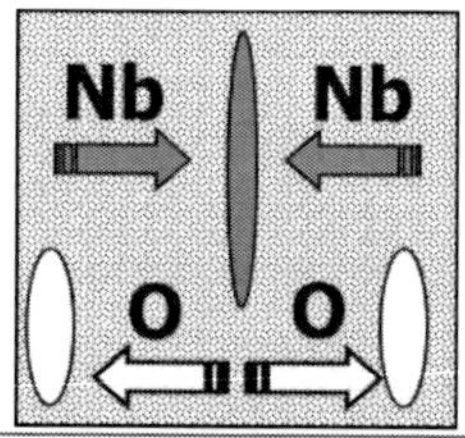
5) Due to a sharp decrease of a temperature, after the forming process is finished, a metallic filament is solidified.	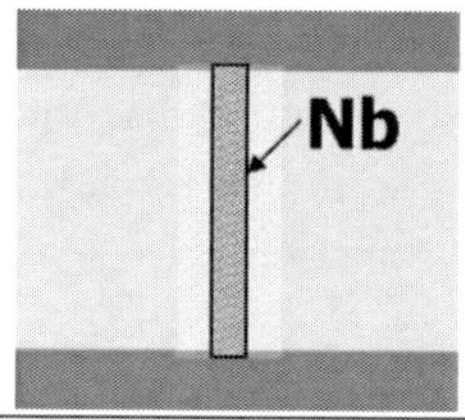

Thus the structure is switched to LRS. The total resistance of a ReRAM cell is determined by a resistance of a metallic filament.

II. The reset process (switching from LRS to HRS)

1) During the reset process a current which flows through the filament becomes a source of electron wind.	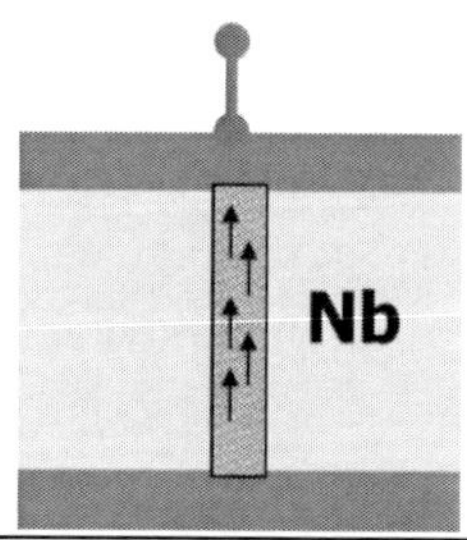
2) Electromigration of metal ions occurs under the action of the electron wind. The migration leads to the rupture of filament in a region close to the cathode. A local domain with a high resistance and electric field is formed. 3) On the border of rupture a part of filament is converted to the oxide due to thermal oxidation under the action of a high electric field.	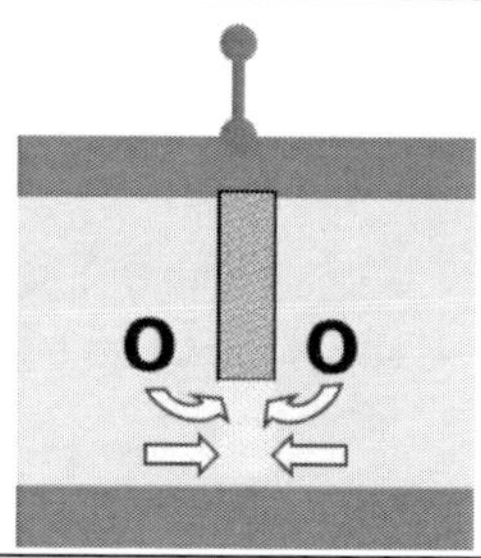

Thus the structure is switched to HRS. The total resistance of a ReRAM cell is determined by a new (reconstructed) section of oxide layer which is created by a rupture of a filament. Note that the resistance of the structure in HRS a significantly smaller than that in an initial state before forming.

III. The set process (switching from HRS to LRS)

1) A dielectric breakdown of a reconstructed section of oxide layer with a required current compliance. 2) A discharge of a capacitor. Release of energy ReRAM as a capacitor structure stored before set process.	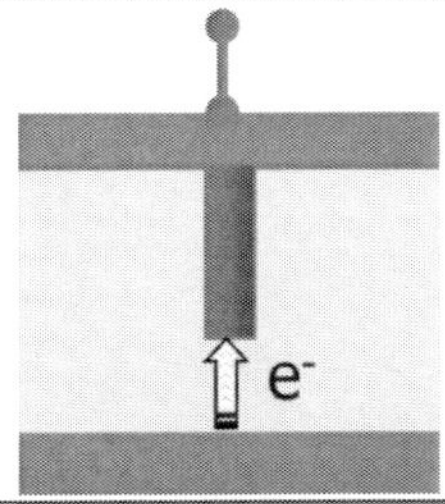
3) Sharp increasing of temperature and, as a result, fast local redox reaction.	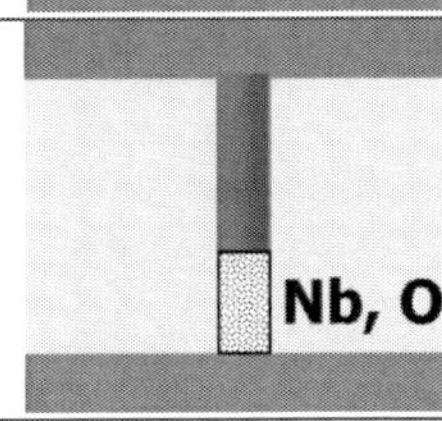
4) Under a gradient of temperature and diffusion process a Soret state is established. Segregation of metal in the center of high temperature region occurs.	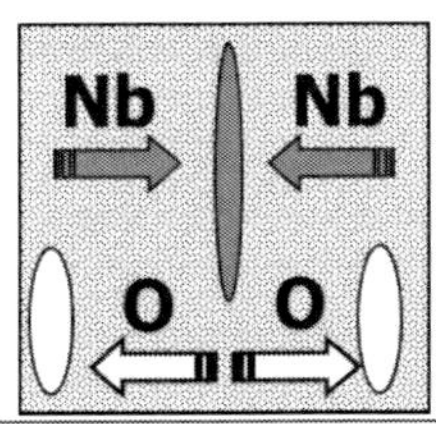
5) Due to a sharp decrease of the temperature after finishing of the set process, the metallic filament is solidified.	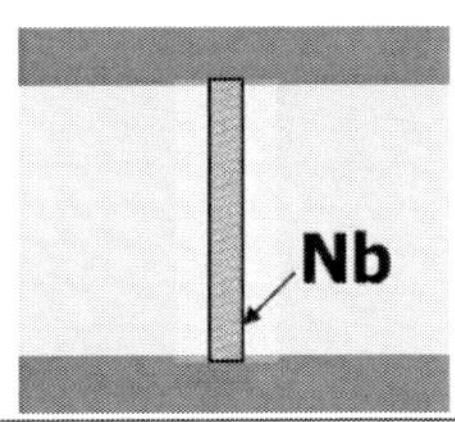

Thus a switching from HRS to LRS is reminiscent the forming process, but it occurs in a smaller volume of oxide structure.

The experimental results show that, first, any polarity of electrical bias of the initial oxide structure with semiconductor type of conductivity induces the growth of the thin filament (or filaments) with metal conductivity inside the oxide matrix (forming).

Secondly, any polarity of electrical bias is able to rupture the metal filament which returns the semiconductor properties to the structure. Further, a transition between HRS and LRS can be repeated many times. The *I-V* curve of the initial oxide structure (Figure 22) during first polarization measured at current controlled regime shows that forming can be classified as irreversible threshold switching with unstable region of the current controlled NDR. These features of the forming allow considering it as a hard breakdown of the insulator oxide. Note that the subsequent LRS-HRS transition can be observed only if adequate compliance current I_C is applied (Figure 23). The electrical bias with high magnitude of the current compliance transfers the structure into LRS which can not be ruptured by the next voltage input. The lowest level of the compliance current is defined by the value of the breakdown threshold current.

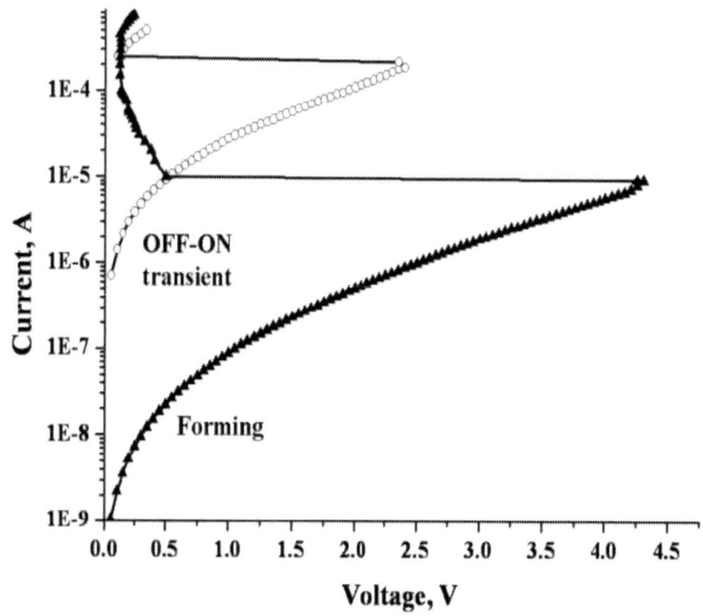

Figure 22. Current-voltage characteristic for forming and OFF-ON transient in current controlled regime of the measurement. [26].

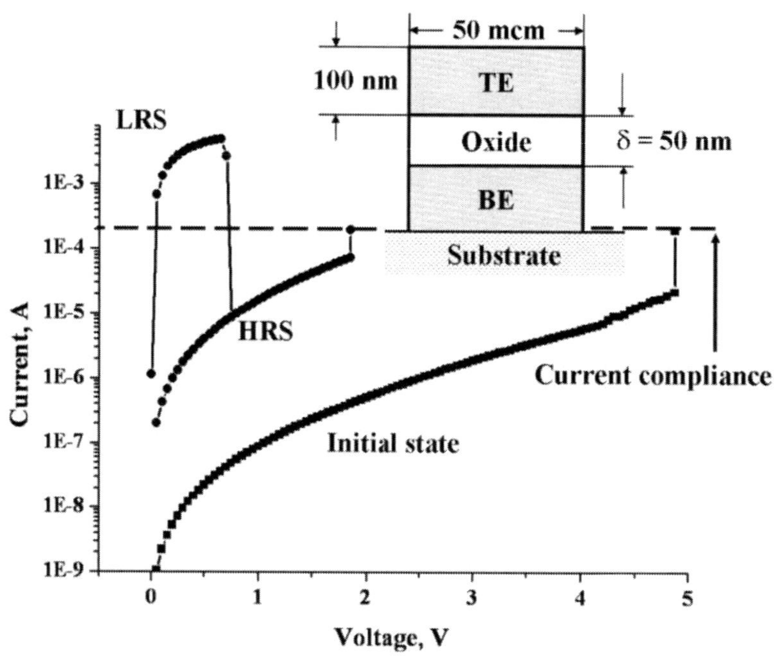

Figure 23. Typical current-voltage characteristic for Pt-NiO-Pt structure with nonvolatile unipolar switching in voltage controlled regime of the measurement. TE and BE are the Pt top and bottom electrodes. [26].

There is no a universal mechanism of thin films breakdown, but all the researchers indicate the two process stages. At the first stage a sudden reduction of the insulator resistance driven by electronic or electrothermal positive feedback mechanism occurs. The NDR region appears in the *I-V* curve, and a narrow conductive channel is formed between electrodes.

During the second stage of the breakdown the permanent conductive filament, whose structure and chemical composition differs from the initial oxide, is formed inside the insulator [15]. Taking into account this universal phenomenological behavior of the thin film insulators, the first breakdown stage is not so important for the presented model. When any electronic or electrothermal instability is initiated and, as a result, the conductive channel is formed, the temperature increases due to the Joule heating of this local region which could result in a local thermochemical modification of the oxide.

There are two approaches to the estimation of the energy dissipation region size. The universal thermodynamic consideration [26] shows that in system with initial uniform current distribution the trend of the current to collect in local domain is governed by the principal of least entropy production. Using approximation which have been developed in [27], the radius a of cross-sectional area of the cylinder conductive channel, which have been formed after first breakdown stage, was calculated to be $a = 5$ nm.

Another approach is based on a strong nonuniform distribution of the current in prebreakdown state of the defect insulator [28, 29]. The statistical model of the electric field enhancement by local geometric thinning of the oxide thickness assumes that the size of the high conductive path after first stage of the breakdown is the same as interface irregularities size – 5 nm [28]. Note also that the other high conductive defect in polycrystalline NiO is the grain boundaries which size have been measured as 5-10 nm [28, 30]. Therefore, the 5 nm as the dimension scale for a is a reasonable estimation.

It is obvious that there are two energy sources for Joule heating of the conductive domain. At first, it is necessary to take into account the action of the direct current trough conductive channel. The density of the dissipated power can be calculated as $P_{DC} = I_C \times V_C / v$, where I_c is the current compliance, V_c is the voltage which corresponds to I_c, and v – the volume of the conductive domain. Obviously that $V_c \leq V_F$, where V_F is forming voltage, because, right after the first breakdown stage, the structure has *I-V* characteristic with current controlled NDR. Accepting a high current domain as cylindrical body with basis radius $a = 5$ nm and oxide thickness as height $\delta = 50$ nm we obtain $P_{DC} = 5\times10^{13}$ W×cm^{-3}.

Before breakdown, the structure is the capacitor with capacitance of C which is charged up to the voltage of V_F. At the second breakdown stage, this energy is liberated by electrical discharge through conductive channel. The storage energy can be written as $E_C = [C \times (V_F - V_C)^2]/2$ and it is equal to 10^{-13} J for the analyzed sample. The capacitance discharge power density P_C changes during energy liberation process but we assume that capacitance discharges occurs with constant rate at characteristic time $\tau_0 = C \times V_C / I_C$ $=10^{-9}$s, and under such an assumption the power density $P_C = E_C /(v \times \tau_0) \approx 10^{16}$ W×cm^{-3}, therefore $P_C >> P_{DC}$. Note that $\tau \leq 10^{-9}$s and it is typical transient time of second breakdown stage for many thin insulator films [15, 29].

For estimations of the temperature space-time distributions we will assume that the heat production is confined to a conductive cylinder with height δ and radius *a*. The temperature will be determined by oxide thermal conduction in both axial and radial direction, and spreading thermal resistance in electrodes. When *a* is enough small as compared to other dimensions of the structure (thickness of oxide, thickness and size of metal electrodes) we can assume that the temperatures of electrodes and oxide matrix volume are equal ambient. Also, for thin conductive path, the cylinder lateral surface is much greater than the basis surface and we can assume that the radial heat flow will dominate. The thermal time constant of the system τ_T can be estimated as $\tau_T = \delta^2 / A$ [31], where $A = N_{NiO} /(c_{NiO}\gamma_{NiO})$ is NiO thermal diffusivity, where K_{NiO} is thermal conductivity, c_{NiO} specific heat, γ_{NiO} density. The time constant τ_T is less then 10^{-10}s and we will be allowed to assume that the steady state conditions are reached during capacitance discharge. The steady-state solution of the heat equation in cylindrical coordinates with heating by a line heat source of a strength $Q_C = P_C / \delta$ along cylinder axis and with $T = 0$ at $r = d$, and $dT/dr = 0$ at $r = 0$ is [31]

$$\Delta T_C = \frac{Q_C}{2\pi K_{NiO}} \ln\left(\frac{d}{r}\right) \tag{1}$$

In practice *d* will not be the sample or electrode size, and in order to obtain a realistic estimation we therefore replace *d* with the oxide thickness δ. Taking nickel oxide thermal conductivity K_{NiO} = 0.71 W/(cm^{0}C), we obtain the rise of temperature on the conductive channel boundary ($r = a$) $\Delta T_C \approx 4000^0$C that is more than the oxide melting point, T_{mNiO} =1990^{0}C.

At last stage of the forming process, when capacitance discharge will be finished, the temperature of the hot region in oxide matrix for the time $\tau_T \le 10^{-10}$ should drop down to values which would be defined by heat production due to current compliance. For estimation of this temperature we can use Equation (1) with the same problem geometry but replacing P_{DC} with P_C, which yields $\Delta T_{DC} \le 100^0$C.

Several processes may develop during high temperature stage of the forming but their importance will be defined by a parity of their time scales to capacitance discharge time. Evidently we should consider melting, thermoreduction of oxide, reoxidation, diffusion and solidification of the component of the reduction-oxidation reaction. Part of these processes will going in parallel and interdependently but their parity can be defined by separate consideration of temporary evolution of each process.

The melting time, t_m, can be evaluated from quasistationary approximation of the Stefan problem of cylindrical body melting due to a line heat source of strength Q_C at $r = 0$. The appropriate solution is given by the equation [32]: $t_m = \pi a^2 \gamma_{NiO} L_{fNiO} / Q_C$ where L_{fNiO} is the oxide latent heat of fusion. Taking L_f = 0.78 kJ/g, tm = 10^{-13}s. We have thus arrive to a conclusion that, during high temperature forming stage, the NiO conductive domain and some region around it should be transformed to melting state.

Second process which should be considered at high temperature forming stage is oxide reduction. Extensive studies of the NiO reduction have appeared in literature and the important result in frames of our consideration is that the reduction of NiO is irreversible, since the equilibrium constant K_{eq} of the reduction reaction reaches 10^3 in high temperature limit [33]. Note that oxide reduction due to direct thermal decomposition is reaction-limited process and we can neglect diffusion of the reaction products for estimation of the reduction time scale. Consider NiO reduction as first-order reaction with respect to Ni we can write solution of the reaction kinetic equation as: $C_{Ni} / C_{NiO} = [1 - \exp(-kt)]$, where C_{Ni} is the Ni concentration, C_{NiO} is the initial NiO concentration, and $k = k_0 \exp(-E_{rmol} / RT)$, where k is the reduction reaction rate constant, E_{rmol} is the molar activation energy and R is the gas constant. Using experimental values: E_{rmol} = 90 kJ/mol and k_0 = 6×10^{13} s^{-1} [34] we can estimate the characteristic time constant of the NiO reduction as $\tau_R = 1/k < 10^{-11}$s. We have to conclude that NiO melting region and nearest solid state region with sufficiently high temperature must be converted to mixture of the Ni and O atoms during capacitance discharge regime.

The presence of the strong temperature gradients can result in temperature gradient-driven diffusion (thermomigration) [35]. Thermomigration in solid is small and therefore it can be usually neglected as compared to concentration diffusion. In a heat flow transient induced by electrical discharge, however, temperature gradient is of the order of 10^8 ^{0}C/cm and thermal diffusion contribution cannot be excluded, especially in the melt state of the oxide. If a homogeneous binary compound is placed in a temperature gradient, a redistribution of the constituents can occur, and one constituent migrates to the cold end of the specimen and other – to the hot end. This phenomenon is called the Soret effect [36]. The direction of the migration and values of the mass flows are defined by the transport heat *f* of the diffusing ions Q*. The values of Q* for Ni and O thermomigration in NiO are unknown. However, we can use the approaches which were developed for liquid conductive compounds [37]. Indeed, in this theory assuming that the liquid is a dense gas and applying the thermo-transport theory in binary gas mixtures, the direction of the diffusion is determined primarily by the mass differences: the lighter component migrates to the warmer end and the heavy component to the cold one. Taking this fact into account, we can assume that Ni ions migrate towards the hot region, whereas oxygen ions diffuse to periphery of the melt region. As a consequence, a temperature gradient drives the establishment of concentration gradients. In the stationary state this concentration gradient depends on the boundary conditions. As melt region are closed for the exchange of oxygen with the surrounding gas phase, the process ends up with zero atom fluxes, defining the so-called Soret state with Ni rich region in the center of the melt. The data given in Figure 24 confirm an opportunity of an establishment of the Soret state at high temperature stage of the forming. The presented results are the SIMS images of the O and Ni distribution near NiO-Pt interfaces for initial oxide structure and after forming. We can see that only O diffuses away from local nonhomogeneous regions of the NiO during forming. Assuming that these local regions have highest conductivity and, as consequence, high temperature due to Joule heating, the atoms redistribution can be defined by thermomigration and Soret state establishment.

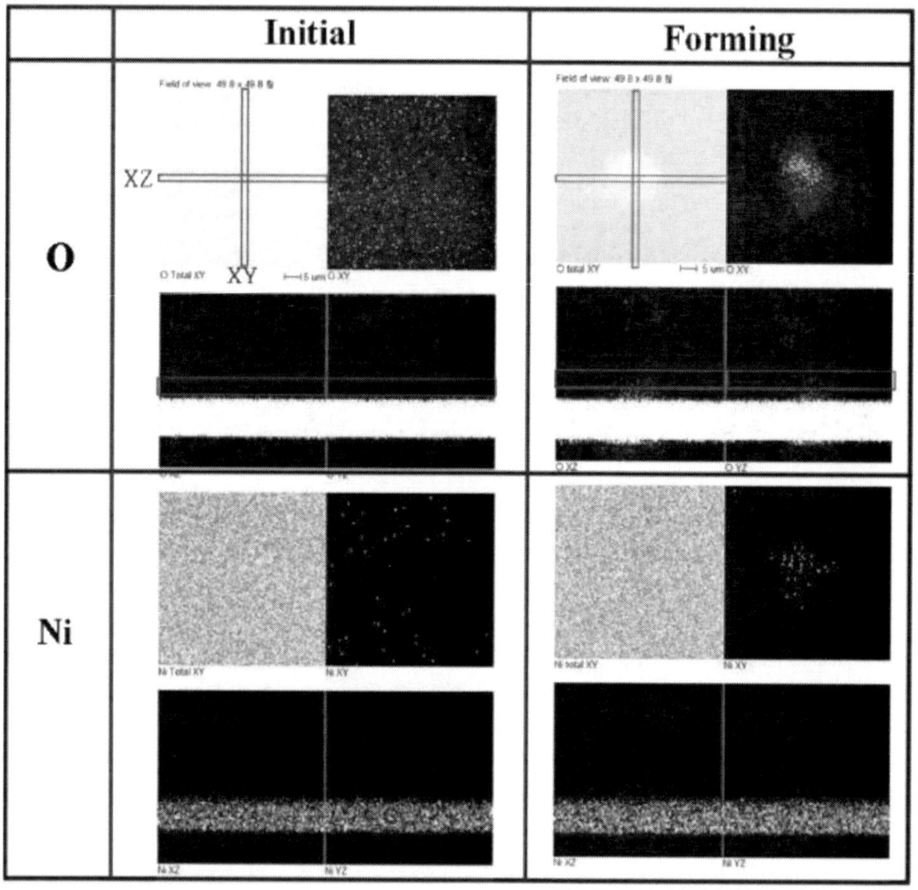

Figure 24. SIMS images of the Ni and O distributions near NiO-Pt interfaces in initial state and after forming [26].

The characteristic time, τ_D of the concentration gradient-driven diffusion can be written as [35] $\tau_D = l_D^2 / D$ where D is diffusion coefficient and l_D is the characteristic distance scale. Accepting for Ni diffusion in melt NiO $D = 10^{-8}$ cm^2/s, and with $l_D = a = 5$ nm, $\tau_D > 10^{-5}$ s, and we can conclude that Fickian Ni diffusion and especially O diffusion, as a slower process, is not crucial for forming.

At the last forming stage, when liberation of the capacitance energy will be finished, the temperature drops down to above estimated low values due to the thermal conductivity, and the solidification of the melting region should occur in time t_s. This time can be estimated from time dependence of the solidification front position $R(t)$. In our case temperature difference between solid and liquid phases near the interface is not so big and we can assume that liquid has melting temperature and temperature profile in the solid is linear. The solution of the appropriate Stefan problem can be written as [31, 32]: $t_S = a^2 \gamma_{Ni} L_{fNi} / 2kNiT_m$. The value of t_s is less than 10^{-11}s and fast solidification should quench the Ni filament inside the oxide matrix.

The low value of the diffusion coefficient for Ni diffusion in NiO and electrode materials (during the final low-temperature stage of electroforming [38]) allows assuming that the influence of Ni diffusion on final Ni filament size is negligible. The oxidation process at the Ni- NiO interface could also be rather slight, because, for this reaction at low temperatures, the oxidation rate is limited by slow oxygen diffusion transport toward the NiO-Ni interface [34].

The strict solution of the problem of the Ni filament size, R_f, is based on considering of the energy conservation equation, but the simple estimations show that heating and heat transfer terms are much less in comparison with melting and chemical reaction terms. Assuming that the volume of the melt is $v = \pi R_f^2 \delta$ and that intensive thermal reduction is going only in melt region, we can write more simple integral energy conservation equation for steady-state regime

$$E_C = Q_{melting} + Q_{reduction} = \gamma_{NiO} v L_{fNiO} + \gamma_{NiO} v \frac{R_{Rmol}}{M_{mol}} \tag{2}$$

where M_{mol} is NiO molar mass. The solution of Eq.(2) is

$$R_f = \sqrt{\frac{E_C}{\pi \delta \gamma_{NiO} (L_{fNiO} + \frac{E_{Rmol}}{M_{mol}})}}, \tag{3}$$

which yields $R_f \sim 7$ nm.

Thus, we can conclude that the Ni melt filament with radius *Rf* is formed inside NiO during energy liberation stage. After discharge the fast solidification of the Ni melt should occur which provides a stable metallic LRS of the oxide structure after the voltage is turned off.

For check of model it is interesting to calculate the dimension of the metallic filament which would have the known resistance (from Figure1 $R_{ON} \approx 50\ \Omega$ for Ohmic section of the *I-V* curve). The temperature dependence of the LRS resistance [38] shows that Mathiesen's rule is valid and the main metal resistivity component, ρ_{Ni}, is defined by intrinsic Ni properties for temperatures more then 50 K. Writing the total resistance as $R = \rho_{Ni} \delta / \pi R_f^2$ and taking $\rho_{Ni} = 6.9 \cdot 10^{-6}$ Ohm·cm, one obtains $R_f \sim 5$ nm, and we can thus arrive at a conclusion that the experimentally obtained filament size and the model estimation coincide satisfactory.

A similar model has been proposed recently in the work [39] where it has been shown that the transition from insulating to metallic conductivity in NiO first results from purely electronic threshold switching, which then causes the formation of a conducting filament by the local high current and high temperature conditions. A set transition time below 1 ns has been evidenced, and the impact of parasitic capacitance has been confirmed by numerical simulations of threshold switching and Joule heating [39]. Also, the TiO2 based sandwich structure studied in [40] has demonstrated behavior resembling the above described processes, i.e. electroreduction and drift process triggered by high electric fields and enhanced by Joule heating [40]. Also, in this work, the results are reported revealing stable rectification and resistive-switching properties of a Ti/TiO2/Pt structure. The oxygen migration and localized conductive filaments play important roles not only in the resistive switching of ReRAM, but also in the process of the rectification of oxide diodes. The rectification properties stable up to 125°C and 10^3 cycles under about 3 V sweep without interference with resistive-switching. This shows a satisfactory reliability of TiO_2 MIM diodes for future 1D1R (one diode – one resistor) ReRAM applications [40].

Now consider briefly reverse transient in which the structure is passed from LRS to HRS with semiconductor conductivity but whose resistance is less on few orders in comparison with initial (before forming) state. Before this transition *I-V* characteristic of LRS follows the Ohmic law that shows the absence any barriers on interfaces or in bulk oxide. Logical base for development of LRS-HRS transient model will be the assumption of the filament rupture by the enough high current passing through structure. In the assumption that all current passes through the metal filament estimation of the current density gives value 10^9 A/cm^2. Such high value of the current density will cause significant filament heating and all other possible processes will be modified by this high temperature. Let's notice, that the time constant for temperature growth remains the same as at a forming stage and has value less then 10^{-10}s that means we can use the steady-state approximation. Using the same geometry and the same arguments for a choice of the basic direction of heat transfer that was applied for calculation of temperature at last forming stage we can use Eq.(1) for temperature estimations. The calculation is shown that filament temperature before switching in HRS $T_f \approx 400$°C and we can conclude that filament has enough high temperatures but Ni melting temperature are not achieved.

The several processes may rupture metal filament but their importance can be checked by a parity of their time scales to experimental time of the transient, which for DC biasing has view microseconds.

One of the probable processes which can break off the metal filament and return structure to HRS with semiconductor conductivity is high temperature oxidation. In situation when Ni filament is surrounded with a thick layer of oxide we can applied the Wagner model of the thermal oxidation [41]. Validity of this approach is proved by absence of the direct atmosphere - Ni interface and absence of the strong electric fields in normal to NiO-Ni interface direction. We can write the Wagner parabolic kinetic equation as $\Delta X^2 = k_p t$, where ΔX is new oxide thickness and k_p is parabolic constant rate. Using the maximal value $k_p = 10^{-10}$ cm^2/s [41] the time for oxidation of the $5 \cdot 10^{-7}$cm Ni specimen (half of the filament diameter) is $2.5 \cdot 10^{-3}$ s. We can conclude that direct oxidation of Ni filament is an important process in transition from metallic to semiconductor state but it does not determine threshold conditions of the ON-OFF switching.

The instability induced by concentration or thermal gradients-driven radial diffusion is ruled out because filament temperature is low as was being shown above.

Summarized presented arguments we can conclude that any radial diffusion mass flux can not be driving force for ON-OFF instability. On the other hand, well known that the most serious and persistent reliability problem in interconnect metallization in VLSI (Very-large-scale integration) circuits is metal atoms electromigration. The typical current density in interconnect lines of this devices achieves values 10^6 A/cm^2. Such current density can cause directional mass transport in the line at the device operation temperature of 100 °C and lead to void formation at the cathode and extrusion at the anode. During ON-OFF switching in NiO with Ni filament size R_f = 10^{-6} cm the current density is about 10^9 A/cm^2 and electromigration may have determining significance in filament rapture process.

There is a high temperature domain inside the Ni filament. In ideal situation this domain should be located in the center of the filament but really its arrangement will be adhered to filament site with the highest resistance (interfaces, geometrical constriction, compositional disordering). Taking into attention the low filament size in comparison with high temperature distribution scale we can assume that filament and the electrodes area adjoining to them has identical temperature $T = T_m$. In this case we can neglect termomigration process owing to small values as termomigration flux and flux divergences together.

In face-centered-cubic metals, such as Ni, atomic diffusion is mediated by vacancies. A flux of Ni atoms driven by electromigration to the anode requires a flux of vacancies in the opposite direction. The diffusion coefficient for Ni self diffusion and Ni diffusion in Pt is match greater then diffusion coefficient for Pt diffusion in Ni and we can neglect Pt diffusion in Ni filament [41]. In this case the vacancy flux will be stopped on cathode interface because there is not the counter atoms flux trough this boundary and vacancy will supply continuously on cathode interface. This conclusion have an experimental support, in [34] shown that destruction of part of the filament is localized in cathode near area.

Now we can assume that if in some part of the filament concentration will fall below this limit there will be a local transition in an insulator state. Also we should note that dimension of this region in current flow direction should be enough high for tunnel non-transparent behavior. Note that if a thin layer with non-conductive characteristics is formed, this leads to the appearance of a high electric field domain in the filament structure, and this part of Ni filament will convert to NiO under the action of thermal oxidation accelerated by electric field.

The reverse process of the Ni filament interruption, and the transition of the structure from LRS back to HRS, is rather more complicated for calculations, and we didn't develop here all these calculations and estimates. At a conclusion it should be noted, however, that all the above described phenomena – electro- and thermo-diffusion, the Soret effect, electronic wind and so on – play an important role, although, all of them are different in their intensity and, thereby, in relative contribution to the mechanism of the LRS – HRS transition. In other words, several processes are involved during this LRS – HRS transition, but their importance (in order to interrupt the Ni-metal filament) will be defined by a parity of their time scales to capacitance discharge time. Evidently, we should consider melting, thermoreduction of oxide, reoxidation, diffusion and solidification of the components of the reduction-oxidation reaction. Part of these processes will be going in parallel and interdependently, but their parity can be defined by separate consideration of temporary evolution of each process. In more detail, all these effects have been considered earlier in the work [42].

The final problem to be discussed is the OFF-ON transition. The phenomenological pictures of the forming and HRS-LRS transition coincide, that allows assumption of the generality of the mechanisms of the two phenomena. Similarly to forming, the OFF-ON transition can be classified as hard breakdown of the insulator NiO layer which was formed near cathode interface during ON-OFF switching. Also we can assume that initial breakdown mechanism is not so important for restoration of the Ni filament and the main processes should be developed the on second stage of breakdown.

5. Flexible ReRAM Structures

A resistive memory devices can be used not only in conventional solid state electronics but also they have advantages in the new developing sectors of electronics: transparent electronics, flexible electronics etc. ReRAM elements with a high ductility will be demonstrated in this section.

Using of thin polymer films and other flexible materials in electronics not only provide a mechanical flexibility of new electronic devices but also reduce the cost significantly. This factor is very important for mass production. Nowadays devices of flexible electronics find increasing application in various fields such as flexible displays, radio frequency identification tags (RFID), electronic paper, solar cells and other devices. Although more wide application of flexible electronics, especially in case of more complex and multifunctional devices, predicted in the future, prototypes of the key electronics elements already exist in science papers: flexible transistor [43-44], diode [45], battery [46] and memory elements [47].

Conventional semiconductor materials are not suitable for flexible memory application because of theirs fragility. Organic and metal oxide compounds are considered as a replacement for silicon. Nevertheless, operation instability and relatively low carrier mobility still delay the development of organic semiconductors devices [48]. Metal oxide dielectrics, in turn, find a successful application in ReRAM devices [49], thereby such oxide becomes suitable for a new flexible ReRAM.

Low temperature fabrication process is a critical condition for flexible electronics devices. The different low temperature methods of deposition can be used: sol-gel method [50], anodic oxidation [17], magnetron sputtering [51].

We have used the method of anodic oxidation to obtain the Kapton/Nb/Nb_2O_5 (Kapton is a polyimide film produced by DuPont) structure at a room temperature, without any destruction of the polymer substrate (Figure 25).

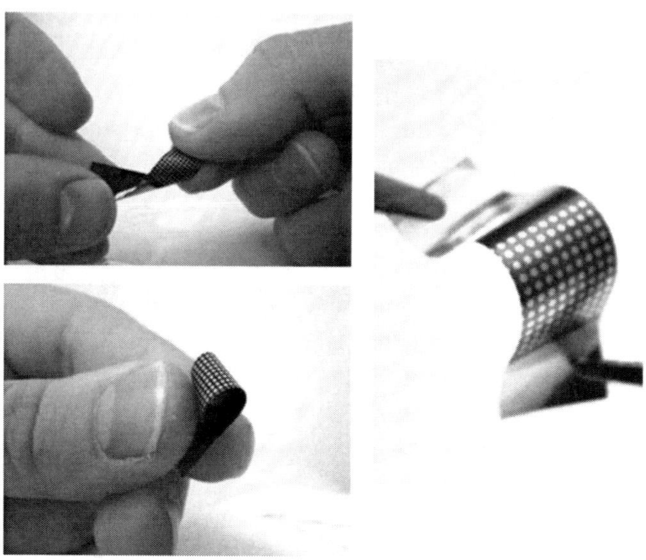

Figure 25. Photographs of flexible ReRAM.

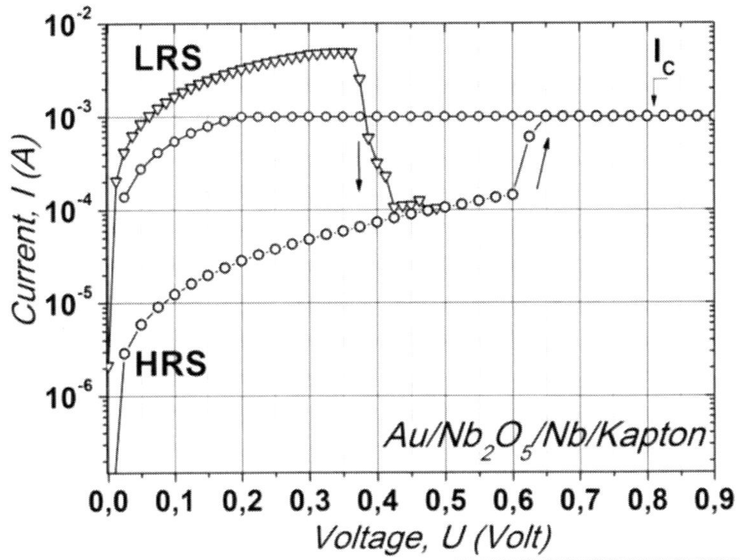

Figure 26. Resistance switching *I-V* characteristics of the flexible Au/Nb_2O_5(75 nm)/Nb/Kapton.

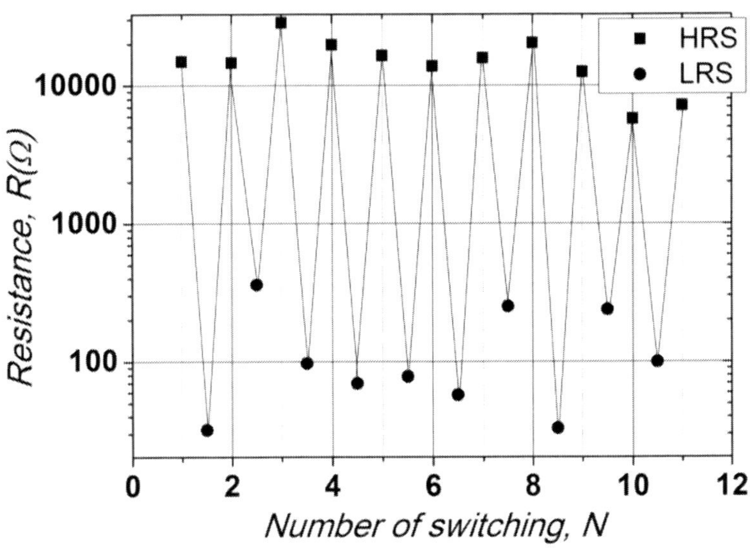

Figure 27. Consecutive switching of flexible Kapton/Nb/Nb_2O_5/Au ReRAM structure.

The fabrication includes the following steps:

1) Polymer metallization. Thin metallic film of Nb is sputtered on kapton substrate by RF magnetron sputtering using a metallic Nb target in an Ar atmosphere.
2) Anodic oxidation of Nb metallic layer. Anodization is performed at room temperature under galvanostatic condition with constant current density of about 1 mA/cm^2 in 0.1 N aqueous solution of H_3PO_4 acid. The thickness of the obtained oxide film is ~75 nm.
3) Deposition of Au top electrodes by thermal vacuum evaporation.

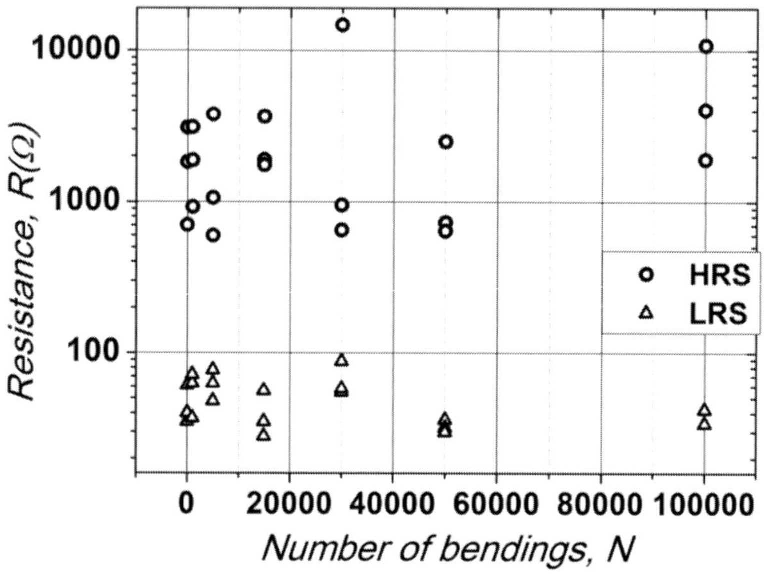

Figure 28. The switching characteristics with continuous bending of the Kapton/Nb/Nb_2O_5/Au structure.

Current-voltage characteristic of a Kapton/Nb/Nb_2O_5/Au structure is a typical characteristics of ReRAM devices that produces unipolar resistive switching between HRS (high resistance state) and LRS (low resistance state) (Figure 26) with set/reset voltage ~ 0.9 V/0.4 V and resistance ration $R_{HRS}/R_{LRS} > 100$. (Figure 27).

In order to confirm the feasibility of obtained ReRAM devices for flexible memory application, mechanical bending tests have been carried out. After several flexing (1000, 5000... and so on, up to 100,000) a low voltage signal of $V = 0–0.1$ V is applied and the *I-V* characteristics, HRS or LRS, of different structures are measured. Calculated from these characteristics resistance of LRS and HRS does not degrade (conserves within an order of magnitude) after numerous bending (Figure 28).

Thus, the memory cells obtained on flexible substrates do not differ from the same cells on solid silicon substrates, as far as its switching characteristics concerns. Since this area of electronics is only at the beginning stage of development, an intensive work is carried out to find the most appropriate materials and technologies which will allow obtaining commercially successful flexible electronic memory devices.

ACKNOWLEDGMENTS

This work was supported by the Strategic Development Program of Petrozavodsk State University (2012 – 2016) and the RF Ministry of Education and Science as a base part of state program № 2014/154 in the scientific field, project no. 1704. The authors also thank A.M. Grishin and S.I. Khartsev (Dept. Condensed Matter Physics, Royal Institute of Technology, Sweden) for discussions and experimental aid and A.K. Vlasova for her assistance in the figure design.

REFERENCES

[1] Gibbons, J. F.; Beadle, W. E. *Solid-State Elect.* 1964, *7*, 785-790.

[2] Fugiwara, K.; Nemoto, T.; Rozenberg, M. J.; Nakamura, Y.; Takagi, H. J. Jpn. *J. Appl. Phys.* 2008, *47*, 8, 6266-6271.

[3] Chang, Y. W.; Lai, Y. C.; Wu, T. B.; Wang, S. F.; Chen, F.; Tsai, M. J. *Appl. Phys. Lett.* 2008, *92*, 022110-1-022110-3.

[4] Kim, S.; Choi Y. K. *IEEE Trans. Electron Devices*, 2009, *56*, 12, 3049-3054.

[5] Kundozerova, T. V.; Grishin, A. M.; Stefanovich, G. B.; Velichko, A. A. *IEEE Trans. Electron Devices*. 2012,*59*, 4, 1144 – 1148.

[6] Kundozerova, T. V.; Stefanovich, G. B.; Grishin, A. M. *Phys. Status Solidi C.* 2012, *9*, 7, 1699-1701.

[7] Park, I. S.; Kim, K. R.; Lee, S.; Ahm, *J. Jpn. J. Appl. Phys,* 2007, *46*, 2172 – 2174.

[8] Wong, H. S.; Lee, H. Y.; Yu, S.; Chen, Y. S.; Wu, Y.; Chen, P. S.; Lee, B.; Chen, F. T.; Tsai, M. J. *Proceedings of the IEEE*, 2012, *100*, 6, 1951 – 1970.

[9] Simmons, J. G.; Verderbert. R. R. *Proc. Roy. Soc. A.* 1967, *301*, 77 - 102.

[10] Zhuang, W. W.; Camas, W. A.; Pan, W.; Ulrich, B. D. *Electron Devices Meeting. 2002. IEDM '02. International.* 2002, 193-196.

[11] Sawa, A. *Materials Today*. 2008, *11*, 28-36.

[12] Waser, R. *Microelectronic Engineering*. 2009, *86*, 1925 – 1928.

[13] Pringle, J. P. *Electrochim. Acta.* 1980, *25*, 11, 1423–1437.

[14] Jung, K.; Kim, Y.; Hyunsik, W. J.; Baeho, I.; Hong, P. G.; Lee, J.; Park, J.; Lee, J. K. *Appl. Phys. Lett.* 2010, *97*, 233509-1 233509-3.

[15] Klein N. *In Advances in Electronics and Electron Physics.* 1969, *26*, 309-424.

[16] Lee, H. D.; Magyari-Kope, B.; Nishi, N. *Phys. Rev. B.* 2010, *81*, 19, 193202-1932206.

[17] Kundozerova, T.; Stefanovich, G. *Applied Mechanics and Materials*. 2013, *346*, 29-34.

[18] Lee, M. J.; Han, S. H.; Jeon, B. H.; Park, B. S.; Kang, S. E.; Ahn, K. H.; Kim, C. B.; Lee, C. J.; Kim, I. K.; Yoo, D. H.; Seo, X. S.; Li, J. B.; Park, J. H.; Lee, Y. *Nano Lett.* 2009, *9*, 1476–1481.

[19] Govoreanu, B.; Kar, G. S.; Chen, Y. Y.; Paraschiv, V.; Kubicek, S.; Fantini, A.; Radu, I. P.; Goux, L.; Clima, S.; Degraeve, R.; Jossart, N.; Richard, O.; Vandeweyer, T.; Seo, K.; Hendrickx, P.; Pourtois, G.; Bender, H.; Altimime, L.; Wouters, D.G.; Kittl, J.A.; Jurczak, M. *Electron Devices Meeting (IEDM). IEEE International.* 2011, 31.6.1–31.6.4.

[20] Yang Y. C.; Pan, F.; Liu, Q. *Nano Lett.* 2009, *9(4)*,1636−1643.

[21] Stoinov Z. B., Grafov B. M. *Electrochemical Impedance*; Nauka; Moscow, Russia, 1991; p.135.

[22] Dearnaley, G.; Stoneham, A. M.; Morgan, D. V.; *Rep. Prog. Phys.* 1970, *33*, 1129-1191.

[23] Fuschillo, N.; Lalevic. B.; Leung, B. *Solid-State Elect*, 1976, *19*, 209-219.

[24] Baek, I. G.; Lee, M. S.; Seo, S. et al, *Tech. Dig.- Int. Electron Devices Meet.* 2004, 587-590.

[25] Seo, S.; Lee, M. J.; Seo, D. H.; Jeoung, E. J.; Suh, D. S.; Joung, Y. S.; Yoo, I. K.; Hwang, I. R.; Kim, S. H.; Byun, I. S.; Kim, J. S.; Choi, J. S.; Park, B. H. *Appl. Phys. Lett.* 2004, *85*, 5655-5657.

[26] Pergament, A.; Stefanovich, G.; Velichko, A.; Putrolainen, V.; Kundozerova, T.; Stefanovich, T. *Journal of Characterization and Development of Novel Materials*. 2012, *4*, 2, 83 – 110.
[27] Ridley, B. K.; *Proc. Phys. Soc.*1963, *82*, 954-966.
[28] Chen, H. L.; Lu, Y. M.; Hwang, W. S. *Surface and Coatings Technology*. 2005, *198*, 138-142.
[29] Ridley, B. K. *J. Appl. Phys*. 1975, *46*, 998-1004.
[30] Sato, H.; Minami, T.; Takata, S.; Yamada T. *Thin Solid Films*. 1993, *236*, 27-31.
[31] Carslaw, H. S.; Jaeger, J. C. *Conduction of Heat in Solid;* Oxford, U.P.: London, 1959; p.517.
[32] Alexiades, V.; Solomon, A. D. *Mathematical Modeling of Melting and Freezing Processes*, Hemisphere Publishing Corporation: Washington; 1993; p.321.
[33] Atkinson, A. *Rev. Mod. Phys*. 1989, *57*, 437-451.
[34] Crank, J. *The mathematics of diffusion*; Clarendon Press: Oxford; 1975; p.411.
[35] Allnatt, A. B.; Lidiard, A. B. *Atomic Transport in Solids*, Cambridge University Press: Cambridge; 1993; p.572.
[36] Bhat B. N.; Swalin, R. A. *Acta Metall.* 1972, *20*, 1387-1391.
[37] Shim, M. T.; Moore, W. J. *J. Chem. Phys.*, 1957, *26*, 802-812.
[38] Kim, M. G.; Kim, S. M.; Choi, E. J. et al. *Jap. J. Appl. Phys*. 2005, *44*, L1301-L1303.
[39] Ielmini, D.; Cagli, C.; Nardi F. *Appl. Phys. Lett*. 2009, *94*, 063511-1 – 063511-3.
[40] Huang, J.-J.; Kuo, C.-W.; Chang, W.-C.; Houa T.-H. *Appl. Phys. Lett.*, 2010, *96*, 26, 262901-1 – 262901-3.
[41] Shatzkes, M.; Lloyd, J. R.; *J. Appl. Phys*. 1986, *59*, 3890-3893.
[42] Stefanovich, G. B.; Lee, M. J.; Kang, B. S.; Ahn, S.-E.; Kim, K. H.; Lee, C. B.; Kim, C. J.; Park Y. S. (2011) *Formation and Rupture of the Nanosized Metal Filament inside Oxide Matrix*. http://arxiv.org:80/abs/1102.3840.
[43] Georgiou, T.; Jalil, R.; Belle B. D. *Nature Nanotechnology*. 2012, *8*, 100-103.
[44] Kuribara, K.; Wang, H.; Uchiyama, N. *Nature Communications*. 2011, *3*, 723, 1-15.
[45] Huang, J. J.; Hou, T. H.; Hsu, C. W.; Tseng, Y. M.; Chang, W. H.; Jang, W. Y.; Lin, C. H. *Jpn. J. Appl. Phys*. 2012, *51*, 04DD09-1 – 04DD09-5.
[46] Koo, M.; Park, K. L. J.; Lee, S. H. *Nano Lett. 2012, 12(9),* 4810–4816.
[47] Kim, S.; Choi, Y. K. *Applied physics letters*. 2008, *92*, 223408-1 – 223508-3.
[48] Kim, Y. H.; Heo, J. S.; Kim, T. H.; Park S.;Yoon, M. H. *Nature*. 2012, *489*, 128-160.
[49] Ha, S. D.; Ramanatha, S. *Journal of Applied Physics*, 2011, *110*, 071101-071101-20.
[50] Jung, S.; Kong, J.; Song, S. et. al. *Applied physics letters*. 2011, *99*, 142110-1 – 142110-3.
[51] Lee, S.; Kim, H.; Yun, D. J.; Rhee, S. W.; Yong, K. *Applied physics letters*. 2009, *95*, 262413-1 – 262113-3.

In: Oxide Electronics and Functional Properties ...
Editor: Alexander Pergament
ISBN: 978-1-63321-499-6

Chapter 2

SOME FUNDAMENTAL POINTS OF TECHNOLOGY OF LITHIUM NIOBATE AND LITHIUM TANTALATE SINGLE CRYSTALS

M. N. Palatnikov*[*] *and N. V. Sidorov
I. V. Tananaev Institute of Chemistry and Technology of Rare Elements and Mineral Raw Materials of Kola Science Centre of RAS, Apatity, Russia

ABSTRACT

In this chapter the results of investigations of single-crystal lithium niobate and tantalate ($LiNbO_3$ and $LiTaO_3$) are aggregated. The chapter describes peculiarities of batch preparation, of $LiNbO_3$ and $LiTaO_3$ crystals growth, their composition and property features as variable phases. It observes the influence of conditions upon output characteristics of the crystal, peculiarities of structural units of cation sublattice of doped crystals and their effect on optical properties. It describes application of laser conoscopy for investigating optical perfection of crystals. Here can be found the investigation of concentration dependencies of Curie temperature data of doped crystals. Raman spectra and concentration dependencies of Curie temperature of crystals of different composition were investigated very thoroughly. This work presents results of studies of stability of electrophysical and optical characteristics of nominally pure and doped crystals of lithium niobate in practically important range of temperatures (300-500 K). Considerable attention is given to photorefractive effect (optical damage) studying, determination of mosaic and radiation-defects in crystals of different composition. Following chapter suggests methods of studying of processes of creation of stable single-domain state of lithium tantalate and evaluation of single-domain uniformity of Raman spectra. Obtained experimental data can be used in development of technology of highly perfected crystals of lithium niobate and tantalate of different composition for optics and acoustoelectronics applications.

[*] palat_mn@chemy.kolasc.net.ru.

1. Structural Features and Some Properties of Lithium Niobate and Lithium Tantalate Crystals

Single crystals and ceramics based on niobium and tantalum oxides are widely used as insulating materials for acoustoelectronics, optoelectronics, communication and automation systems, and optical storage media. The most important of them are ferroelectric single crystals of lithium niobate $LiNbO_3$ (LN) and lithium tantalate $LiTaO_3$ (LT), possessing a fortunate combination of electrooptical, pyroelectric, piezoelectric, and nonlinear-optical characteristics. The large-scale application of these compounds and their attractiveness as test objects are due to these characteristics. Recently, the preparation of stoichiometric LN and LT single crystals of high structural perfection for various optical applications has become important [1-3].

In order to design optical-quality lithium niobate and lithium tantalate single crystals, the following is important: strictly standardized precursor preparation; well-developed schedules for feed synthesis, doping, crystal growth, and post-growth processing; and efficient quality control at each stage. This highlights the importance of fundamental research into the following fields: LN and LT crystals of various compositions and various extents of structural perfection, disordered crystalline phases based on niobium and tantalum compounds as test structures, and order-to-disorder transitions. These investigations are of great practical value; structural imperfection largely governs the quality of the physical parameters of the materials.

It is essential to note that the physical parameters of materials based on LN and LT single crystals, especially optical parameters, are largely controlled by defect formation in various sublattices both during feedstock preparation and during the growth and post-growth processing of single crystals. The primary feature of LN and LT single crystals is a loose cation sub-lattice; this allows the accommodation of (doping with) extremely different ions. For high-quality crystals, defects controlled by subtle features of cation order (governed by minor fluctuations in the matrix composition R = Li/Nb) or small dopant amounts are important for optical characteristics [1, 2]. Comparative studies of the fine structural features of various sublattices in nominally pure crystals (as a function of their chemical composition) and in crystals doped with cations whose ionic radii are close to the Li^+ or Nb^{5+} (Ta^{5+}) ionic radius are currently of great interest. Such dopants readily substitute for Li^+ and Nb^{5+} ions and are incorporated into vacant octahedral interstices, thus locally changing the extant order in cation arrangement along the polar axis. Even when the dopant concentration is on the level of a tenth or hundredth fraction of weight percent, a crystal can substantially change its dielectric and optical properties, e.g., its sensitivity to laser damage.

Roughly, lithium niobate and lithium tantalate are isomorphous. Fragments of the ideal crystal structure of lithium niobate are depicted in Figure 1. The structure is built of slightly distorted oxygen octahedra O_6 linked through shared faces and edges.

The oxygen framework is the closest hexagonal packing. Octahedral voids are arranged along polar axis z, and only two-thirds of them can be populated with cations (Li^+, Nb^{5+}, impurity cations), while the others are vacant.

From this standpoint, a near-ideal structure can, potentially, exist in stoichiometric (R = 1) single crystals of high perfection.

In lithium-deficient crystals (R < 1), in crystals of congruent composition (R = 0.946) among them, the cation sublattice is substantially disordered [1, 2].

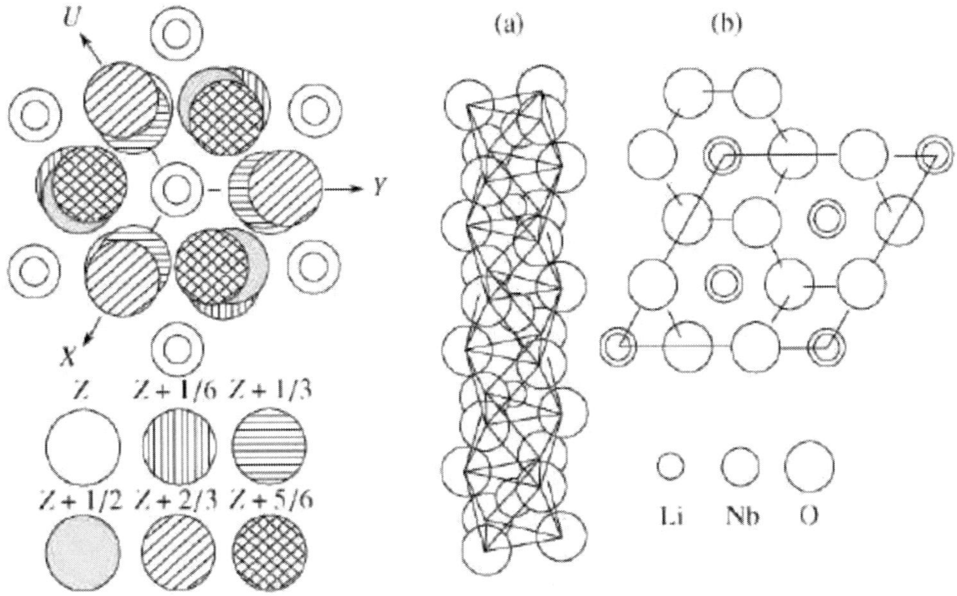

Figure 1. Projection of the crystal structure of LN on plane [0001] [9].

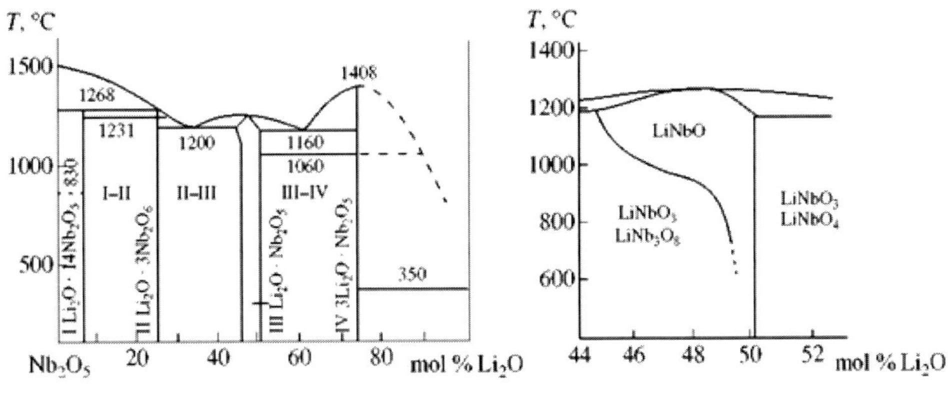

a b

Figure 2. Panel (a): Li_2O–Nb_2O_5 phase diagram [1]. Panel (b): a fragment of this phase diagram [7].

Lithium niobate and lithium tantalate are phases of variable composition distinguished by extensive homogeneity regions in the phase diagrams. For lithium niobate, the homogeneity range is 44.5-50.5 mol% Li_2O at 1460 K and 49.5-50.5 mol% Li_2O at 293 K [1, 7, 8]; for lithium tantalite, 46-50.4 mol% Li_2O [8] (Figure 2a) [1-4, 7, 8-10]. The congruent freezing point, at which the melt composition corresponds to the composition of the growing crystal, for these crystals does not coincide with their sstoichiometry [1-4, 8, 9]. Such structures are usually distinguished by a significant three-dimensional inhomogenity and a complex spectrum of point and extended defects, which create a complex, hardly modeled structure disorder [1-3, 8, 9, 11-16].

There is no consensus on the congruent melting point of nominally pure LN. The position of the congruent melting point in the phase diagram varies from 48.3 to 48.65 mol % Li_2O [1, 2, 4, 8, 14, 17].

There are many reasons for this discrepancy. Some of them are considered in [18]. An uncontrolled oxygen deficit in the precursor niobium pentaoxide associated with off-stoichiometry can be one such reason [14]. This uncontrolled deficit introduces an uncertainty into the ratio R = Li/Nb even at the stage of feed preparation. Another reason can be the different volatilities of the matrix components; this can change R depending on the thermal history (the feed synthesis schedule and melt-exposure time) [17-20]. This matter is not clear.

In [17, 19, 20], only lithium losses are taken into account; in [20], R changes toward a niobium deficit upon long melt exposures under oxidative conditions. Therefore, given that all other conditions are equal, researchers using niobium pentaoxide of different grades and purchasing it from different sources can create differing results.

Many parameters (the Curie point, critical synchronism angle and the temperature of the SHG phase matching of laser radiation, line widths of NMR and vibrational spectra, position of the fundamental optical absorption edge, luminescence, and photorefractive properties) within the homogeneity region of nominally pure crystals are significant functions of the chemical composition of a crystal, above all, of R [1, 2, 4, 8, 21-34]. These functions were used to develop methods for controlling the homogeneity and stoichiometry of LN crystals [17, 19, 24, 25, 28, 29, 32, 35].

These methods, based on measurements of changes in some physical parameter of a crystal, are circumstantial and must lean on straightforward chemical and physicochemical measurements. Strictly speaking, the congruent melting composition is uniquely defined only by the dystectic ordinate on the Li_2O-Nb_2O_5 phase diagram. Moreover, the applicability of some methods is greatly limited by the essential dependence of physical properties on structural perfection and on the presence of uncontrolled impurities. Therefore, a method can give divergent results for different crystals even though their R values are the same. For example, the position of the optical absorption edge is largely dictated by an oxygen deficit and by impurities that generate optically active energy sublevels in the bandgap [28, 32]. The adequacy of the holographic determination of R = Li/Nb is also affected by photorefractive impurities [29, 36]. The same refers to SHG methods.

In addition, the properties of lithium niodate and its extent of homogeneity are strongly affected by the thermal history of a crystal. When temperature drops below 910°C, the solubility interval abruptly narrows (Figure 2b) [1, 2, 4, 7, 8; 37], and a new phase can freeze. Below 910°C, the solubility interval can be as narrow as 0.5 mol % on each side of the stoichiometric point. This means that the congruent melting composition falls outside the homogeneity region. However, since equilibration at temperatures below 910°C requires hundreds of hours, nonequilibrium compositions, e.g., a congruent melting one, can be obtained by rapid cooling. Nonetheless, crystals having identical compositions but differently annealed postgrowth can differ in homogeneity.

In addition to point defects, cluster-type density inhomogeneities are observed in the cation sublattice of LN; these defects, like point defects, spoil the translational invariance of the structure without changing the overall symmetry of the unit cell [38, 39]. Therefore, the adequacy of composition determination from Raman line broadening [19, 25] (which is observed when the translational symmetry of the lattice is spoiled [4]) is also to some extent controlled by the growth parameters and the thermal history of a sample. The compositional homogeneity of a crystal along the growth axis (determined by, e.g., holography or judged from the constancy of the synchronism angle [24, 29]), likewise, cannot be regarded as unambiguous evidence that the crystal has the congruent melting composition.

The homogeneity of single crystals can be appreciably improved by employing special growth and post-growth processing schedules [40, 41]. In [40], growth in electric fields appreciably improved the homogeneity of crystals with strongly incongruent compositions. In [41], a similar result was achieved through long-term anneals near the melting point in a weak electric field. We may conclude that for real crystals the feed ratio R = Li/Nb corresponding to the congruent melting composition is, likely, dictated not only by the physico-chemical and thermodynamic properties of a system but also, to a large extent, by process parameters.

Clearly, rather homogeneous incongruent crystals with comparatively small sizes can be grown from a large melt bulk at low growth rates that provide the diffusion of an excess component into the melt and melt enrichment near the freezing interface. For example, stoichiometric crystals can be grown from a melt containing about 58 mol % Li_2O [1].

Qualitatively, DTA can aid in finding the deviation of the lithium niobate feedstock from the congruent melting composition. When thermoanalytical curves are recorded for crystals with various R values, a single liquidus peak must be observed for samples with congruent melting compositions. When the sample has an incongruent composition (whether it is lithium poor or lithium rich compared to the congruent melting composition), extra peaks appear on exothermal or endothermal DTA curves (otherwise, the curves become noticeably skew). Therefore, the most symmetrical thermoanalytical curves of endothermal and exothermal events must correspond to a dystectic point. When we studied a feedstock produced at the Institute of Rare-Element and Mineral Chemistry and Technology, the most symmetrical thermoanalytical curves were recorded for an LN sample containing about 48.6 mol % Li_2O. Compositions with $[Li_2O]$ = 48.7 or 48.5 mol % gave skew cooling peaks. As the composition departs from 48.6 mol % Li_2O, the heating and cooling thermoanalytical curves become progressively more skew; for compositions with $[Li_2O]$ = 48.4 or 48 mol %, DTA heating curves display extra peaks. Likely, the dystectic ordinate in the phase diagram for the LN feedstock we studied lies near 48.6 mol % Li_2O. Most of the studies reviewed in [1] give consistent results. The feedstock was prepared from high-purity reagents (Li_2CO_3, 11-2 high-purity grade; Nb_2O_5 from the Institute of Rare-Element and Mineral Chemistry and Technology, total cationic impurities $< 1.5 \times 10^{-3}$ wt %).

Measurement errors might mainly arise from the uncontrolled oxygen nonstoichiometry of the niobium pentaoxide. Lithium niobate single crystals grown from a melt having the congruent melting composition have a disordered structure, since they cannot be free of defects that provide the electrical neutrality of a crystal [1]; as a result, the crystals are sensitive to laser damage, which limits their application in optics. Crystals richer in lithium, e.g., stoichiometric crystals, have more ordered lattices and are more resistant to optical damage. However, it is difficult to grow large crystals; a significant compositional inhomogeneity along the boule length generated during the growth usually leads to cracking of a crystal and to a scatter in its physical parameters. Nonetheless, physicochemically, there is no fundamental difference between crystals of congruent melting composition and stoichiometric crystals. They differ only in their internal defectiveness.

The investigation of internal defects, in particular, subtle features of structure-unit order in the cation sublattice (associated with fluctuations in chemical composition and with the thermal history), and the investigations of their effects on the physical and physicochemical parameters are important for the following reasons: these investigations highlight tendencies in the properties of real crystals and promote progress in the technology of single crystals of high homogeneity and structural perfection.

Lithium niobate is a representative congruent melting nonstoichiometric phase of variable composition. Its phase diagram is characterized by the fact that liquidus and solidus maxima are very flattened and that the position of the dystectic point differs from the stoichiometric composition (Figure 2).

Property-composition diagrams for such phases have no unique concentration points. Properties vary monotonically across the homogeneity region; there is no well-defined composition within the homogeneity region that would be characterized by maximal order in the arrangement of dissimilar atoms or ions [42].

A stoichiometric crystal has no specific properties. Extensive investigations of LN over a wide range of concentrations, covering the homogeneity range, showed no unique points on property-composition diagrams near the concentration corresponding to R = Li/Nb = 1 [1–4, 8, 10, 18, 23, 25, 27, 30].

In an ideal lithium niobate crystal, the order of cation alternation along the polar axis must be the following: Li^+, Nb^{5+}, and a vacant octahedron [1]. In this context, an absolutely perfect structure should have belonged to crystals with R = Li/Nb = 1, i.e., to stoichiometric crystals in which a maximum occupancy of lithium and niobium positions in the ideal structure can potentially exist.

However, even for the stoichiometric composition, single-crystal X-ray diffraction in real crystals shows that the unit cell dimensions better fit the structure in which niobium ions can in part substitute for lithium ions and reside in vacant octahedra and in which some niobium-site vacancies exist [1].

This is in part due to both the nonequilibrium crystallization of real crystals and mainly due to the fundamental features of structure formation in phases of variable composition [1].

An extensive homogeneity region necessitates that a free energy versus composition curve have an extended, gently sloping portion near an extreme point [43]. The curve has this shape given high levels of various types of structure defects, i.e., given a high extent of intrinsic disorder (vacancies, antisite defects in cation sublattices, and other defects) [1]. This means that a situation in which all sites in the ideal structure are occupied by proper cations does not exist in these systems even when this is in principle possible (R = Li/Nb = 1).

The extent of structural perfection in similar phases of variable composition, which have a developed defect structure, must be controlled by the amount of intrinsic defects leading to the greatest disorder. Such defects in LN are, supposedly, niobium ions arranged along the polar axis in lithium ion sites [1, 2, 4, 8].

This supposition fully agrees with the defect-structure models currently considered for LN: the Li-site vacancy model and the Nb-site vacancy model. The former is described by the formula $[Li_{1-5x}Nb_x(Li_v)_{4x}][Nb]O_3$, where (Li_v) stands for a Li-site vacancy in the ideal structure; the latter is described by $[Li_{1-5x}Nb_{5x}][Nb_{1-4x}(Nb_v)_{4x}]O_3$, where (Nb_v) stands for a Nb-site vacancy in the ideal structure.

Charge neutrality in these crystals is conserved through the generation of antisite defects Nb_{Li} and, accordingly, cationic vacancies. The appearance of Nb_{Li} defects is accompanied by the perturbation of translational invariance along the polar axis. Similar cations in a crystal occupy structurally nonequivalent positions; in the cation sublattice, cluster-type density inhomogeneities appear (several antisite cations and (or) vacancies are clustered).

2. The Search for Homogeneity of $LiNbO_3$ Crystals Grown of Charge with Different Genesis

The interest towards $LiNbO_3$:Mg crystals doped by "threshold concentrations" (5-5.5 mol %) is caused by high optical damage resistance and by the possibilities to use such crystals at optical transducers based on periodically poled structures [44, 45]. But the methods of obtaining of the Mg doped defect-free crystals with homogeneous dopant distribution in the bulk of the boule are yet not perfect [46-50]. The influence of genesis of the initial components on the optical quality and the dopant concentration homogeneity is usually not taken into account. In paper [51] at a boron (B) example a technique of homogeneous doping of $LiNbO_3$ by addition of boron to the re-extract at the stage of clean Nb_2O_5 obtaining is first described. The optical homogeneity and optical damage resistance for crystals grown from Nb_2O_5:B charge were higher than for the crystals grown by usual method.

In this paper the comparative analysis of optical and structural homogeneity was carried out for Mg doped lithium niobate crystals grown by Czochralski method of usual charge synthesized by adding MgO to the Nb_2O_5 and of Nb_2O_5:Mg charge synthesized at homogeneous Mg doping during the extraction of Nb_2O_5.

During the Nb_2O_5 extraction the extractant contained 35% carboxylic acids dimethylamides $C1_0$-$C1_3$ fractured, 30% octanol-1, 35% Eskaid thinner. At the 16-step cascade of extractors of "mixer-settler" type were obtained re-extracts containing 50-60 g/L Nb_2O_5 and 40-50 g/L F. The strictly measured amount of pure MgO was added to the re-extract. MgO completely dissolved at this. A batch of Nb_2O_5:Mg containing 0.947 wt. % Mg was prepared due to the scheme in Figure 3. The niobium hydroxide was precipitated by 25% solution of NH_4OH from the solution that contained magnesium. To precipitate magnesium oxide completely together with precipitation of niobium hydroxide by NH_4OH one needs to keep high pH (>11.5) of the solution. So the necessary concentration of the OH~ ions is provided. At such pH the loss of the magnesium is less than 0.3 wt. %. Filtrates and wastes contained less than 0.5 mg/l. We concluded that at such condition almost all magnesium goes into the niobium hydroxide and on, to the solid charge Nb_2O_5:Mg.

The precipitate was washed with deionized water 3 times by decantation at solid:liquid volume ratio about 3÷5. Then the mixture was heated at 1000 °C. The concentration of admixtures in Nb_2O_5:Mg was less than $5 \cdot 10^{-4}$ wt%. The Mg amount in the Nb_2O_5:Mg, in the lithium niobate charge and in the $LiNbO_3$:Mg crystals were determined by mass spectrometry with inductively coupled plasma at ELAN 9000 DRC-e (MC ISP). The additional concentration control was accomplished by the Curie temperature measurements by DTA method. The method error was ± 0.5° C. Due to the X-ray analysis heated at 1000°C Nb_2O_5: Mg contained $MgNb_2O_6$ phase along with the Nb_2O_5:Mg phase. Additional heating at 1250°C provided a single-phase compound, with X-ray diffraction was similar to Nb_2O_5 one. Nb_2O_5: Mg (method 1) and Nb_2O_5 (method 2) were used to synthesize charges by (method 1) and by (method 2) due the method described in [9].

The powdered Li_2CO_3–Nb_2O_5–MgO (method 1) or Li_2CO_3 – Nb_2O_5:Mg (method 2) were thoroughly mixed in a fluoroplastic mixer with fluoroplastic rods. After that the mixtures were heated at 1250±5 °C and a granulated charge with high bulk density (3.4 g/cm^3) was obtained. This charge enables to complete the fusion in the crucible at one stage.

The phase compound was detected by the X-ray fluorescence analysis. The admixtures concentrations was controlled by spectral analysis method. The Li/Nb ratio in the charge corresponded to the congruous compound (Li/Nb = 0.946). The magnesium concentration was 0.84 and 0.85 wt.% for the charges prepared by (method 1) and (method 2), respectively. The preparation of the melt before growing of all crystals from charges prepared by (method 1) and (method 2) was identical. The melt was overheated to 100° above the melting temperature for 2 h. All final crystals had the same size and close magnesium concentrations. Thermal conditions were also identical: the rotational speed, growth speed, the temperature gradient at the phase boundary were the same.

All $LiNbO_3$:Mg crystals had length of the cylindrical part 25 mm and diameter 30 mm. The crystals were grown 2 mm/h at (001) direction by Czochralski method, the speed of rotation was governed by the condition of flat solid-melt interface and was 12 rpm. For all crystals 25% of the melt in crucible turned into the crystal.

Three homogeneously doped $LiNbO_3$:Mg crystals were grown by (method 1) - crystals A, B, C, and one usually doped crystal was grown by (method 2) - crystal D. To escape thermoelastic stress all crystals were heated at 1195°C for 20 h and then were put into the single-domain state by high-temperature electrodiffuse annealing (HTEA) at 1238°C and aftercooling at current until 980°C. To determine magnesium concentration and to search defect structure and the Curie temperature, the plates were cut from the cylinder bottom, from the part of the cone going into the cylinder and from basic X-plate along the growth axis of each boule after heating and HTEA. Optical quality of the crystals was determined by the amount of the light-scattering centers per volume unit.

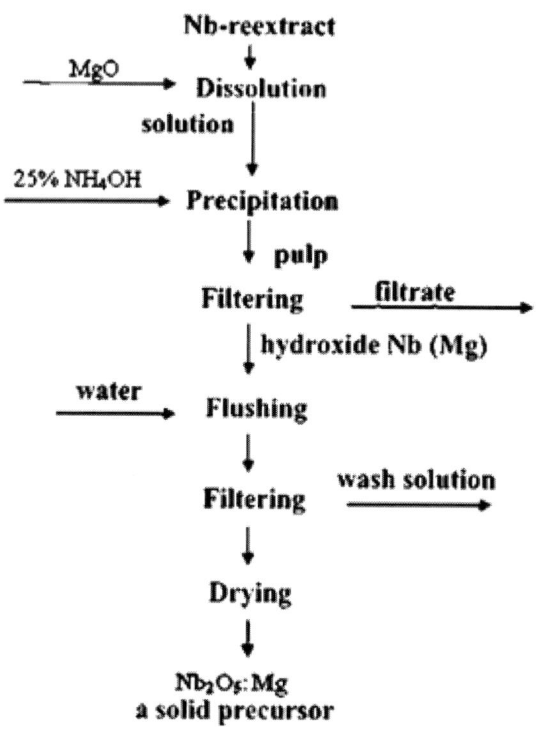

Figure 3. The scheme of obtaining of homogeneously doped Nb_2O_5:Mg solid charge (method 1).

For this method He-Ne laser LG-112 was used, wavelength 632.8 nm, beam diameter 0.05 cm. The defect structure was searched by the system of image analysis "Thixomet". Crystal plates were smoothed, polished and acid-etched in a mixture of mineral acids $HF:HNO_3=1:3$. To evaluate the crystals quality a method of counting of defects was developed.

The method includes building of a panorama image of the searched object using software system (Thixomet-PRO). As the result of analyzing of the panorama we obtain the following parameters: the average diameter of defects (d, μm), the density of defects (p, mm^{-2}) and the ratio between the area of all defects and the total searched area (S%, %).

The Raman spectra of the powders made from the searched crystals were excited by 514.5 nm line of argon laser Spectra Physics and were registered by spectrometer T64000 by Horiba-Jobin Yvon with resolution 0.5 cm^{-1} at the "reflection" geometry.

The data from Table1 confirm that magnesium concentration in the bottom part of the boule (C_{bottom}) differs from magnesium concentration in the top part of the boule (C_{top}) to ΔC¼C_{top}-C_{bottom} r3% for the $LiNbO_3$:Mg crystals grown of the charge prepared by (method1). This means that the magnesium is distributed homogeneous in the homogeneously doped $LiNbO_3$:Mg crystal (method 1). For the crystals grown of the charge prepared by method 2, the value of ΔC is ~ 6.5%.

The Curie temperature for different parts of all grown boules is shown in Table 1. Due to Table 1 the Curie temperature decreases as the magnesium concentration decreases in the crystal along the polar axis.

Note that the Curie temperature of the central part of the boule agree within the error with the arithmetic mean of Curie temperatures of the top and bottom part of the boule. This could mean that the magnesium concentration changes smoothly through the crystal along the polar axis (Table 1).

The Curie temperature of different samples cut from one slice of the boule coincided within experiment error which shows that the distribution of magnesium in the crystals at directions perpendicular to the growth axis is absolutely homogeneous.

The method of counting of defects revealed a higher optical quality of $LiNb0_3$:Mg crystals grown of charge (method 1) after the HTEA in comparison with the crystal grown of charge (method 2). The light scattering centers were absent from the crystals A, B, C and the amount of light scattering centers in the crystal D were 7 cm^{-3} (Table 2).

By optical microscopy the most typical micro-and macrostructure of $LiNb0_3$:Mg was searched before and after HTEA. After heating and after HTEA in the x-cut planes of $LiNb0_3$: Mg samples A, B and C (method 1) were free of growth bands and other micro- and macrodefects (Figures 4a, 5a).

Table 1. The magnesium concentration in homogeneously doped (method 1) and usually doped (method 2) crystals

Crystal type	C_{top} (mol%)	T_C (°C), top of the crystal	C_{bottom} (mol%)	T_C (°C), bottom of the crystal	T_C (°C), middle of the crystal	ΔC (mol%)	C_{top} - C_{bottom}/ C_{top} × 100 %
Method 1, crystal A	5.3	1209±0.5	5.13	1205±0.5	1207±0.5	0.2	3.75
Method 1, crystal B	5.32	1209±0.5	5.24	1208±0.5	1208±0.5	0.08	1.5
Method 1, crystal C	5.32	1209±0.5	5.17	1206±0.5	1207±0.5	0.15	2.8
Method 2, crystal D	5.36	1210±0.5	5.01	1203±0.5	1207±0.5	0.35	6.5

At the same time crystal D (method 2) had growth bands after heating (Figure 4b). Despite the general homogeneity of the crystal D along the growth axis (Figure 6c) after HTEA, it had vestigial domains along the Z-axis (Figure 6d).

Moreover, Figure 6c (the x-cut) reveals a highly etched surface that is a circumstantial evidence of strong local inhomogeneity. Therefore, the data obtained from the method of optical microscopy revealed the less defectivness of $LiNb0_3$:Mg A, B and C (method 1) compared to D (method 2) crystals (Figures 2-4, Table 2). For example, crystals A, B and C (method 1) had no registered microdefects after HTEA (Figure 5c and d, Table 2).

The $LiNb0_3$:Mg structure order was searched by Raman-spectra and the data confirm results of optical microscopy method. Figure 7 presents Raman spectra of the powder obtained by grinding of crystal B and crystal D. The characteristics of Raman-spectra were different for crystals B and D. At the same time Raman spectra of samples cut of different parts of one crystal B coincide within the error both along the boule and across the boule (sample B in Table 1).

The low frequency area of Raman spectra (20-300 cm^{-1}) specify the vibrations of cations located in the octahedron interstice. The area of 300-500 cm^{-1} is well known to correspond to deformational vibrations and in the area 500-950 cm^{-1} is associated with stretching vibrations of the oxygen octahedrons. The basic parameters of the spectral bands are shown in Table 3.

Figure 5 and Table 3 show that differently obtained $LiNb0_3$:Mg crystals have sharp distinctions. The widths and the intensities of some bands of samples B and D are quite different. It is caused by cations order difference and by the fact that oxygen octahedrons deformation is different.

Table 3 shows that the intensity of band 276 cm^{-1} corresponding to the Li^+ ions totally symmetric vibrations in the octahedrons [44] is maximum for the sample D and noticeably decreases for sample B. This shows that Li^+ ions order in the homogeneously doped sample is higher than in sample D. The 869 cm^{-1} band corresponds to the valence-bridge vibrations (VBV) of oxygen atoms in the $B0_6$ octahedrons (B is Nb or the doping element) along the polar axis. The parameters of the band 869 cm^{-1} are considered to determine the dipole moment and respectively spontaneous polarization of the $LiNb0_3$ crystals [44].

So the 869 cm^{-1} band intensity can be used to evaluate the dipole ordering of cation sublattice of $LiNbO_3$ and other crystals and solid solutions with oxygen-octahedron structure [44, 52, 53]. This band reveals at the Raman spectra of the ferroelectric phase of lithium niobate and it is absent from the Raman spectra of paraphase [54-56]. The more ordered are cations along the polar axis, the higher dipole moment of the unit cell is and the more 869 cm^{-1} band intensity is [44, 52, 53].

The fact that increase in the intensity of the band corresponding to VBV of oxygen in oxygen octahedrons BO_6 corresponds to cation sublattice ordering has a good correlation with the decrease in the width of the bands corresponding to the basic lattice vibrations [44]. This means that homogeneously doped $LiNbO_3$:Mg (sample B) has bigger dipole moment of the unit cell and bigger spontaneous polarization than $LiNbO_3$:Mg sample D due to the fact that the intensity of the 869 cm^{-1} band is higher and the width is less for the sample B (Figure 7, Table 3). So, the optical properties of crystals with oxygen-octahedron structure can be evaluated by intensity of the band corresponding to VBV of oxygen in oxygen octahedrons BO_6 and by the width of the bands corresponding to the basic lattice vibrations.

Table 2. The microdefect structure of z-cut of $LiNbO_3$:Mg crystals grown from different charge after HTEA

The doping method	Homogeneous (method 1) samples A, B and C	Usual (method 2) sample D
The amount of light scattering centers (cm^{-3})	The centers are absent	7
d (μm)	Microdefects are absent	3.5
p (mm^{-2})		13
$S_{\%}$(%)		0.02

The disorder in cation sublattice along the polar axis in the pure and doped lithium niobate crystals leads to the multimode regime of the band corresponding to VBV of oxygen in oxygen octahedrons BO_6.

The widths of the lines that correspond to totally symmetric and doubly degenerate vibrations of the oxygen octahedrons (618 and 658 cm^{-1}) are also different for differently doped $LiNbO_3$:Mg crystals (Table 3).

The widths of 618 and 658 cm^{-1} lines are less for sample B so the oxygen octahedrons are less distorted than for sample D. The correlation between octahedral geometry and cations order along the polar axis is observed: the better cations are ordered, the less octahedrons are distorted.

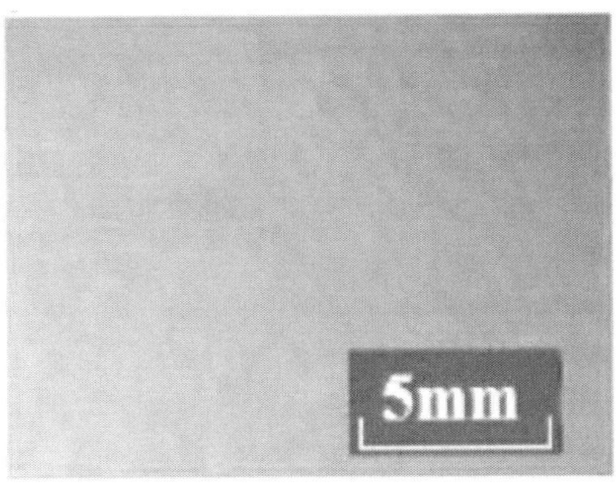

a

Figure 4. (Continued)

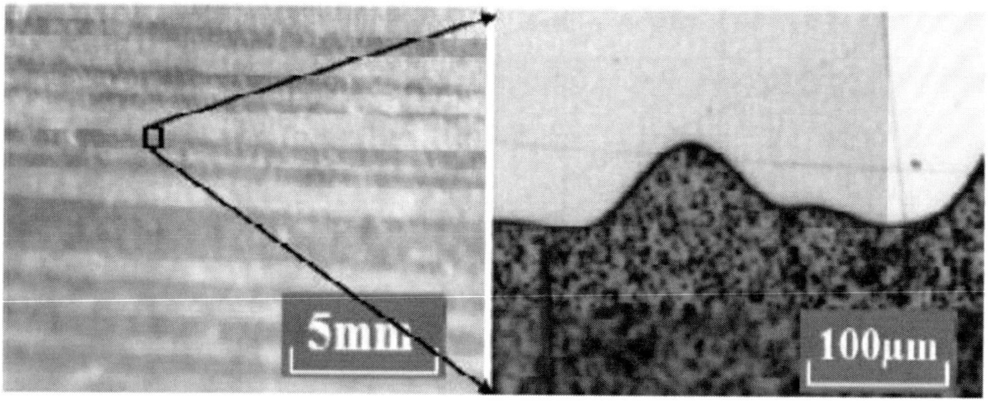

b

Figure 4. The macrostructure of $LiNb0_3$:Mg: a - obtaining homogeneously doped charge containing Nb_20_5:Mg, the x-cut, after heating (method 1); b - usually doped, the x-cut, after heating (method 2).

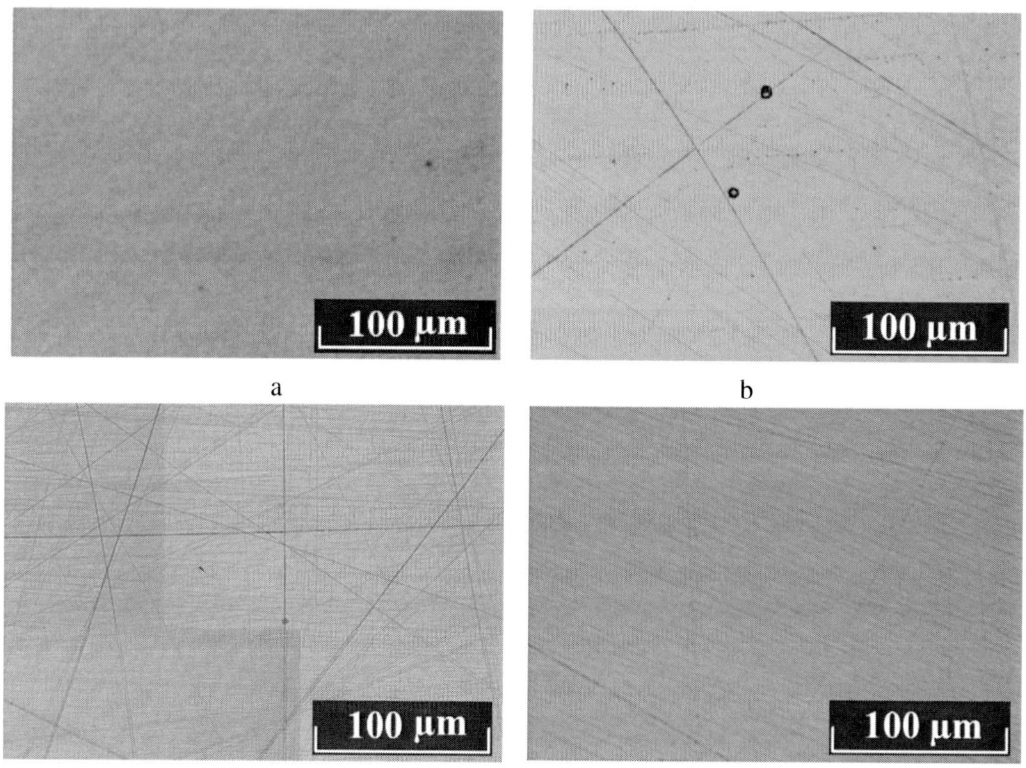

Figure 5. The microstructure of $LiNbO_3$ crystal grown of homogeneously doped Nb_2O_5:Mg solid charge (method 1): a – x-cut, b – z-cut after heating, c – x-cut, d – z-cut after HTEA.

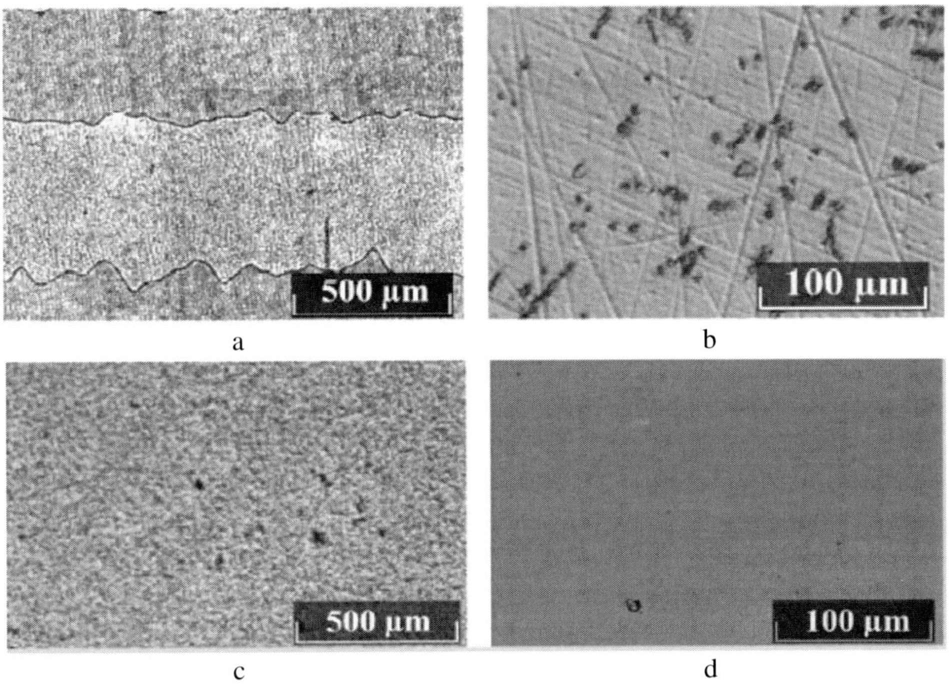

Figure 6. The microstructure of $LiNbO_3$ crystal grown by usual method (method 2): a – x-cut, b – z-cut after HTEA.

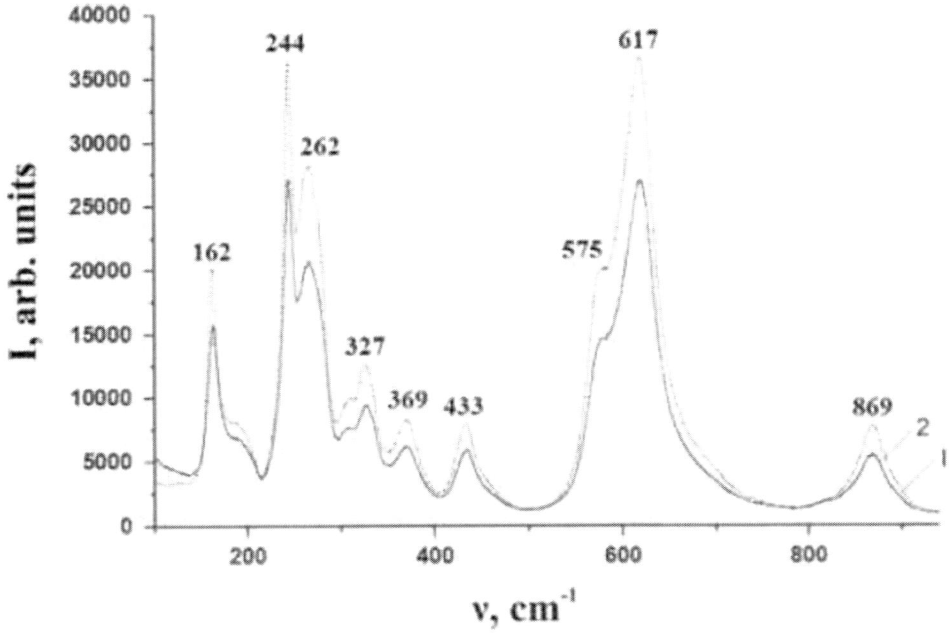

Figure 7. Raman spectra of homogeneously doped $LiNbO_3$:Mg powder (method 1), and of usually doped $LiNbO_3$:Mg powder (method 2). The exciting radiation wavelength λ¼ = 514.5 nm, power – 300 mW.

Table 3. The values of frequency (V, cm^{-1}), linewidth (S, cm^{-1}), intensity (I, arb. units) of Raman spectra lines of homogeneously doped $LiNbO_3$:Mg (sample B) and of usually doped $LiNbO_3$:Mg (sample D)

Sample D			Sample B		
V	S	I	V	S	I
163	13	15,782	163	12	11,186
186	35	6876	186	31	3825
244	13	27,165	244	14	21,630
262	17	20,791	262	17	19,760
276	26	18,852	277	18	16,872
304	29	7664	304	23	5178
328	25	9477	328	26	7403
370	28	6222	370	29	4625
433	32	5910	434	24	4018
577	31	14,730	574	30	12,366
618	47	27,038	618	39	26,641
658	85	7099	658	69	4018
868	43	3415	869	39	5584

So the less 276 cm^{-1} line width is the less 618 and 658 cm^{-1} lines width are (Table 3). Note that lines in the low-frequency area of the spectra (186 and 304 cm^{-1}) are slightly narrower for the homogeneously doped sample B (Table 3). The isolated line 433 cm^{-1} that probably corresponds to the E-type deformational vibrations of oxygen atoms in symmetrical bridge Nb-O-Nb [1] is substantially narrower for sample B (Figure 7 and Table 3).

So the Raman spectra demonstrated that homogeneously doped lithium niobate (B) has more ordered structure than lithium niobate doped by usual method (D).

We assume that magnesium is distributed more homogeneously at the re-extract. Therefore the clusters in melts obtained of charges (method 1) and (method 2) will have different structure and size. This means that even if the conditions for melts (method 1) and (method 2) will be identical, the crystallization will differ and the optical quality of the final crystals will differ. The problem was described in the papers of Sobol and Voronko [54-56], where authors associate the structure and the size of the clusters with the compound and the thermal history of the melt. They also suppose that clusters attach to the growing crystal during crystallization.

To synthesize new homogeneously doped charge for magnesium doped lithium niobate a new method was developed to obtain Nb_2O_5:Mg. The method is based upon addition of dopant (MgO) into the re-extract at the stage of obtaining of niobium oxide by extraction. The micro- and nanostructure search by optical microscopy proved that homogeneously doped $LiNbO_3$:Mg (method 1) has significantly less defects than crystal obtained by (method 2). The data obtained by Raman spectroscopy agree with the data on optical microscopy and reveal that crystals A, B and C have more perfect structure than crystal D. Using Nb_2O_5:Mg charge during growth of doped lithium niobate crystals allow to obtain more optically perfect and structurally homogeneous samples than the usual technology. The results can be explained by the difference in structure and size of the clusters in the melt due to the different genesis of the charge. It means that the mechanisms of crystallization of the melt are different and the optical properties of the crystals will change.

3. Formation of a Stoichiometric Layer and New Polar Phase Upon Exposure of $LiTaO_3$ Single Crystals to Lithium Vapor

The physical properties of lithium tantalate single crystals can be effectively tuned by varying the Li/Ta ratio in the crystals. At room temperature, the spontaneous polarization of lithium tantalate, $LiTaO_3$, is ~ 60 μC/cm^2, but the electric field needed for complete switching in congruent lithium tantalate is rather high, ~ 210 kV/cm [57-60]. The coercive (switching) field of stoichiometric crystals is considerably lower. However, melt-grown stoichiometric lithium tantalate crystals are inhomogeneous, and their optical quality is not very high. The Li : Ta ratio in relatively thin plates can be raised through vapor transport equilibration (VTE): prolonged high-temperature annealing of nonstoichiometric single-crystal lithium tantalate in a lithium-saturated atmosphere. The objectives of this study were to study the switching behavior and kinetics of layers of different phase compositions and stoichiome-tries, to find out whether saturated dielectric hysteresis loops corresponding to the complete switching of lithium tantalate after VTE processing can be obtained in relatively low fields and at moderate temperatures, and to investigate the phase transition in the polar (ferroelectric) structure produced by VTE processing in the surface layer of lithium tantalate.

Z-cut plates (with their faces normal to the *Z* crys-tallographic axis) 14 × 16 × 1.2 mm in dimensions were prepared from a lithium tantalate single crystal with T_C = 628°C (which corresponds to 48.71 mol % Li_2O [44]).

The plates were oriented with an accuracy of 30' or better and annealed in a closed system (in a "crucible" fabricated from a 50% $LiTaO_3$ + 50% Li_3TaO_4 mixture) at 1200°C for ~ 220 h. Pt electrodes were deposited by magnetron sputtering onto specimens 7 × 8 × 1.2 mm in dimensions, prepared from *Z*-cut VTE $LiTaO_3$ plates. The specimens were then stored at room temperature for 24 h. Dielectric loop measurements were performed at a frequency of 0.01-0.02 Hz in a sinusoidal electric field of 12.5 kV/cm peak, using a classic Sawyer-Tower circuit. The measurements were made at room temperature and during heating or cooling in the temperature range 18-205°C. Stoichiometry-depth profiles were obtained by Raman spectroscopy, using the known relationship between the width of the line at 140 cm^{-1}, due to vibrations of symmetry *E* [5], and the Li/Ta ratio.

The spectra were measured on a Ramanor U1000 spectrometer equipped with a confocal microscope, which enabled laser beam scanning over the sample surface in 0.1-mm steps. The experimental procedure was described in detail elsewhere [17, 19, 25, 44].

In the initially polydomain VTE $LiTaO_3$ samples studied, a pyroelectric effect may be due only to a slight natural unipolarity Because P_s is a weak function of temperature far away from the Curie point, the pyroelectric current proper should be negligible.

On the other hand, any changes in the charge or dipole moment distribution, independent of their localization and origin, should produce a temperature-dependent spontaneous current density, j_s, in the external circuit. If a phase transition occurs at a temperature T_0, $j_s(T)$ should have a well-defined anomaly in the vicinity of T_0 and, if the transition is reversible, the magnitude and sign of j_s will depend on those of the temperature scan rate, $dT/d\tau$.

Figure 8 shows a typical $j_s(T)$ curve, which was obtained in a heating-cooling cycle at a rate $dT/d\tau$ = ± 1 K/min. Quantitatively similar results were obtained for the other samples studied, in particular in subsequent measurement cycles.

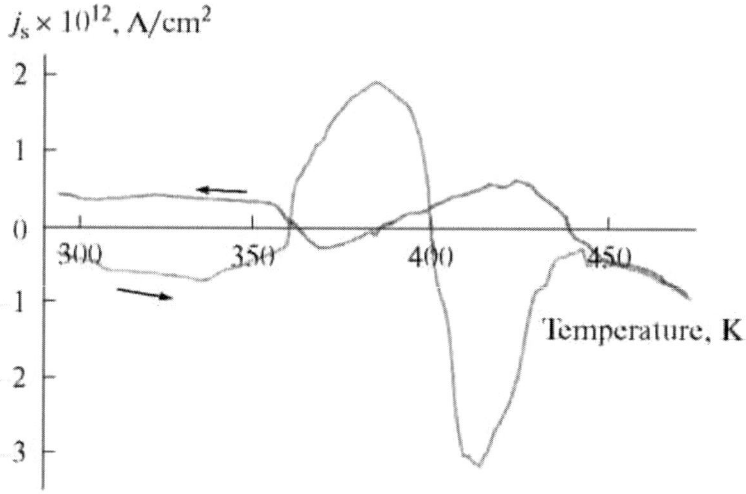

Figure 8. Temperature dependence of the spontaneous current density for *Z*-cut VTE $LiTaO_3$.

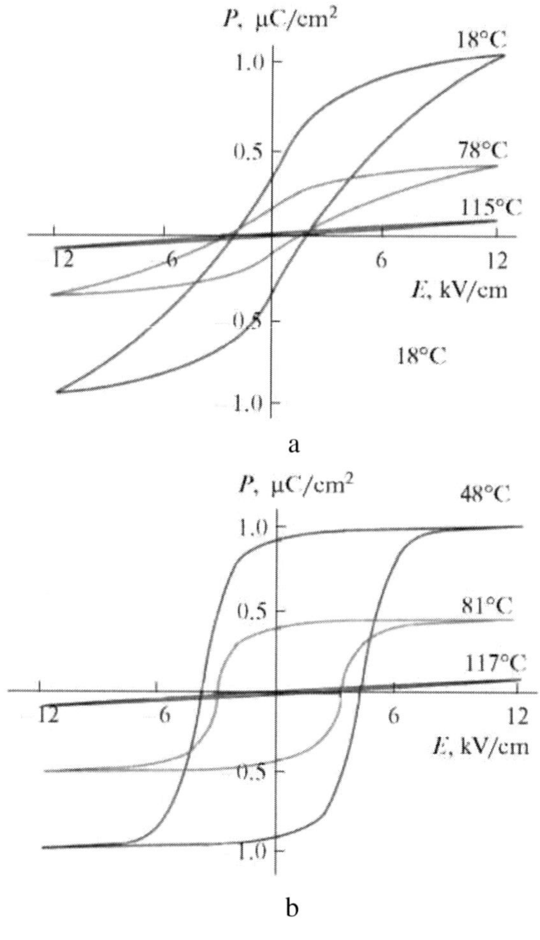

Figure 9. Quasi-static dielectric hysteresis loops of *Z*-cut single-crystal VTE $LiTaO_3$ (f = 0.02 Hz): (a) first heating, (b) first cooling.

The shape of the $j_s(T)$ curve attests to a phase transition. Interestingly enough, both anomalies correspond to roughly the same value $\int j_s dx \sim 0.6\ \mu C/cm^2$, which is close to the room-temperature residual polarization of the samples (Figures 8, 9).

The room-temperature hysteresis loops are clearly unsaturated, with partial switching. With increasing temperature, the switchable polarization decreases (Figure 9a). For $t >$ 117°C, there is no dielectric hysteresis, and P is an almost linear function of E, like in the case of a transition to a paraelectric state. The sample was then heated to 120°C, held there for 10 min, and cooled. During cooling, the switchable polarization increased again, and the $P(E)$ plot assumed the form of a typical dielectric hysteresis loop with a clear tendency for the polarization to saturate (Figure 9b). Upon cooling to room temperature, the switchable polarization returned to its original level (~ 1 $\mu C/cm^2$).

The results obtained in subsequent dielectric loop measurements were qualitatively similar to those above: the residual polarization (P_r) was ~ 15% lower (Figures 9, 10).

The polarization was also found to decrease with increasing temperature (Figure 10). The dielectric hysteresis loops were unsaturated, like those in Figure 9a.

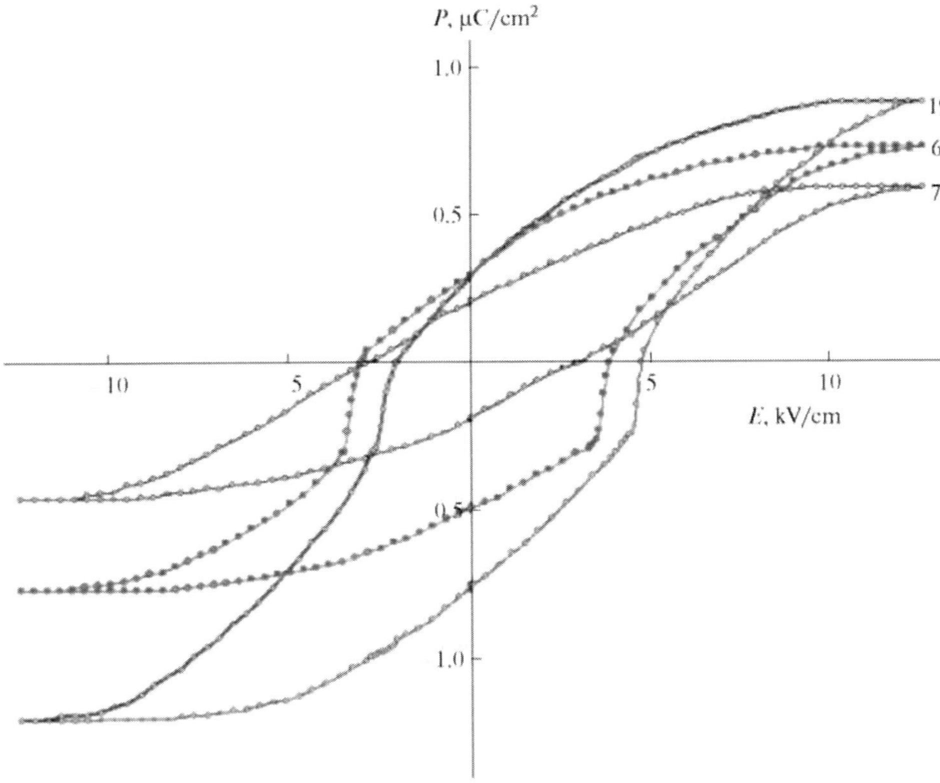

Figure 10. Quasi-static dielectric hysteresis loops of Z-cut single-crystal VTE $LiTaO_3$ at different temperatures.

Therefore, we deal with residual polarization, P_r, rather than with spontaneous polarization, P_s. However, in contrast to the data in Figure 9, the dielectric hysteresis here does not disappear at ~ 120°C (Figure 11a). Moreover, P_r increases in subsequent loop measurement cycles at constant temperature (Figure 11), whereas the coercive field E_c varies insignificantly (the curves are numbered according to the number of the measurement cycle).

In successive loop measurement cycles, the switching process involves an ever increasing volume, that is, the stoichiometric layer proper of stoichiometric VTE $LiTaO_3$. It seems likely that the anomalies in the spontaneous current and the switching processes represented in Figures 8 and 9 arise only from the surface layer, which differs in properties from the bulk of the material. After a relatively thin (~30 μm) surface layer was removed on the side that had been exposed to lithium ions during the VTE processing, no low-temperature switching processes and no anomalies in the spontaneous current were detected (Figures 8, 9).

At higher temperatures, an ever increasing volume of the sample is involved. The following features are seen in Figure 11:

1 The dielectric hysteresis loop becomes ever more saturated as the number of measurement cycles increases.
2 With an increase in the number of cycles, P_r increases. (Clearly, the switching kinetics play a significant role: with increasing temperature, the field "sways" the polydomain structure more rapidly in each subsequent measurement cycle. It seems likely that, at a very large number of cycles, high P_r values can be obtained at a lower temperature.)
3 At positive fields, all of the loops have one more step in polarization ("frozen" domains), which shifts downfield as the number of cycles increases, which also points to an increase in the volume of VTE lithium tantalate involved in switching.

These effects are well-defined throughout the temperature range studied (Figure 11). During field cycling, P_r increases steadily. The asymptotic value is probably the P_b ~ 60 μC/cm^2 reported for lithium tantalate far away from its Curie temperature [57-60]. The data in Figure 4 are precisely for this temperature range. The coercive field E_c in this range is constant at ~ 3 kV/cm, which is tens of times lower than that in congruent lithium tantalate single crystals [1-4]. As the temperature is raised to 200°C, the dielectric hysteresis loop gives P_s ~ 60 μC/cm^2, known for lithium tantalite single crystals (Figure 12).

Further raising the temperature (t > 200°C) has no effect on the shape of the dielectric hysteresis loop, which gives P_s ~ 60 μC/cm^2. At 200°C, saturated dielectric hysteresis loops correspond to the switching of the entire stoichiometric lithium tantalate layer obtained upon VTE processing (Figure 12). Since the measurements were made under quasi-static conditions, field cycling led to drift of the loop as a whole, without increasing the polarization. Because of this, no measurements were made at considerably higher temperatures. Raman spectra (Ramanor U1000, excitation with the 514.5 nm argon laser line) demonstrate that the VTE $LiTaO_3$ samples contain layers differing in Li/Ta ratio. Just beneath the surface layer, there is an ~ 0.5-mm-thick layer of constant Li/Ta ratio. Judging from the coercive field in this layer, it has a nearly stoichiometric composition.

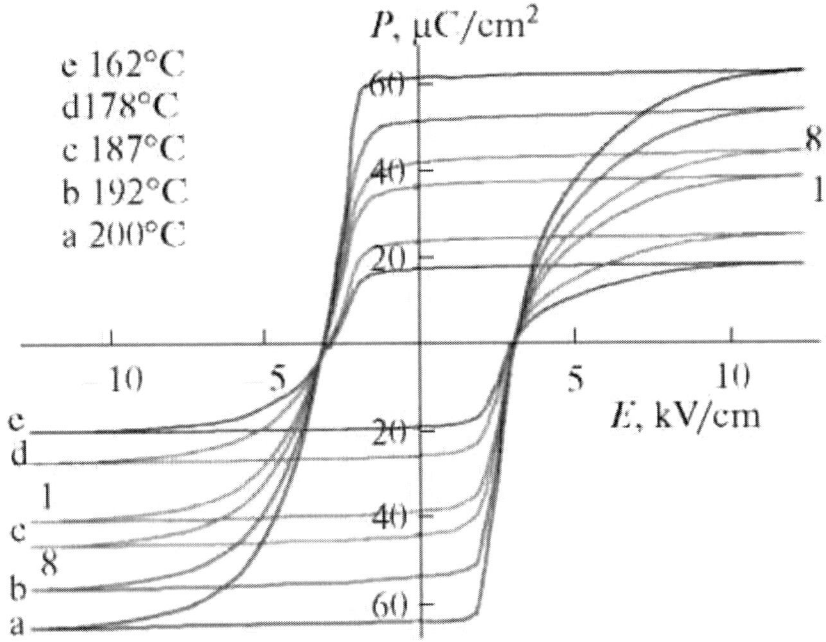

Figure 11. Quasi-static dielectric hysteresis loops of Z-cut single-crystal VTE $LiTaO_3$ at (a) 122, (b) 148, and (c) 156°C. The curves are numbered according to the number of the measurement cycle.

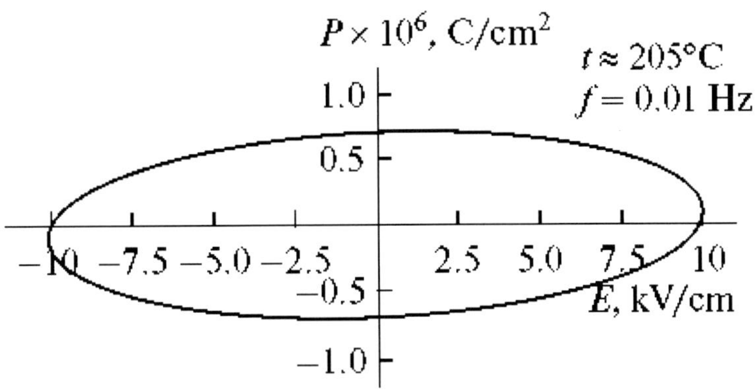

Figure 12. Quasi-static dielectric hysteresis loops of Z-cut single-crystal VTE $LiTaO_3$ (the same sample as in Figure 4) at higher temperatures. The 187°C curves are numbered according to the number of the measurement cycle.

Table 4. Width of the 140 cm^{-1} Raman line (*S*) as a function of distance from the surface (*L*) for VTE $LiTaO_3$

L, mm	0.1	0.2	0.4	0.5	0.6	0.7	0.8	1.0	1.0	1.1
S, cm^{-1}	10.3	10.3	10.3	10.3	10.7	12.5	12.6	12.7	12.7	12.7

That the Li/Ta ratio is constant is evidenced by the fact that the width (S) of the 140-cm^{-1} Raman line, *E(TO),* in the spectrum of VTE $LiTaO_3$ is independent of the distance from the surface in the range L = 0.1–0.5 mm (table 4).

Beneath a relatively thin (~100 μm) transition layer *(L = 0.6 mm, S = 10.9 cm^{-1})*, there is a layer with an almost constant Li /Ta ratio, which seems to be similar in composition to the parent congruent lithium tanta-late crystal, which had a rather high coercive field.

After grinding off the layer with a new, polar (ferroelectric) structure (~30 μm) and the stoichiometric layer (~500 μm in thickness), the sample behaves as a standard linear dielectric with low conductivity. The 205°C *P(E)* plot in Figure 12 has the shape of an ellipse whose parameters correspond to weak-field static values: $G_s = 1.3\times10^{10}$ S/m, $6'_s$ — 82 at the same temperature.

VTE processing of lithium tantalate single crystals produces layers tens to hundreds of microns thick which differ in phase composition and stoichiometry. Stoichiometric layers, up to 500 μm in thickness, can be used to create integrated optic devices (optical converters and minilasers) with a regular domain structure. After VTE processing, a thin (within 30 μm) surface layer of crystalline material has a new, polar (ferroelectric) structure with a phase transition temperature of ~ 120°C. Thus, VTE of single-crystal lithium tantalate produces not only stoichiometric layers but also new, polar phases. At t < 120°C, we observe partial switching with well-defined nonlinearity and pronounced hysteresis. The switchable polarization decreases as the temperature approaches ~ 120°C. Concurrently, dielectric hysteresis disappears and P becomes an almost linear function of E, which means a transition of the new polar (ferroelectric) phase produced by VTE to a paraelectric state. Upon cooling, the switchable polarization rises systematically to its original level.

All these findings lead us to conclude that VTE produces a new, polar (ferroelectric) phase with a Curie temperature of ~ 120°C.

4. Vibrational Spectrum and Defect Structure in Lithium Niobate Crystals

The physical properties of crystalline solids are frequently controlled by both their atomic and molecular composition and, to a significant extent, by crystal-structure features, in particular, fine ordering features. Structure parameters precisely affect the most important parameters of crystals, meaning, the response of a crystal to external actions. The description of the structure and structure evolution in response to external actions, together with the development of original methods for studying fine structural features, is a way to attack problems of paramount importance, namely, to govern the physical parameters of crystals.

Three-dimensional invariance is intrinsic to crystals; its absence is a signature of a noncrystalline substance. The actual situation in crystals is much more intricate: spontaneous perturbations were found in several crystal structures, either periodically (spatially) inhomogeneous or with a period indivisible by the main structure period (e.g., incommensurable structures).

Given this, either the structure loses three-dimensional periodicity, or the spatial period becomes so large that a conventional description in terms of group theory is highly in efficient. More adequate in this regard might be the addition of three variables, describing this perturbation, to the three spatial coordinates. To account for such new degrees of freedom requires that symmetry groups in >3D spaces be invoked in symmetry description. Therefore, the description of a real crystal structure is an immeasurably more complex problem.

In lithium niobate and lithium tantalite, the type and amount of intrinsic defects that spoil the ideal structure symmetry are varied and depend on many factors, such as off-stoichiometry, the thermal history of a crystal, crystallization kinetics, nonequilibrium crystallization conditions, frozen high-temperature disorder, inherited seed-crystal defects, and mechanical stresses. The basic reasons for the existence of intrinsic point defects in a crystal are off-stoichiometry and nonequilibrium crystallization.

Irregularities in the arrangement of small cation amounts along the polar axis of an LN crystal caused by off-stoichiometry practically cannot be detected using diffraction methods, in particular, single-crystal X-ray diffraction; observed diffraction peaks refer to a unit cell and are averaged over a crystal, containing many such cells. The dynamic properties of a crystal are far more responsive, than static properties, to similar structure defects [18, 30, 38].

Local structure perturbations along the polar axis in nominally pure and doped LN crystals can be detected in vibrational spectra, in the frequency ranges of the full-symmetry vibrations of octahedral ions. These vibrations are Raman active and are usually the strongest in the two-phonon excitement frequency region. Local structure perturbations can distort the spectra of two-phonon states through changing selection rules for the overall wave vector of quasiparticles. Lines due to local vibrations are expected to appear in the spectrum, along with a notable broadening of lines from the oxygen framework and the appearance of forbidden lines, as a result of the fact that incorporated impurity cations deteriorate the symmetry of oxygen octahedra.

Note also that, apart from applied (materials science) aspects, phonon spectra for a set of single crystals with compositions changing from ordered to disordered structures are of interest for fundamental research. LN and LT crystals with various extents of ordering doped with lanthanides and other elements are suitable test systems; fundamental vibrational spectra for nominally pure LN and LT crystals have been studied in detail and reliably assigned in the ideal structure approximation.

The rhombohedral unit cell of a lithium niobate crystal has space group $R3c$ and contains two formula units [9, 44]. $4A_1 + 9E$ fundamental vibrations are Raman and IR active. These vibrations, being polar, are split into longitudinal (LO) and transverse (TO) components. In addition, $A_1 + E$ acoustic and $5A_2$ optical vibrations exist; these vibrations are forbidden in Raman and IR spectra for wave vector $k = 0$ [61-63]. Therefore, a total of 26 lines associated with fundamental phonons must appear in Raman spectra given that phonons propagate along the principal crystallographic axes and in view of LO-TO splitting. In polycrystalline samples, only 13 lines are expected to appear due to A_1 and E fundamental phonons.

Vibrational (Raman and IR) spectra for the ferroelectric phase of mono- and polycrystalline LN have been studied carefully [61-78].

Those studies aimed at the assignment of fundamental phonon lines to symmetry types and to the LO or TO type. To this end, the Raman spectra of oriented single crystals were measured with polarized light using various scattering geometries [61-69], including angular dependences of frequencies for mixed LO-TO (anisotropic, oblique) phonons [70-72] and bulk polaritons [73, 74]. In [62, 70, 75, 76], IR reflection and absorption spectra were studied. Polarized IR absorption spectra have barely been studied because of the difficult preparation of thin, oriented single-crystalline samples.

When the results of these experiments were interpreted, in most cases the subtle features of the complex internal structure of the real crystals were ignored; rather, the ideal structure approximation for the stoichi-ometric ($R = 1$) composition was used. The test samples had, however, the congruent melting composition (R = 0.946) or a near-congruent one; these crystals are usually grown in the industry. Results obtained by different authors, in fact, refer to crystals with different internal structures and are controversial.

Attempts were made to account for the chemical composition of LN crystals in Raman spectroscopic studies [18, 30, 38, 64, 66, 68, 72, 77, 78].

In [18, 30, 38, 64, 72, 77], the Raman spectrum evolution as a function of chemical composition is directly assigned to structure ordering.

Figure 13 shows Raman spectra for a congruent crystal measured with scattering geometry *X(ZZ)Y* (A_1(TO) phonons are active) and scattering geometry *X(ZX)Y* (E(TO) and E(LO) phonons are active) [38]. The spectra of A_1 and E fundamental vibrations differ significantly from one another: lines due to E vibrations have lower intensities and are narrower than lines of A_1 vibrations. Because symmetry type A_1 corresponds to ionic vibrations along the polar axis and because type E corresponds to ionic vibrations normal to the polar axis, the above facts prove a strong anisotropy of the elastic constants of the crystal and a higher disorder level in the direction of the polar axis.

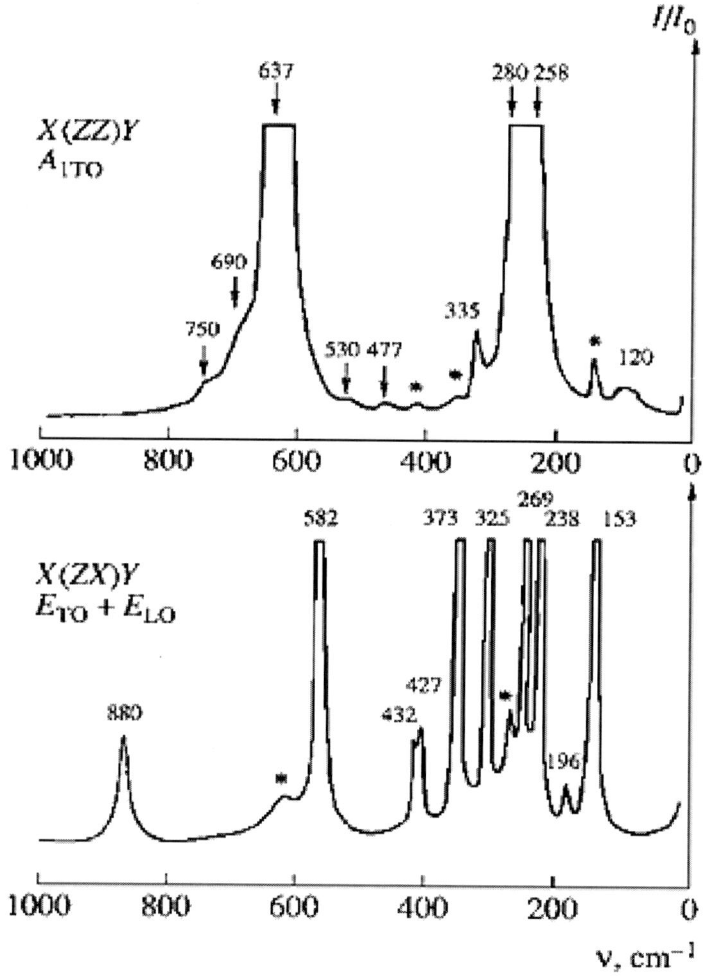

Figure 13. Raman spectra (77 K) for LN single crystals of congruent melting composition using scattering geometries *X(ZZ)Y* and *X(ZX)Y*.

The spectra shown in Figure 13 also display lines other than those of A_1 and E fundamentals. With scattering geometry *X(ZZ)Y* (where the selection rules allow only A_1 phonons to appear), several extra weak peaks appear along with the fundamental (the strongest) lines; these extra peaks are forbidden by the selection rules with reference to LO-TO splitting for the taken scattering geometry [72].

Some of these peaks are due to errors in polarization measurements and the photorefraction effect. These lines in Figure 13 are marked by asterisks. Their intensities are notably higher for congruent crystals, especially for the ones doped with photorefractive dopants [72]. However, this explanation does not apply to some other low-intensity lines; these lines are unob-servable in congruent crystals in other scattering geometries or in stoichiometric crystals, but they are well defined in polycrystals with various degrees of disorder (Figure 4) [38, 64, 72, 77]. These lines in Figure 13 are marked by arrows. Their frequencies are independent of the angle formed by phonon wave vectors and the polar axis, which circumstantially proves that they are not fundamental [72].

The authors of [18, 30, 38, 64, 72, 77] assign these lines to fine ordering features in lithium niobate crystals. Various types of defects, ones characteristic of the lithium niobate crystal structure among them, can be manifested in various ways in a vibrational spectrum. The defects responsible for local perturbations of translational symmetry in cationic and anionic sublattices fall, according to their spectral manifestations, into two main categories: defects arranged randomly or with some order.

Randomly arranged defects are understood well as regards to their effect on the vibrational spectrum. They exist in all crystals regardless of their chemical composition and structural features. In cases where randomly arranged defects only insignificantly perturb the crystal structure, line broadening is, as a rule, the only change in the vibrational spectrum.

When only such defects exist in the LN structure, fundamental vibrational spectra for disordered nonstoichiometric crystals must correspond, in the number of lines, to the fundamental spectrum of a high-order stoichiometric crystal, which displays no extra lines [38, 64, 72, 77].

However, in the Raman spectra of nominally pure real lithium niobate crystals with $R < 1$ (containing no impurity phases or extraneous ions), low-intensity extra lines appear that are not allowed by the selection rules for space group C_{3v}^{6} (*R*3*c*) [18, 30, 38, 64, 72, 77].

The frequencies of these extra lines are fixed and, unlike the fundamental lines, are independent of the chemical composition of a crystal [18, 30, 38, 64, 72, 77]. X-ray crystallography shows that the space group of a crystal also remains unchanged.

It is unlikely that extra lines arise from chaotic perturbations of the structure order induced by defects. Strong local perturbations induced by antisite ions or groups of ions can, under certain conditions, deteriorate the vibrational symmetry, which is manifested as frequency shifts and the appearance of new lines.

The entire fundamental spectrum is, as a rule, broadened and distorted. However, there is abundant experimental evidence that, although line widths in the fundamental spectra of LN crystals within the homogeneity region change appreciably (more than twofold), the spectrum as a whole conserves its individuality [38, 48, 72, 77].

An essential fact is that extra lines are observed mostly when the scattering geometry is such that it allows A_1 phonons to appear, corresponding to ion vibrations along the polar axis of a crystal, such as geometries *X(ZZ)Y* and *Y(XZ)Z*, Figure 14. At the same time, spectra for

E vibrations coincide nicely with fundamental spectra. This fact proves that cation order in an LN crystal is significant for a vibrational spectrum.

It is difficult to achieve an almost full and adequate match of the structural order in complex crystals, such as LN, with the vibrational spectrum. Such crystals, even though grown under identical conditions, frequently differ in their chemical compositions and the defect state [1, 2].

The formula unit $LiNbO_3$ only conventionally fits the chemical composition of an LN crystal. It characterizes the composition of an ideal (defect-free) crystal, which actually does not exist. The actual chemical composition of crystals having such a wide homogeneity range can appreciably differ from the composition given by this formula unit [1, 2, 11, 13, 14, 25, 56, 78, 79]. The space group remains the same, but unit cell parameters can vary within small ranges [1].

Numerous studies show that Raman spectra are very responsive to compositional variations in a nominally pure lithium niobate crystal within the homogeneity range [19, 25]. In particular, line widths and intensities change substantially as a function of R = [Li]/[Nb]. A notable correlation exists between the crystal composition and line width [19, 25, 30, 38] (Figure 15). This correlation holds for Li_2O–Nb_2O_5 melts [54-56].

Currently, there is no dominant view on how the intrinsic defect structure affects the physical, in particular, the optical, parameters of an LN crystal.

The fact that optical absorption (crystal color) does respect, nominally pure LN crystals differ little from perovskites like $BaTiO_3$, in which nonstoichiometry is usually associated with color centers and an increased electrical conductivity.

The associated increases in the density and unit cell volume in the Li_2O deficit region ($R < 1$) suggests that in such LN crystals some of the excess Nb^{5+} cations can occupy Li^+ sites or other extra positions of the unit cell.

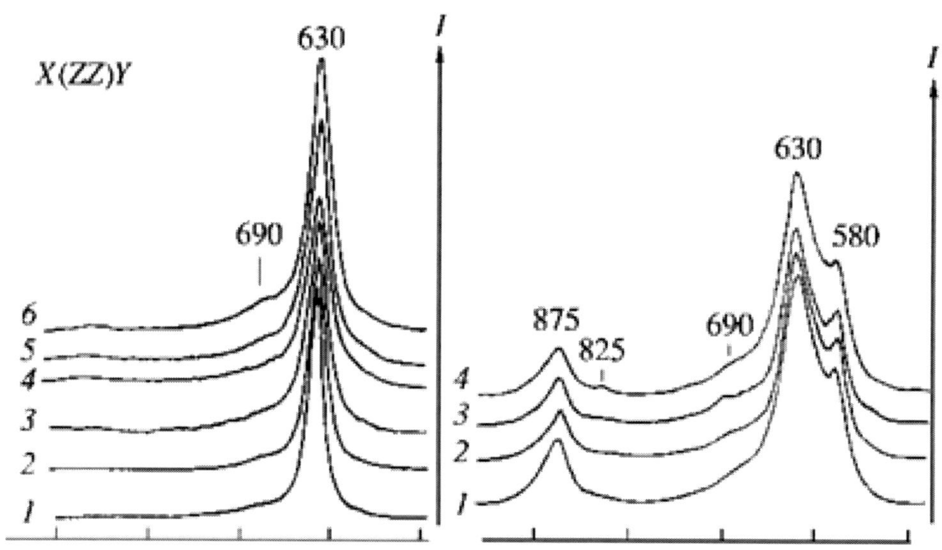

a b

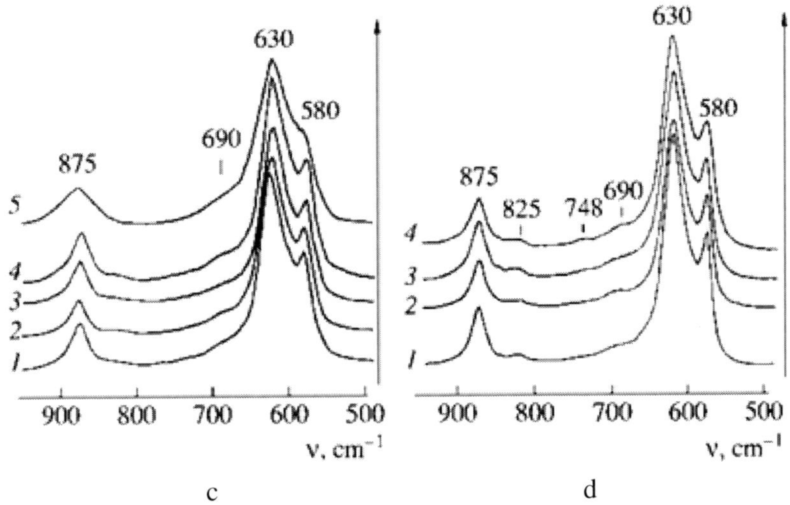

Figure 14. Fragments of Raman spectra ($T = 293$ K) for LN in the frequency ranges in which the vibrations of oxygen octahedra appear.

For electroneutrality to hold, not change significantly as a function of R within the homogeneity range proves that the charges of lacking Li^+ ions (for $R < 1$) or Nb^{5+} ions (for $R > 1$) are not compensated for by color centers, e.g., electrons located in oxygen vacancies [1-4].

At the same time, a significant photo-refraction effect proves that there are enough energy levels from which electrons can migrate over a crystal exposed to laser radiation followed by their localization on deep trapping sublevels in the bandgap [1]. In this it is required that a proper amount of Li- or Nb-site vacancies form.

The structural perfection of nominally pure phases of variable composition, like lithium niobate crystals, having a developed defect structure must be primarily controlled by the concentration of intrinsic defects, which lead to maximal structure disorder [1-4, 11, 13, 14, 18, 30, 38, 64, 72, 77, 79]. In the cation sublattice of stoichiometric crystals (potentially, the most ordered crystals), Li^+-site vacancies are experimentally observed and Nb^{5+} ions can substitute for Li^+ ions and reside in vacant octahedra [1]. The majority defects in a congruent LN crystal are, likely, excess Nb^{5+} ions in lithium sites.

Therefore, charge neutrality in this crystal is conserved by antisite defects Nb_{Li}. An increased density of niobium-rich crystals cannot be interpreted in terms of the oxygen vacancy model. This type of defect, likely, plays only an insignificant role in electroneutrality conservation [56].

Therefore, the generation of intrinsic and impurity cation defects in lithium niobate is accompanied by substantial perturbations of translational invariance in the cation sublattice along the polar axis. Structural disorder can be very intricate: above all, similar cations can appear in crystallographically nonequivalent positions (in sites of other cations or vacancies). Density inhomogeneities in the form of clustered cations and vacancies can form in the cation sublattice [1, 11, 79, 15].

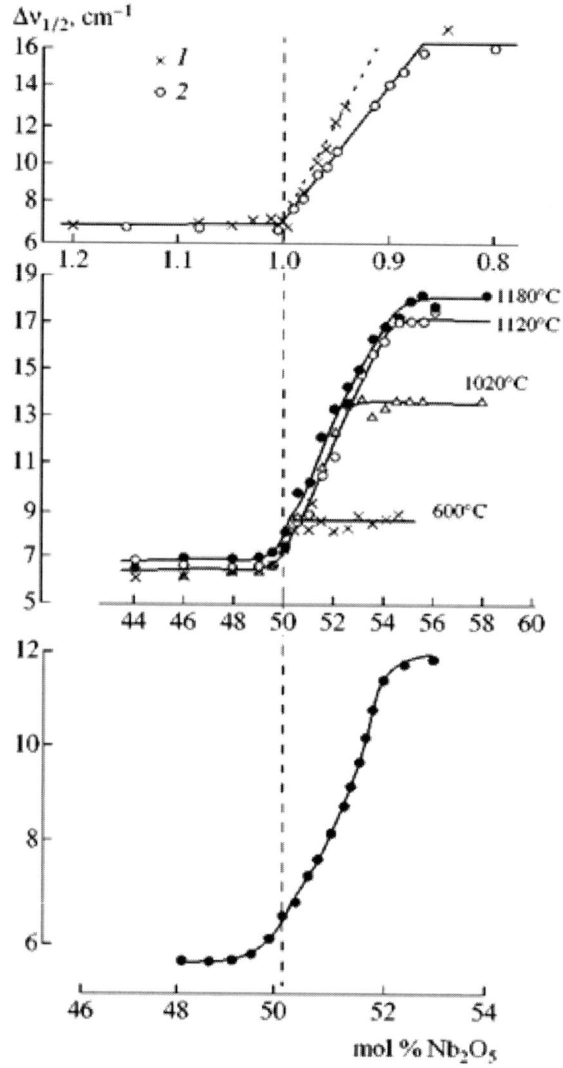

The upper and middle panels show data taken from [19] and [25], respectively. The lower panel presents data from [26].

Figure 15. Plot of the half-width of the line 152 cm^{-1} *E*(TO) vs. the chemical composition of an LN crystal.

These clusterlike regions can be extensive and reach 5-10 translation periods; their concentration in a nominally pure crystal of a congruent composition can be significant (>10^{20} cm^{-1}) [15, 39]. Microinclusions of an ilmenite-like structure can exist in lithium niobate crystals; their dimensions are also several translation periods [76]. These microinclusions can appear as a result of the distortion of the matrix crystal structure in the vicinity of Nb_{Li}, intrinsic defects most characteristic of $R > 1$ crystals [39, 76, 80, 81]. Complex cluster defects can form as a result; they include, together with ilmenite inclusions, charged centers Nb^{4+}, V_{Li+}, V_{Nb5+} , and V_0 [11, 79]. In doped crystals, molecular complexes are also formed in their cation sublattices [1]. The value and direction of spontaneous polarization in such clusters can strongly differ from their crystal-average quantities; cluster defects per se can form ordered sublattices in the LN structure rather than being randomly distributed over the crystal. This issue was separately considered in [38, 64, 72, 77].

Thus, cation disorder patterns in the cation sublat-tice of lithium niobate crystals are diversified and are governed by many factors whose constancy is hardly achievable in crystal growth runs.

Moreover, lithium niobate is not necessarily the only product of high-temperature solid-phase synthesis in $Li_2CO_3 : Nb_2O_5 = 1 : 1$ mixtures [1, 82-85]. Other niobates, above all, Li_3NbO_4 and $LiNb_3O_8$, can form in addition. At high temperatures, they react with one another and with feed components. Synthesis, thus, can involve several solid-phase chemical reactions yielding intermediates.

With regard to the aforesaid, recall that the properties of lithium niobate crystals and their structure perfection are strongly affected by the thermal history of a crystal. At room temperature, LN crystals with $R < 1$ are metastable; under certain conditions, their structure degrades to segregate other phases [1, 2, 7, 85]. The segregation is possible because $R < 1$ compositions at room temperature lie outside the homogeneity region of the Li_2O - Nb_2O_5 system (Figure 2b).

From the above, we suggest that, in a certain concentration interval, the crystal perfection must be improved by the incorporation of dopant cations that compete with Nb^{5+} cations for lithium sites and, accordingly, decrease the formation probability of anti-site defects Nb_{Li}. This was observed in [18, 30, 38, 64, 72, 77, 86, 87]. In [18, 30, 38], it was shown that, for Li^+ ions substituting for Nb^{5+} ions, this ordering effect must go beyond the stoichiometric composition; the tendency toward structural perfection likely holds for $R > 1$ compositions.

The Raman spectra of lithium niobate ceramics [18, 30, 38] confirmed that an increase in Li^+ concentration positively affects structural ordering within the range $[Li_2O] = 47$-52 mol %. Figure 16 shows fragments of Raman spectra for ceramic LN samples, differing in $R = [Li]/[Nb]$, in the frequency range in which the fundamentals of ions residing in oxygen octahedra appear. The line 277 cm^{-1} notably increases its intensity when the lithium concentration of the ceramics rises. This fact can serve to support that the line 277 cm^{-1} is due to the full-symmetry fundamental vibrations (A_1) of Li^+ ions along the polar axis. In this case, the line 255 cm^{-1} can be assigned to similar vibrations of Nb^{5+} ions.

An increase in $R = [Li]/[Nb]$ must improve the order of the cation sublattice because Li^+ ions occupy more of their own sites. Antisite defect (Nb_{Li}) formation becomes progressively less probable. This is also manifested in Raman spectra: the line widths corresponding to the A_1 vibrations of Li^+ and Nb^{5+} ions notably decrease (Figure 6) [18, 30, 38]. Thus, when the Li^+ content of a nominally pure LN crystal rises, the structural perfection is improved because of the better ordering of Li^+ and Nb^{5+} cations and vacant octahedra along the polar axis.

The Raman line width versus Li_2O concentration dependence for ceramic lithium niobate samples is found in [18, 30, 38], which is fully consistent with the data in [19, 25] (Figure 15), proves that the state of the system corresponds to the equilibrium of the sintering temperature (1323 K). A notable decrease in the line width observed up to $[Li_2O] = 51$ mol % (Figure 15) supports the inference that, likely, the amount of Li^+ cations residing in proper lithium octahedra increases even when the lithium concentration is higher than 50 mol %. The crystal structure, in particular, the cation sublattice, becomes markedly ordered. The weaker dependence of Raman line widths on the Li_2O concentration when $[Li_2O] > 50$ mol % suggests that this process slows down but does not stop. An insignificant growth in the line intensity at 277 cm^{-1} supports this suggestion (Figure 16) [18, 30, 38].

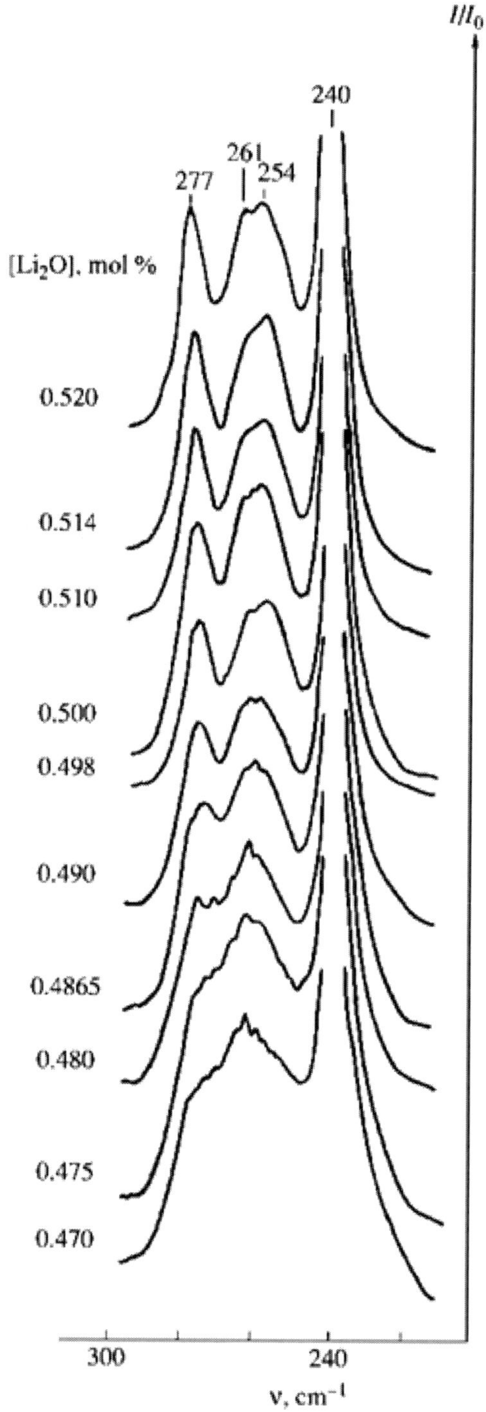

Figure 16. Evolution of the Raman spectra (293 K) of ceramic LN samples in the frequency region of the full-symmetry (A_1) fundamental vibrations of Li^+ and Nb^{5+} ions as a function of chemical composition.

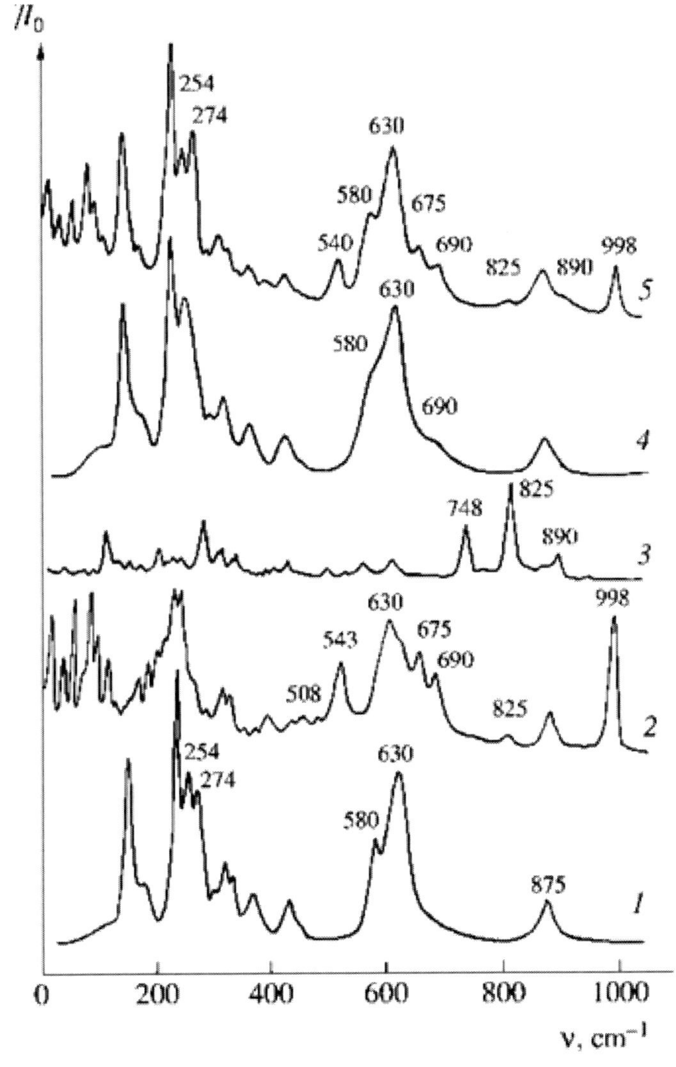

Figure 17. Raman spectra of ceramic samples measured at 293 K for (*1*) LN of congruent melting composition, (*2*) $LiNb_3O_8$, (*3*) Li_3NbO_4, (*4*) LN of congruent composition thermally processed for 20 h at 1500 K, and (*5*) LN containing $LiNb_2O_8$ and Li_3NbO_4 phases.

First, the evolution of Raman spectra as a function of the stoichiometry and impurity composition of a lithium niobate crystal was considered in [19, 25, 66, 88, 89].

For the majority of real LN crystals, Raman lines, as a rule, broaden in response to a change in R = [Li]/[Nb], while R remains within the homogeneity range; the reason for this is an increased defect concentration associated with spatial, chaotic, local perturbations of translational symmetry [66, 68, 78].

The fundamental vibrational spectrum, on the whole, corresponds to the fundamental spectrum of a stoichiometric crystal. $R < 1$ crystals, however, display low-intensity (extra) lines, forbidden by the selection rules for space group C_{3v}^{6} ($R3c$) [38, 64, 72, 77, 88] (Figures 13, 17).

One reason for the aforesaid is the following: impurities or intrinsic defects, under certain conditions, are located in the lattice so as to stabilize a superstructure that slightly differs from the structure of a stoichiomet-ric crystal in the matrix lattice [76, 80, 90-92]. Such a complex crystal can substantially differ in its vibrational spectrum from a stoichiometric crystal. Defect-induced chaotic perturbations of the translational symmetry, unaffecting space group, usually cause vibration dephasing [93]. Such the chaotic phonon dephasing on defects can statistically be modeled by spatial-damping waves with the damping factor $c = 1/L$ (L is an average defect–defect distance) [93]. Damping leads to line broadening as a result of the spoiled interference conditions during scattering.

From Figure 5, it is seen that a change in the Li_2O concentration from 47 to 52 mol % causes the line width at 150 cm^{-1} to change more than twofold.

Heavier perturbations can cause the Brillouin zone to open; not only limiting ($k = 0$) optical frequencies but also other frequencies in the Brillouin zone (dictated by the scatter in wave vector k) become observable in the spectrum; their intensities are proportional to the defect concentration [93]. In view of the low optical dispersion, rather narrow extra lines, not allowed by the selection rules for the space group of an ideal crystal, can appear in the spectrum [93]. However, the origin of particular extra lines is more complex and needs special investigations involving particular structure modeling.

If we take that Brillouin zone opening as a result of defect-induced perturbation of the translational invariance of the cation sublattice is the only reason for extra Raman lines in nonstoichiometric LN crystals, we oversimplify the interpretation of the Raman spectrum. This explanation is too general. It is primarily based on the chaotic defect distribution, and it ignores many features of the complex internal structure of an LN crystal and the features of defect distribution.

Recent precision investigations of subtle features of structural order in an LN crystal [11, 13, 18, 30, 38, 64, 72, 77–79, 86, 87, 91-94] suggest that not only are defects distributed chaotically over the crystal lattice (this is primarily manifested as line broadening), but, under certain conditions, impurity or intrinsic defects, clusters, or other entities (e.g., molecular complexes) are located in a nonstoichiometric crystal so as to stabilize a superstructure (defect sublattice) in the matrix structure. Defects in such a crystal are arranged in a definite order rather than randomly.

The resulting sublattice has a structure that, in general, differs from the highly ordered structure of a stoichiometric crystal. The vibrational spectrum of such a crystal can substantially differ from the highly ordered stoichiometric crystal. Up to now, several theoretical structural models have been created for crystals with a similar type of disorder; an ordered location of defects is discussed in these models [11, 15, 79].

Because experimental results on Raman line assignment for real LN crystals are discrepant and because the chemical composition of a crystal was ignored in the assignment, in [64, 72, 77] polarized spectra from stoichiometric and congruent single crystals were comprehensively studied and vibrations were classified according to their symmetry (LO or TO) types. Using the dependence of the LO and TO phonon frequency on the angle formed by the phonon detection direction and the polar axis, lines from fundamental vibrations were separated out, and LO and TO phonons were separated in pairs of lines corresponding to one branch.

No differences in the frequencies of lattice fundamentals were found between stoichiometric and congruent crystals. However, significant differences were found in the number of lines (Figure 14) [64, 72, 77].

The defect structure of complex crystals such as LN can vary as a function of the chemical composition and thermal history; its spectral manifestations are also diverse. For example, under some conditions, impurity atoms and (or) intrinsic defects can be located in the structure so as to form an extra ordered impurity (defect) sublattice. In order to distinguish lines due to lattice fundamentals from lines that can arise from the defect sublattice, the Raman spectra of high-purity LN single crystals of congruent and stoichiometric compositions were studied in comparison to crystals having the same Li/Nb melt ratio but having various types of artificially generated impurity defects [64, 72, 77].

The congruent crystals showed fewer low-intensity extra lines [64, 72, 77] than other authors observed. This is evidence of a higher perfection of the congruent crystals used in [64, 72, 77]. Some of the extra lines disappear upon doping [64, 77, 87]. In stoichiometric crystals, no extra lines were found. Only lines due to lattice fundamentals were observed here. Almost all extra lines were found in the crystals in which various types of structural disorder were generated artificially, through changing a chemical composition, cation doping, or thermal processing (Figures 14, 17) [38, 64, 77]. In such disordered crystals, all lines associated with lattice fundamentals are appreciably broadened compared to the spectral lines from stoichiometric crystals, the latter potentially having the highest cation order (Figures 14, 18). No differences were observed in the fundamental frequencies [64, 72, 77, 86, 87].

Nonstoichiometric crystals are specific in that, when their nonstoichiometry increases, the lines 254 and 274 cm^{-1}, due to the full-symmetry A_1(TO) fundamentals of cations residing in oxygen octahedra, change first; at higher nonstoichiometry levels (upon doping of a congruent crystal), the lines 580 E(TO) and 630 A_1(LO) cm^{-1} also broaden; these lines are due to the vibrations of oxygen octahedra (Figure 8) [64, 72, 77].

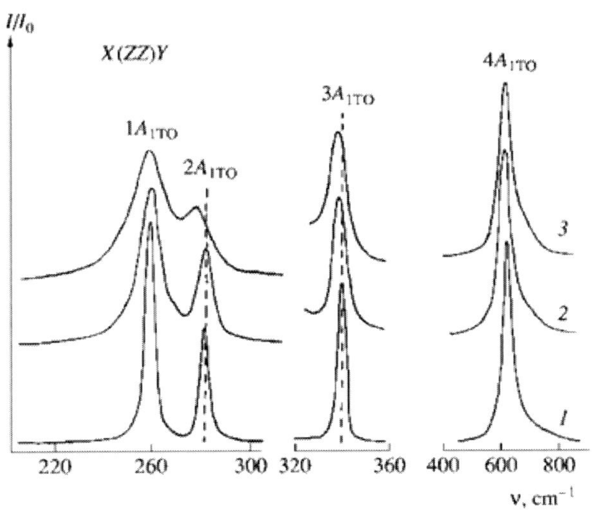

Figure 18. Raman spectra of A_1(TO) phonons for LN single crystals of various chemical composition: (*1*, *2*) stoichiometric and congruent compositions and (*3*) a congruent composition doped with Gd^{3+} (0.23 wt %) and Mn^{3+} (0.51 at %). $T = 77$ K.

This is especially pronounced in crystals in which the Li/Nb melt ratio corresponds to the congruent composition and which are doped comparatively heavily with cations whose ionic radii are close to the Li^{+} or Nb^{5+} radius and whose charges are between the charges of these ions (e.g., Mg^{2+}, B^{3+}, Gd^{3+}, or some others; Figure 8) [38, 64, 72, 77, 87].

5. Photorefractive and Raman Light Scattering in Lithium Niobate Ferroelectric Single Crystal

A lithium niobate ferroelectric single crystal has a variable composition and a strongly imperfect structure and exhibits a clearly pronounced photorefractive effect, which considerably depends on the composition and the degree of imperfection of the crystal structure [44, 95]. Information on the photore-fractive properties of the $LiNbO_3$ single crystal and on its photorefractive light scattering is very important for solving problems on the creation of materials for holographic information recording, for generation and frequency conversion of laser radiation, and for laser-assisted controlling the properties of materials. A special role in the formation of the photorefractive effect and photorefractive scattering is played by intrinsic and impurity defects with localized electrons and defects induced by laser radiation [44, 96].

The photorefractive effect arises in an illuminated region of a ferroelectric crystal as a result of the spatial transfer of electrons under the action of light and their subsequent capture on deep energy levels with the formation of a field of a nonequilibrium space charge, which changes the refractive index [44, 95-98]. As the laser radiation propagates through this inhomogeneous single crystal, it experiences a random modulation, which manifests itself in the structure of the scattered light, which also makes it spatially inhomogeneous. The laser radiation scattered by inhomogeneities interferes with the pumping radiation to form a complicated pattern of intensity minima and maxima of photorefractive scattering, the speckle pattern [99]. That is, in the course of the irradiation of the crystal, in the spatial region of the propagation of the laser beam, instability develops in the system under strongly unsteady-state conditions and structures with conspicuously pronounced self-organization are formed. The photorefractive scattering negatively affects the information recording and the transformation of laser radiation.

In the region of the propagation of the laser beam and in a certain vicinity near it (whose size can reach a few millimeters) both the refractive index of the crystal and its structure noticeably change, with these changes being preserved for a long period of time after the action of the laser radiation [95-97]. Despite the fact that these distortions have been examined in a series of solid publications (a review is given in [44, 95, 97-99]), their subtle features in relation to the composition of the $LiNbO_3$ single crystal still remain to be clarified. It is most important to study the fluctuating and static micro- and nanodefects induced by the laser radiation and characteristics of radiation scattered by them. Laser-induced defects in single crystals doped by pho-torefractive[1] cations arise because the charge state of these cations changes [44, 96-98].

[1] Photorefractive (variable-valence) cations change their charge in a crystal under the action of light and enhance the photorefractive effect. Nonphotorefractive cations do not change their charge under the light and, under certain conditions, can reduce the photorefractive effect.

In the literature, the photorefractive effect and the photorefractive light scattering were mainly studied in congruent lithium niobate crystals (R = Li/Nb = 0.946) doped with Fe and Rh photorefractive cations, which considerably enhance the photorefractive effect [44, 95-99].

At present, the nature of fluctuating and static defects that are induced by laser radiation in nominally pure $LiNbO_3$ single crystals of a stoichiometric composition (Li/Nb = 1), which exhibit a stronger photorefractive effect compared to congruent single crystals (Li/Nb = 0.946), is absolutely unclear and has not been investigated at all. The difference between the Nb-O and Li-O bond strengths, as one of the reasons for the inadequacy between the composition of the congruent melt and the stoichiometric composition, gives rise to a comparatively easy formation of lithium vacancies in the crystal. The number of these vacancies does not decrease due to the heterovalent isomorphism process, i.e., the replacement of lithium with niobium in the cationic sublattice (because the ionic radii of Li^+ and Nb^{5+} are close to each other). An inevitable consequence of this process is the formation of new vacancies at lithium sites. The main result of the mentioned isomorphism is the disordering of the structure of the cationic sublattice of the crystal, which, among other things, is related to a partial reduction of Nb^{5+} ions and formation of intrinsic cluster charged defects, governing the character of the photorefractive effect and photorefractive scattering in the crystal. The large role played by intrinsic defects with localized electrons in the formation of the photo-refractive effect in these crystals is evident [44, 95].

We study the characteristics of speckle structures (the scattering indicatrix and the distributions of speckle fields and intensities) and the Raman spectra of a stoichiometric lithium niobate single crystal that was grown by the Czochralski method from a melt with 58.6 mol % of Li_2O. Stoichiometric $LiNbO_3$ single crystals are promising materials for holographic information recording (because of their relatively strong photorefractive effect) and for nonlinearly active laser media with periodically polarized submicron domains (because the coercive field strength in them is considerably lower compared to congruent crystals [100]).

Experiments on photorefractive scattering were performed using radiation from an argon laser (Spectra Physics; λ_0 = 514.5 nm) and a MLL-100 Y:Al garnet laser (λ_0 = 530.0 nm) with a power of up to 160 mW. The speckle structure of photorefractive scattering was observed on a semitransparent screen placed behind the crystal and was recorded by a digital videocamera. Frames were selected using a special program and the opening angle of the photorefractive scattering indicatrix was determined. In more detail, the experimental technique was described in [99]. Raman spectra were excited by radiation at 514.5 nm (P ~ 200 mW) from a 2018-RM argon laser (Spectra Physics) and were recorded with a spectrograph of an original design [101]. Stoichiometric single crystals were grown by the Czochralski method on a Kristall-2 setup from a melt that contained 58 mol % Li_2O. Single crystal samples for investigations were cut as parallelepipeds with an overall dimension of ~7 × 6 × 5 mm with their edges that coincide with the X, Y, and Z crystallographic axes (Z is the polar axis of the crystal). The faces of the parallelepiped were thoroughly polished.

As a photorefractive $LiNbO_3$ single crystal is irradiated by visible laser light, a speckle structure is formed (Figure 19). At the very first moment of irradiation of the crystal, the scattered light has a shape of a single circular central spot with a small opening angle of the indicatrix (Figure 20a; $t = 1$ s). Then, in the course of time, the opening angle of the speckle structure increases, and three layers of the structure can be observed (Figure 20; $t = 30$ s).

The indicatrix of photorefractive scattering that is opened upon laser irradiation of the single crystal is not a single formation, but, rather, has three types of speckles, which are arranged successively one by one. The indicatrix of photorefractive scattering is opened as a figure-of-eight that is oriented along the polar axis of the crystal such that a larger lobe of the figure lies in the positive direction of the polar axis, while the a smaller lobe is aligned along negative direction of the axis. The central layer of the speckle structure is a bright spot with a highest intensity, the intensity of the brightness of the second layer is lower, and the third, peripheral, layer has a clearly pronounced granular speckle structure (Figure 19). Each layer of the speckle structure is shown in more detail in Figure 21.

As the time and power of irradiation increase, the shape, contrast, and intensity of the speckle structure change, and the opening angle of the indicatrix of photorefractive scattering increases because the refractive index changes (Figures 20 and 4). In this case, the peripheral third layer experiences the most considerable changes with increasing power and action time of the laser radiation (Figures 20a and 20b).

Therefore, all the three layers of the speckle structure of the $LiNbO_3$ single crystal are opened stage by stage. The central spot of the indicatrix of photorefractive scattering appears nearly instantaneously at a rate that is close to the velocity of propagation of the electromagnetic wave. Then, the second layer, which corresponds to the photorefractive scattering by static defects induced by the laser radiation [96], is opened.

And only after that, the third layer, which corresponds to the photorefractive scattering by fluctuating laser-induced defects, is opened.

It is likely that, with increasing power of the excitation radiation, each layer of the speckle structure can be observed separately.

At low powers of the laser radiation, one should observe only the central spot. An increase in the power leads to the successive appearance of the second and, then, of the third layer of the speckle structure.

The shape of the scattering indicatrix depends on the structure of the crystal, the polarization of the radiation, and the geometry of experiment. The opening angle of the indicatrix of photorefractive scattering attains a steady-state value considerably faster at high pumping powers compared to low powers (Figures 20, 22).

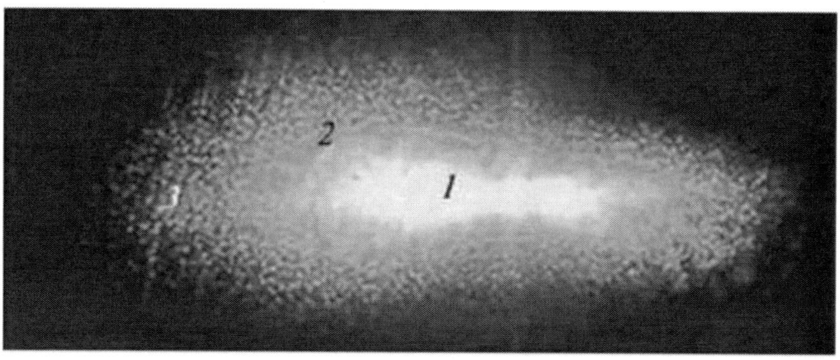

Figure 19. Three-layer speckle structure of photorefractive scattering in stoichiometric lithium niobate single crystal grown from melt with 58.6 mol % of Li_2O: (*1*) central layer, (*2*) second (static) layer, and (*3*) third (fluctuating) layer.

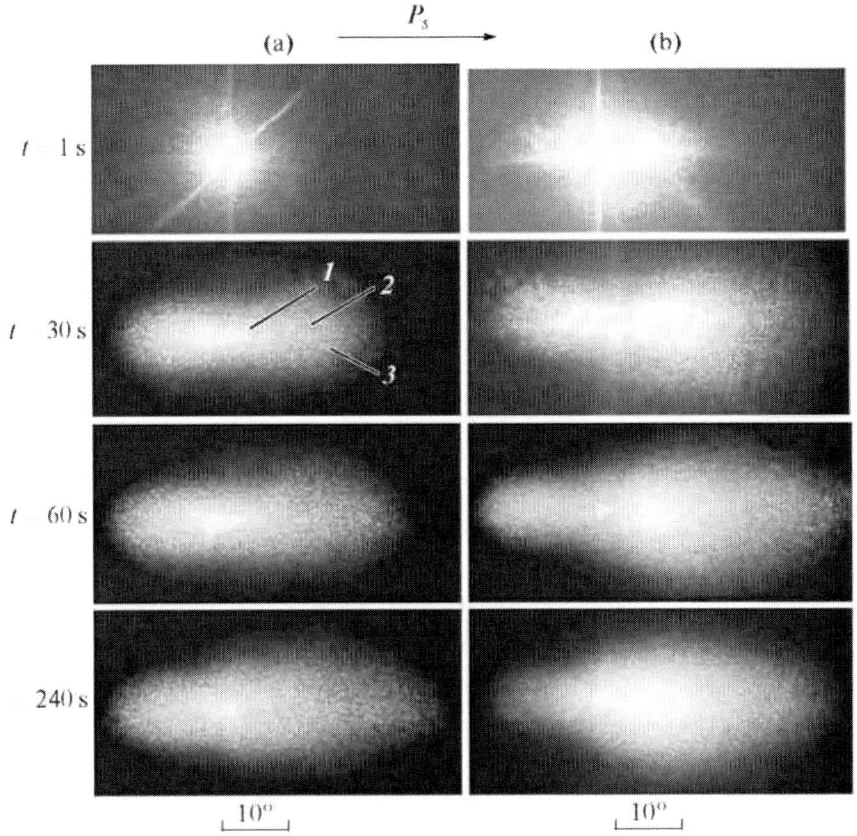

Figure 20. Indicatrix of photorefractive scattering in stoichiometric $LiNbO_3$ single crystal upon excitation by MLL-100 Y:Al garnet laser (λ_0 = 530.0 nm) with power of (a) 35 and (b) 160 mW: (*1*) central layer, (*2*) second layer, and (*3*) third layer.

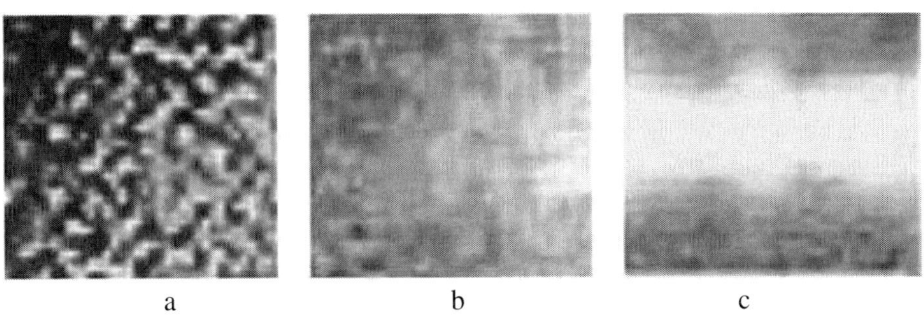

Figure 21. Structures of speckle layers obtained upon laser irradiation of stoichiometric lithium niobate crystals: (a) first layer showing photorefractive scattering by micro-structures of crystal with fluctuating refractive index; (b) second layer showing photorefractive scattering by microstructures with changed (static) refractive index; (c) third layer showing photorefractive scattering by lowest track.

The dynamics of the development of the photorefractive effect in a ferroelectric $LiNbO_3$ single crystal becomes clear from the presented results. The photo-refractive effect is also developed in three stages. Initially, in the propagation region of the laser beam in the single crystal, bright dots appear that are caused by the scattering of radiation by intrinsic micro- and nan-odefects and by micro- and nanodefects (fluctuating and static) induced by the laser radiation. With increasing irradiation time, as well as with increasing power of the laser radiation, the number of induced defects increases and they are gradually transformed into a laser track in which the refractive index differs from the refractive index of the single crystal not exposed to the action of the radiation (Figure 23). However, near the track, where the action of the radiation on the crystal is significantly lower, laser-induced micro- and nanoinhomogeneities of the structure can be clearly seen as static or fluctuating micro- and nanostructures (Figure 24). In this case, the distribution of static defects over distances from the center of the laser spot has several clearly pronounced maxima (Figure 25).

After irradiation, the laser track can occur in the crystal for a very long period of time (up to one year in the dark), which is determined by the time of the Maxwell relaxation.

The occurrence of the track indicates that this material can be used for information recording. In this case, the photorefractive scattering is the factor that impedes the information recording. In the literature, the laser track was observed only in single crystals doped by photorefractive cations. We were the first to observe the laser track in a stoichiometric $LiNbO_3$ single crystal (Figure 23).

In contrast, the laser track in congruent crystals has not been detected.

The shape of the studied three-layer speckle structure (Figure 19) is characteristic of $LiNbO_3$ single crystals both nominally pure (stoichiometric and congruent) and doped with photorefractive (e.g., Fe and Rh) or nonphotorefractive (Zn^{2+}, Mg^{2+}, Gd^{3+}, etc.) cations [99, 102, 103]. At the same time, the speckle structure of photorefractive scattering in different crystals has specific subtle features, by which one can study the structure crystals and their homogeneity at the micro- and macrolevel.

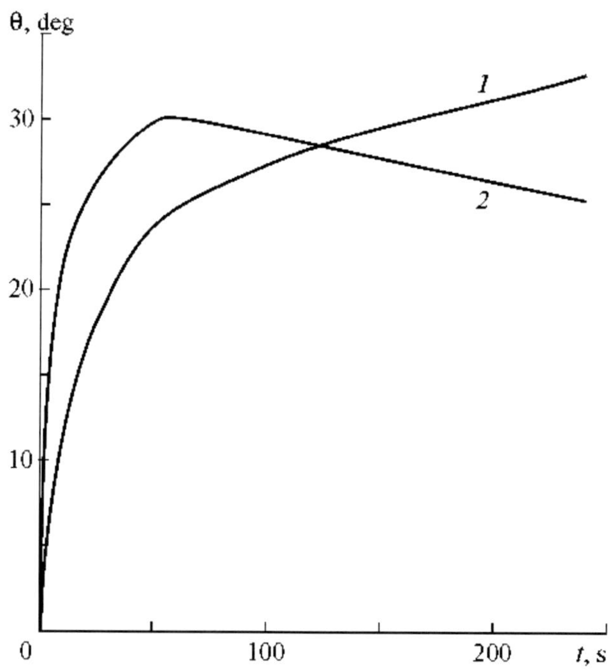

Figure 22. Time dependences of photorefractive scattering angle in stoichiometric $LiNbO_3$ single crystal upon excitation by radiation from MLL-100 Y:Al garnet laser ($\lambda_0 = 530.0$ nm) with power of (*1*) 35 and (*2*) 160 mW.

Further studies of speckle structures in lithium niobate crystals of different composition, which differ in the ordering of structural units of the cationic sublattice and state of imperfection of the oxygen and cationic sublattices, are of undoubted interest for creating materials with predetermined photorefractive characteristics.

Finally, we should emphasize the following. In the Raman spectrum, the photorefractive effect, among other things, manifests itself in a significant depolarization of the excitation laser radiation and in the occurrence of lines in the spectrum that are forbidden by the selection rules for the given examined geometry of scattering [44]. Furthermore, it is believed in the literature that the intensity of forbidden lines gradually increases with opening of the scattering indicatrix [88]. The results obtained show that the intensities of forbidden lines in the Raman spectrum increase to a maximal level almost instantaneously (just like the photorefractive effect) because the refractive index changes under the action of light at a rate of the motion of electrons in the crystal. This is evidenced by the almost instantaneous occurrence of the central layer of the speckle structure (Figure 20; t = 1 s). To verify this assumption, we measured the Raman spectra of a stoichiometric $LiNbO_3$ single crystal with a rather strong photorefractive effect within 30 min with a step of 1 s. The spectra were measured on a multichannel spectrograph of an original design [101], which made it possible to record the entire Raman spectrum of lithium niobate within ~ 0.1 s.

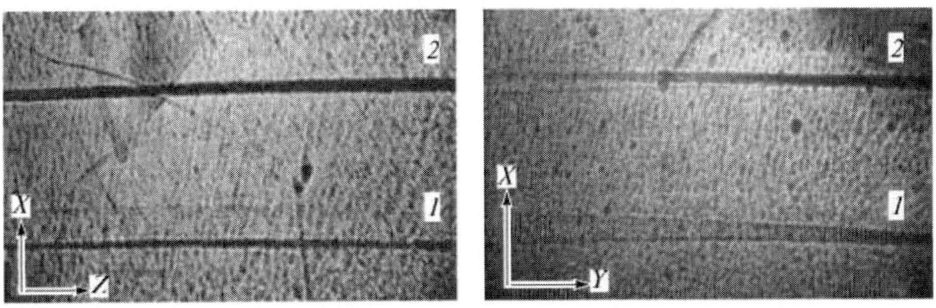

a b

Figure 23. Images of laser beam in stoichiometric lithium niobate crystal in (a) *ZX* and (b) JY planes obtained in *(1)* 5 and *(2)* 12 min; vector E of laser radiation coincides with polar axis.

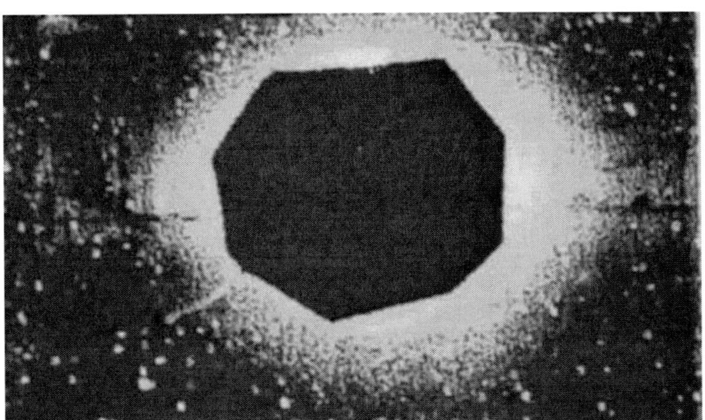

Figure 24. Photograph of illuminated region near laser beam in photorefractive stoichiometric lithium niobate single crystal. Polar axis and laser beam are perpendicular to plane of figure.

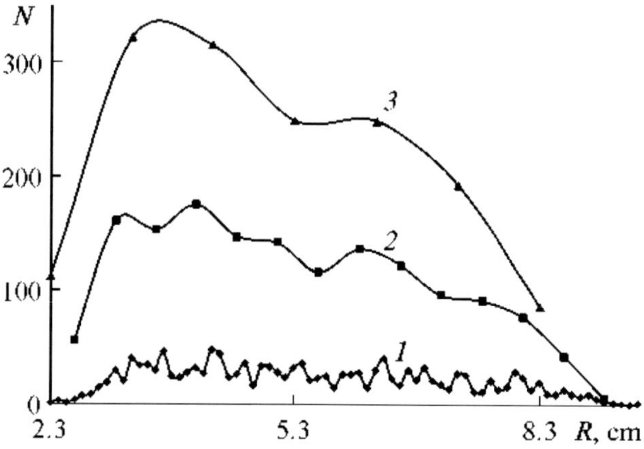

Figure 25. Distributions of dots over distances from center of laser track: curves *1*–*3* show number of dots contained in concentric ring whose radius increases at each step by (*1*) 1, (*2*) 5, and (*3*) 10 mm, respectively.

This allowed us to conduct a detailed study of the dynamics of spectral changes for 30 min, since the moment of simultaneous excitation of the photorefractive effect and the Raman spectrum. The results of these measurements are presented in Figure 26, which shows the spectra measured within the first 30 s using the *X(YZ)X* scattering geometry.

According to the selection rules, only lines that correspond to vibrations that belong to the *E(*TO*)* symmetry species should be observed in this scattering geometry, whereas lines of other symmetry species (A_1(TO), A_1(LO), *E*(LO)), which are seen in Raman spectra of the lithium niobate single crystal recorded in these scattering geometries [44], should be absent.

It can be seen from Figure 26 that, within the entire irradiation time of the crystal by the laser, the spectra do not differ from each other.

Since the first second of the excitation of the photorefractive effect in the crystal, the Raman spectrum exhibits lines (e.g., the line at 630 cm^{-1}, which corresponds to vibrations of the A_1(TO) symmetry species) that are forbidden in the Raman scattering according to the selection rules for the used scattering geometry but that are observed in this geometry due to the occurrence of the photorefractive effect.

In the literature, the line at 630 cm^{-1} (A_1(TO)) is commonly used as an analytical line in studies of the photorefractive effect via changes in Raman spectra [44].

Therefore, our results convincingly demonstrate that the intensities of forbidden Raman lines buildup to their maxima almost instantaneously (just like photorefractive effect).

All subsequent subtle changes observed in the Raman and photorefractive scattering are caused by the formation of laser-induced static and dynamic defect structures, which determine the dynamics of the development of the second and third layers of the indicatrix of photorefractive scattering and by the energy transfer from layer to layer.

These structures exhibit the property of self-similarity on different scales and can be identified as fractals.

The general characteristic of these structures is the fact that they are formed far from the thermodynamic equilibrium at a certain magnitude of the supercritical action; i.e., these are dissipative structures, which arise at high external energy flows and are products of self-organization in the open system.

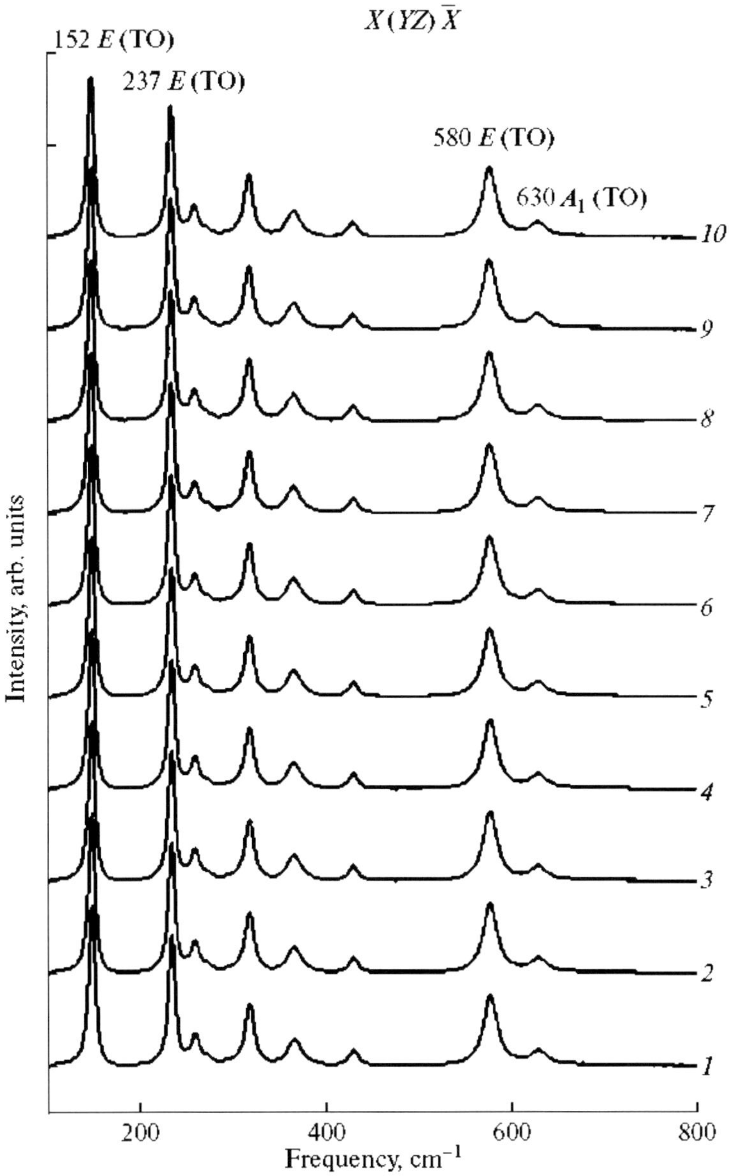

Figure 26. Raman spectra of lithium niobate single crystal recorded with one-second step in (*1*) 3, (*2*) 6, (*3*) 9, (*4*) 12, (*5*) 15, (*6*) 18, (*7*) 21, (*8*) 24, (*9*) 27, and (*10*) 30 s after beginning of laser irradiation of crystal.

As visible laser light propagates through a $LiNbO_3$ single crystal, due to the photorefractive effect in it, local micro- and nanostructures with the fluctuating refractive index are initially formed in the region of propagation of the beam.

Following an increase in the irradiation intensity over time, more and more of these structures are formed; then, they are transformed into static micro- and nanoformations, which subsequently are converted into a continuous laser track. However, near the track, where the action of the radiation on the crystal is significantly lower, laser-induced static and fluctuating micro- and nanoinhomogeneities of the structure can be clearly seen.

The speckle structure of the photorefractive scattering in the $LiNbO_3$ crystal has the shape of an asymmetric figure eight and three layers (the central layer, the layer of static defects with the changed refractive index, and the layer of defects with the fluctuating refractive index). The Raman lines that are forbidden by the selection rules for the given scattering geometry but that are observed in this geometry due to the occurrence of the photorefractive effect buildup to their maximal intensity nearly instantaneously, as well as the photorefractive effect does. All subsequent subtle changes observed in the Raman and photorefractive scattering are caused solely by the formation of laser-induced static and dynamic defects, which determine the dynamics of the development of the second and third layers of the indicatrix of photorefractive scattering and by the energy transfer from layer to layer. We have shown that stoichiometric lithium niobate single crystals grown from a melt with 58.6 mol % of Li_2O exhibit a fairly strong photorefractive effect for their use as materials for information recording and storing. However, the photorefractive scattering, which occurs in these crystals, is the factor that limits the practical application of the crystals as optical materials. At the same time, congruent single crystals, in which the photorefractive scattering is considerably lower, are incapable of information recording with laser radiation.

6. Effects of the Ordering of Structural Units of the Cationic Sublattice of $LiNbO_3$:Zn Crystals and Their Manifestation in Raman Spectra

A ferroelectric lithium-niobate crystal ($LiNbO_3$) has a unique combination of piezoelectric, electrooptical and nonlinear optical properties, which makes this crystal one of the most demanded electronic and optical materials [1, 44, 104]. Lithium niobate has a wide homogeneity range on the phase diagram and, being a phase with a variable composition, is characterized by a highly defective structure [44, 104]. Many physical characteristics of this crystal essentially depend on stoichiometry, impurity composition, and the state of defectness of the structure [1, 44, 95, 105]. The presence of the photorefractive effect (optical damage), which leads to a distortion of the wavefront of the laser beam propagating through the crystal, is a factor that substantially restricts the application of lithium niobate in electrooptical, nonlinear optical, and laser devices [44, 95, 105]. One of the methods to increase the stability of a congruent crystal (Li/Nb = 0.946) to optical damage is doping it with "nonphotorefractive" (optical-damage resistant) cations (Zn^{2+}, Mg^{2+}, Gd^{3+}, In^{3+}, etc.), which do not change their charge state in the crystal under the action of the laser radiation [44, 95].

These cations are capable of significantly suppressing the photorefractive effect.

Experimental and calculation data show that, upon doping with nonphotorefractive cations (Zn^{2+}, Mg^{2+}, Gd^{3+}, In^{3+}, etc.), the ordering of structural units of the cationic sublattice along the polar axis and deformations of oxygen octahedra NbO_6 change nonmonotonically, and the state of defectness of the structure of the crystal on the whole also changes [44].

Furthermore, the concentration dependences of physical characteristics exhibit clearly pronounced anomalies at certain concentrations, which indicates that doping cations enter into the crystal structure in a thresholdlike manner [44, 106-108]. In the most general case, the following regular feature is observed: an increase in the ordering of structural units of the cationic sublattice along the polar axis (i.e., lowering of the potential energy of the crystal) in nominally pure crystals leads to an increase in the defectness of their structure on the whole, i.e., to an increase in the entropy factor and to an enhancement of the photorefractive effect. In this case, a special role in the formation of the photorefractive effect is played by intrinsic and impurity defects with electrons localized on them [44].

Doping of a congruent lithium-niobate crystal with Zn^{2+} cations leads to a change in the polarizability of oxygen octahedra NbO_6, parameters of the lattice of the crystal, and electrooptical characteristic [44, 106-114]. The mechanism by which impurities enter the crystal has a threshold character and is determined by the concentration of Zn^{2+} ions [44, 106, 107]. The coefficients of the linear electrooptical effect in a $LiNbO_3$:Zn single crystal are smaller than in a congruent crystal, and they exhibit a minimum at concentrations of Zn^{2+} of ~ 2-3 mol % and a maximum at ≈ 7 mol % [44, 107]. At a concentration of Zn^{2+} higher than 7 mol %, the electrooptical effect is weak and, upon further increase in the concentration of zinc, these coefficients almost do not change [44, 107]. In this case, there are no Nb_{Li} defects at all in the $LiNbO_3$:Zn crystal, while Zn^{2+} cations occupy basic positions of Li^+ and Nb^{5+} cations in certain proportions [44, 107].

It is significantly interesting to investigate subtle features of the concentration rearrangement of the structure of the $LiNbO_3$:Zn crystal below the first threshold concentration of Zn^{2+} ions, i.e., in the concentration range of Zn^{2+} of 0-2 mol % [7], in which the photorefractive effect changes (decreases) most [44]. In the concentration range of Zn^{2+} ions of 0 - 3 mol %, the electrooptical effect decreases from 3.1×10^2 to 6.6×10^2 W/cm^2, while, in the concentration range of Zn^{2+} of 5-7 mol %, its magnitude changes from 7.1×10^2 to 9.8×10^2 W/cm^2 [44]. Therefore, a maximal change in the photorefractive effect is observed in the range of the first concentration threshold, whereas, a weakest electrooptical effect is observed in the range of the second concentration threshold.

Obtaining optically perfect lithium-niobate crystals with a weak photorefractive effect by doping a congruent crystal with small concentrations of Zn^{2+} ions (up to 2 mol %) is also interesting economically, because, in this case, technological regimes of the crystal growth almost do not differ from the growth regimes of nominally pure congruent crystals, which are well developed in industry.

Raman light-scattering spectroscopy is well known to be an informative method for the study of subtle features of the crystal structure, the state of its defectness, and doping-induced changes [44]. Raman spectra are very sensitive to changes in interactions between structural units of the crystal, as well as to the occurrence of intrinsic defects and defects induced by laser radiation [44, 115]. At present, Raman light-scattering spectroscopy is the only method that is capable of simultaneous investigation of the photorefractive effect and changes in the crystal structure caused by it. A significant merit of Raman spectroscopy is that, by studying Raman spectra of a photorefractive crystal at different powers of the excitation radiation, it makes it possible to clearly distinguish changes in the structure of the crystal that are caused by its doping from changes that are caused by the photorefractive effect proper. In particular, if the power of the excitation radiation is small, the electrooptical effect is almost zero and changes in the spectrum of the crystal are mainly caused by changes in its composition.

By measuring Raman spectra of (i) nominally pure stoichiometric lithium-niobate crystals (Li/Nb = 1) that were grown from a melt with 58.6 mol % of Li_2O ($LiNbO_3$(stoich)), (ii) congruent crystals (Li/Nb = 0.946, $LiNbO_3$(congr)), and (iii) congruent crystals doped with Zn^{2+} cations ($LiNbO_3$: Zn) in the concentration range 0-1.59 mol %, we comparatively investigate subtle features of the structure of these compounds.

The Raman spectra of congruent and stoichiometric lithiumniobate crystals were previously studied in [44, 53, 116, 117], and the spectra of congruent crystals doped with Zn^{2+} ions were examined in [109, 118].

Stoichiometric crystals are a promising material for information recording and as active nonlinear laser media with periodically polarized domains of micron and submicron dimensions [44, 105, 119, 120], whereas $LiNbO_3$:Zn crystals, in which the photorefractive effect is weak, are promising for nonlinear laser media that are used for the transformation of broadband and coherent optical radiation [1, 44, 105, 120]. All single crystals were grown in air atmosphere by the Czochralski method on a Kristall-2 setup in accordance with the unified technique. We used an original lithiumniobate granulated batch with a high apparent density, which was synthesized at the Tananaev Institute of Chemistry and Technology of Rare Elements and Mineral Raw Materials, Kola Scientific Center of the Russian Academy of Sciences, and which makes it possible to obtain absolutely colorless (water white) nominally pure lithium-niobate single crystals [121]. The dopant was introduced as CuO oxide (high-purity grade). The crystal-growth technique and batch-preparation procedure were described in detail in [122].

Since the photorefractive effect in nominally pure lithium-niobate crystals is determined both by intrinsic defects, with electrons localized on them, and by trace amounts of impurity multiply charged cations (Fe, Rh, Cu, etc.) [44, 95, 105], Table 5 lists concentrations of cationic impurities in the crystals under study, which were determined by the spectral-analysis method. It can be seen from Table 5 that crystals are characterized by a high homogeneity along the growing axis with respect to both the composition of impurities and the content of the basic components (R = [Li]/[Nb]). The values of the Curie temperature (T_C), which is a function of the ratio R = [Li]/[Nb] in a nominally pure crystal, for the upper and lower parts of the crystal boule were the same. The specimens for investigations had a shape of a parallelepiped with dimensions of ~5 × 4 × 3 mm with their edges parallel to the crystallographic axes *X, Y,* and *Z.* The *Z* axis coincided in direction with polar axis P_s of the crystal. Faces of parallelepipeds were thoroughly polished. Raman spectra were excited by an Ar–Kr laser (Spectra Physics; λ_0 = 514.5 nm) and were registered with a resolution of 1 cm^{-1} using a T64000 spectrograph (Horiba Jobin Yvon), which was equipped with a confocal microscope. Spectra were recorded using *Y(ZX)Y* and *Y(ZZ)Y* scattering geometries, in which the electrooptical effect and structural distortions caused by it manifest themselves maximally in the Raman spectrum, because the electrooptical effect is predominantly induced by the laser radiation that is polarized along the polar axis of the crystal (*Z* axis) [44]. In order for the magnitude of the electrooptical effect caused by laser radiation in a specimen to be minimal, spectra were excited by laser radiation of small power on the specimen (~3 mW).

In this case, differences in spectra of crystals with different values of the ratio Li/Nb and different concentrations of Zn^{2+} ions will mainly be determined by differences in the crystal structure that are caused by doping rather than by the photorefractive effect. At this radiation power on the specimen, we have not observed photorefractive (photoinduced) light scattering in $LiNbO_3$(congr) and $LiNbO_3$:Zn crystals, and only insignificant circular scattering has been observed [103], which indicates that the photorefractive effect is weak.

However, the $LiNbO_3$(stoich) crystal, in which the photorefractive effect is stronger than in $LiNbO_3$(congr) and $LiNbO_3$:Zn crystals, exhibits clearly pronounced photorefractive light scattering [102].

Processing of contours of complex spectral lines and determining of their basic parameters (frequencies, widths, intensities) were performed using the programs LabSpec 5.0, Origin 8.0, and Bomem Grames/386 (version 2.03). The determination accuracies of line frequency ν, width *S*, and intensity *I* were ± 1.0 cm^{-1}, ± 2.0 cm^{-1}, and 5%, respectively.

Figure 27 presents Raman spectra of $LiNbO_3$(sto-ich), $LiNbO_3$(congr), and $LiNbO_3$:Zn ([Zn] = 0-1.59 mol %) single crystals that were measured in the *Y(ZX)Y* and *Y(ZZ)Y* scattering geometries (in which fundamental phonons of the *E*(TO) and A_1(TO) symmetries, respectively, are active [44]). Changes in the main parameters of lines (frequencies, widths, and intensities) in relation to the composition of the crystal are given in Table 6. It can be seen from this table that the frequencies of the majority of lines barely change as the composition of the crystal is varied, which indicates that the quasi-elastic constants of vibrations remain unchanged. However, the widths of lines noticeably change.

The widths of all the lines are minimal in the spectrum of the stoichiometric crystal, because its cationic sublattice is most ordered.

It is important to note that the concentration dependences of the widths of many lines in the spectrum of the $LiNbO_3$:Zn crystal have a minimum in the concentration range of 0.05-1.12 mol % (Figure 28). It can be seen from Figure 28 that, with an increase in the concentration of Zn^{2+} ions in the $LiNbO_3$:Zn crystal, the widths of some lines change nonlinearly; namely, in the concentration range of Zn^{2+} ions 0-0.94 mol %, they decrease and then, in the concentration range of Zn^{2+} of 0.94-1.59 mol %, they increase. This minimum is especially clearly pronounced for the concentration dependences of the widths of lines with the frequencies at 156, 240, 268, 371, 434, 576, and 876 cm^{-1} (*E*(TO)) and 254 and 274 cm^{-1} (A_1(TO)), which correspond to vibrations of Nb^{5+} and Li^+ cations in oxygen octahedra and internal vibrations of oxygen octahedra. A decrease in the widths of the lines with the frequencies at 254 and 274 cm^{-1} (A_1(TO)), which correspond to totally symmetric vibrations of Nb^{5+} and Li^+ ions along the polar axis, unambiguously indicates that, in the concentration range of Zn^{2+} ions of 0.05 - 1.12 mol %, the cationic sublattice of the lithium niobate crystal is ordered along the polar axis. In this case, oxygen octahedra become more perfect. This is evidenced by a decrease in the width of the line with the frequency at 626 cm^{-1}, which corresponds to totally symmetric (A_1(TO)) vibrations of oxygen octahedra (Figure 28).

Table 5. Results of spectral analysis of plates cut from upper and tail parts of nominally pure congruent and stoichiometric lithium-niobate crystals

Impurity element	Impurity concentration, wt %	
	upper part	tail part
Zr	$<1 \times 10^{-3}$	$<1 \times 10^{-3}$
Mo	$<1 \times 10^{-3}$	$<1 \times 10^{-3}$
Ca	$<5 \times 10^{-3}$	$<5 \times 10^{-3}$
Fe	$<1 \times 10^{-3}$	$<1 \times 10^{-3}$
Ti	$< 1 \times 10^{-3}$	$<1 \times 10^{-3}$
Si	$<1 \times 10^{-3}$	$<1 \times 10^{-3}$
Pb, Ni, Cr, Co	$<1 \times 10^{-3}$	$<1 \times 10^{-3}$
Al	$<5 \times 10^{-4}$	$<5 \times 10^{-4}$
Cu	$<5 \times 10^{-4}$	$<5 \times 10^{-4}$
Mn, V, Mg, Sn	$<5 \times 10^{-4}$	$<5 \times 10^{-4}$
T_C of the $LiNbO_3$ crystal, °C	1142.0	1142.0

Therefore, at [Zn] ~ 0.05-0.94 mol %, $LiNbO_3$:Zn crystals have a region of a more ordered structure such that the order of sequence of basic ions, impurity ions, and vacancies along the polar axis of the cationic sub-lattice is more perfect, while the oxygen octahedra are close to ideal. A maximal ordering in the structure of the $LiNbO_3$:Zn crystal is observed at concentrations [Zn^{2+}] ~ 0.05–0.94 mol %. In this case, the widths of lines in the Raman spectrum of the $LiNbO_3$:Zn crystal ([Zn] ~ 0.05-0.94 mol %) are smaller than in the spectrum of the $LiNbO_3$(congr) crystal, and they approach the widths of lines in the spectrum of the $LiNbO_3$(stoich) crystal (Figure 28). This indicates that the degree of ordering of structural units of the cationic sublattice of the $LiNbO_3$:Zn crystal ([Zn] ≈ 0.05 - 0.94 mol %) is high and that it approaches the degree of ordering of the stoichiometric crystal.

Previously, we obtained similar results by studying Raman spectra of congruent lithium-niobate crystals doped with Mg^{2+} and Gd^{3+} ions [44, 86, 87].

Our spectroscopic data for $LiNbO_3$:Zn crystals ([Zn] = 0-1.59 mol %) correlate well with the concentration dependence of parameters of the unit cell determined by the X-ray diffraction analysis [44, 123].

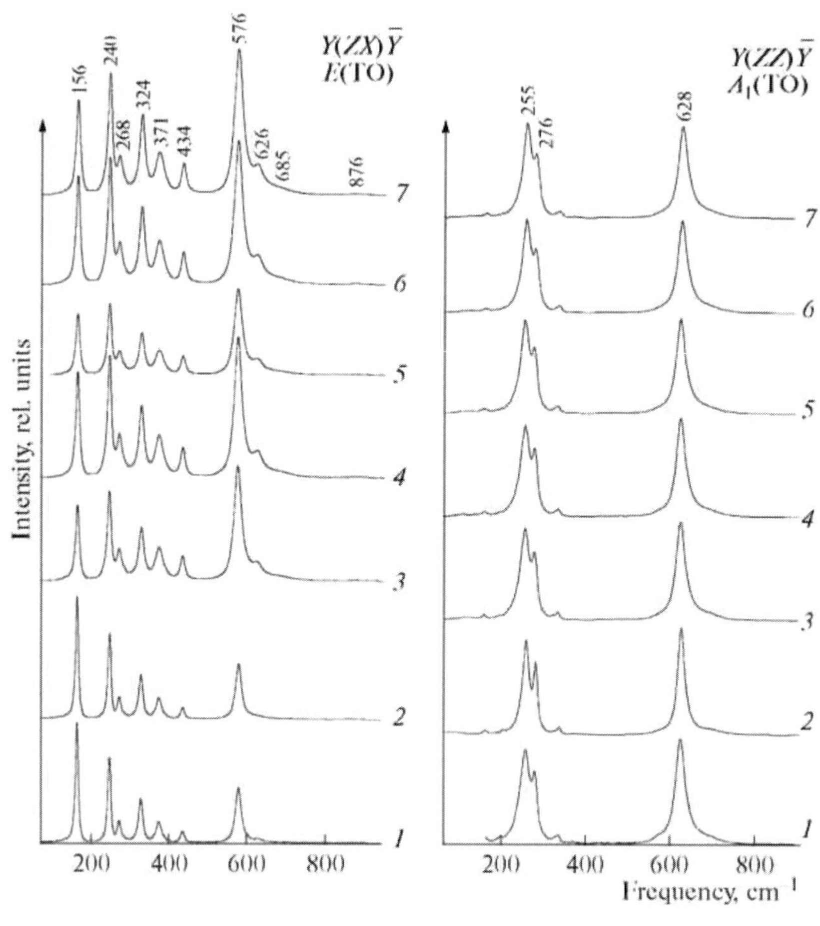

Figure 27. Raman spectra of crystals: (*1*) $LiNbO_3$(stoich), (*2*) $LiNbO_3$(congr), and (*3*) $LiNbO_3$:Zn ([Zn] = (*3*) 0.03, (*4*) 0.05, (*5*) 0.94, (*6*) 1.12, and (*7*) 1.59 mol %) measured in the *Y(ZX)Y* and *Y(ZZ)Y* scattering geometries.

In the concentration range of Zn^{2+} ions of 1-2 mol %, the concentration dependence of the parameter *c* of the unit cell exhibits a minimum [3, 28], whereas, in accordance with the Vegard law, the parameter *c* should increase if the ionic radius of the impurity cation increases compared to the ionic radius of the replaced cation of the matrix. The ionic radii of the Zn^{2+}, Li^{+}, and Nb^{5+} ions are 0.74, 0.68, and 0.68 Å, respectively [44].

The spectra of the $LiNbO_3$congr and $LiNbO_3$:Zn crystals contain a low-intensity line at a frequency of 682 cm^{-1} (Figure 27). In accordance with the data of [116], the manifestation of this line in Raman spectra is caused by the activity of biphonons. However, according to the data of [54-56], this line refers to fundamental phonons of the A_1(TO) symmetry.

In the Raman spectrum of the $LiNbO_3$(stoich) single crystal, we have not observed a line at a frequency of 682 cm^{-1}, as well as a line with a frequency of 120 cm^{-1}, which corresponds to two-particle states of acoustic phonons, the total wave vector of which being zero.

It can be seen from Figures 28 and 29 that, in the concentration range 0-1.12 mol %, the width and intensity of the line at a frequency of 682 cm^{-1} monotonically increase with increasing concentration of Zn^{2+} ions, whereas the frequency of this line, conversely, decreases. However, as the concentration of Zn^{2+} ions is further increased to 1.59 mol %, there is significant decrease in the width of this line (by 9 cm^{-1}; Figure 28, Table 6), which points to the manifestation of ordering effects. However, in this case, both the frequency and the intensity of this line, conversely, increase. In accordance with [117], an increase in the intensity of the line with a frequency of 682 cm^{-1} corresponds to an increase in the concentration of defects Nb_{Li}. However, this contradicts the results of [44, 106-108, 118, 123], in which it was unambiguously shown that, in $LiNbO_3$:Zn crystals, with an increase in the concentration of Zn^{2+} ions, the amount of Nb_{Li} defects decreases. In this case, if the concentration of Zn^{2+} ions is in the range 0–5 mol %, the mechanism of displacement of antisite Nb_{Li} defects by Zn^{2+} cations predominates. Therefore, the line with a frequency of 682 cm^{-1}, which is observed in the spectra of $LiNbO_3$:Zn crystals and which is absent in the spectrum of the $LiNbO_3$(stoich) crystal, has a high sensitivity to the concentration of Zn^{2+} cations in the crystal structure, and the behavior of its main parameters (Figures 28 and 29) can evidence that, at concentrations of Zn^{2+} cations in the range ≈0.94-1.12 mol %, these cations enter into the crystal structure in a thresholdlike manner.

The width of the line at a frequency of 876 cm^{-1} (*E*(TO)), which corresponds to stretching bridge vibrations of oxygen atoms along the polar axis, shows the behavior similar to that of the width of the line with a frequency of 682 cm^{-1}. The occurrence of this line in the spectrum of the centrosymmetric paraelectric phase of the lithium-niobate crystal is forbidden by selection rules [54-56]. A change in the character of the behavior of the line at a frequency of 876 cm^{-1} (as well as of the line with a frequency of 682 cm^{-1}) is observed at a concentration of Zn^{2+} cations of 1.12 mol % (Figure 28).

In the literature, the intensity of the line with a frequency of 876 cm^{-1} is used to evaluate the quality of crystals with an oxygen-octahedral structure [44]. The higher the intensity of this line, the more is the cationic sublattice is ordered along the polar axis and the higher the spontaneous polarization of the crystal is [44].

Figure 29 also shows the dependence of the relative intensity of the line with a frequency of 626 cm^{-1} (A_1(TO)) on the concentration of Zn^{2+} ions in $LiNbO_3$:Zn crystals. This line corresponds to totally symmetric vibrations of oxygen octahedra and is forbidden by selection rules in the Raman spectrum for the *Y(ZX)Y* scattering geometry. The intensity of this line is maximal in spectra measured in the *Y(ZZ)Y* scattering geometry [44, 116] (Figure 27).

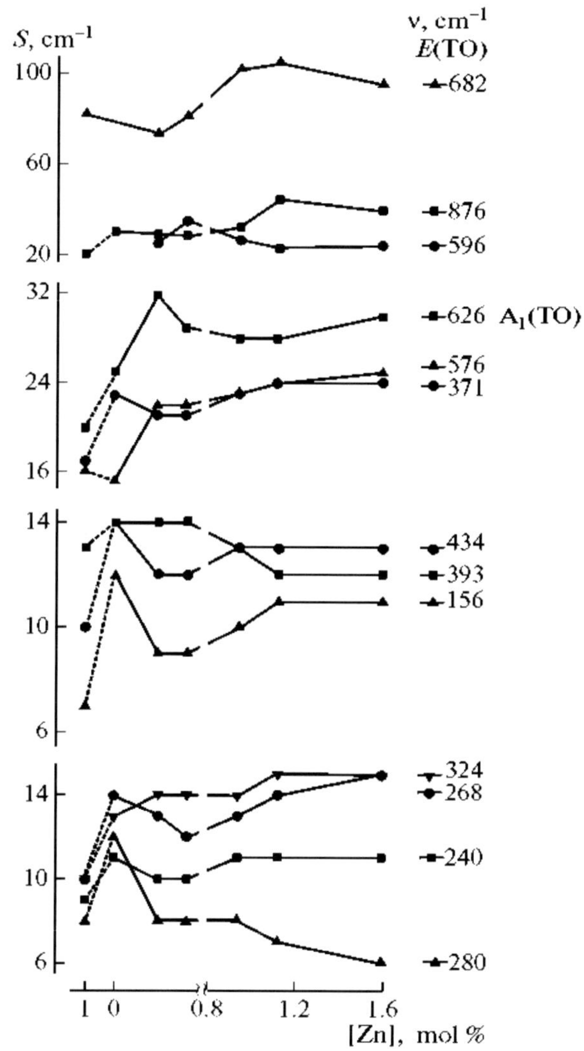

Figure 28. Dependences of the widths of lines in the Raman spectra of $LiNbO_3$:Zn crystals on the concentration of Zn^{2+} cations. Dashed lines show changes in the widths of lines that occur upon passage from the $LiNbO_3$(stoich) crystal to the $LiNbO_3$(congr) crystal.

Table 6. Main parameters of lines that correspond to vibrations of the E(TO) and A_1(TO) symmetries in the Raman spectra of the $LiNbO_3$(stoich), $LiNbO_3$(congr), and $LiNbO_3$:Zn [0,03 ÷ 1,59 mol. %] single crystals

$LiNbO_3$(stoich)		$LiNbO_3$(congr)		$LiNbO_3$: Zn														
				[Zn] = 0,03			[Zn] = 0,05			[Zn] = 0,94			[Zn] = 1,12			[Zn] = 1,59		
E(TO)																		
v	*s*	*v*	*s*	*v*	*s*	*Irel*	*v*	*s*	*Irel*	*v*	*s*	*Irel*	*v*	*s*	*Irel*	*v*	*s*	*Irel*
156	7	156	12	156	9	71,77	155	9	81,78	155	10	75,75	155	11	80,95	155	11	68,83
240	9	240	11	240	10	84,59	240	10	95,83	240	11	90,39	240	11	95,81	240	11	89,94
268	10	268	14	268	13	29,03	268	12	32,1	268	13	30,18	268	14	30,23	268	15	28,4
280	8	280	12	280	8	11,47	280	8	13,28	279	8	12,89	280	7	12,55	279	6	12,74
324	10	324	13	324	14	47,96	324	14	52,85	324	14	50,36	324	15	55,67	324	15	57,18
371	17	371	23	371	21	28,9	371	21	30,59	371	23	29,81	371	24	30,74	370	24	29,76

$LiNbO_3$(stoich)		$LiNbO_3$(congr)		$LiNbO_3$: Zn														
				[Zn] = 0,03			[Zn] = 0,05			[Zn] = 0,94			[Zn] = 1,12			[Zn] = 1,59		
E(TO)																		
393	13	393	14	393	14	9,89	393	14	10,27	394	13	9,46	394	12	10,14	394	12	10,13
434	10	434	14	434	12	22,73	434	12	22,66	434	13	23,73	435	13	23,96	435	13	23,36
576	16	576	15	576	22	100	576	22	100	576	23	100	576	24	100	576	25	100
-	-	-	-	596	25	26,53	597	35	28,11	598	26	26,63	598	23	27,22	598	24	28,42
-	-	-	-	682	73	5,15	682	81	5,5	666	102	6,5	662	104	7,9	668	95	7,64
876	20	876	30	876	29	1,35	874	29	1,54	873	32	1,39	876	44	1,7	874	39	1,67
A_1(TO)																		
v	*s*	*v*	*s*	*v*	*s*	*Irel*	*v*	*s*	*Irel*	*v*	*s*	*Irel*	*v*	*s*	*Irel*	*v*	*s*	*Irel*
255	18	255	26	255	25	-	255	24	-	255	23	-	255	23	-	255	23	-
276	11	276	14	276	14	-	276	14	-	276	15	-	276	15	-	276	16	-
626	20	626	25	626	32	17,07	625	29	19,25	625	28	19,63	625	28	21,1	625	30	21,81

v (cm^{-1}) – is the frequency and *s* (cm^{-1}) is the width of a spectral peak and I_{rel} is its intensity in percent with respect to intensity of a peak with a frequency of 580 $см^{-1}$ in percent, c_{Zn} (mol %) is the concentration of Zn^{2+}.

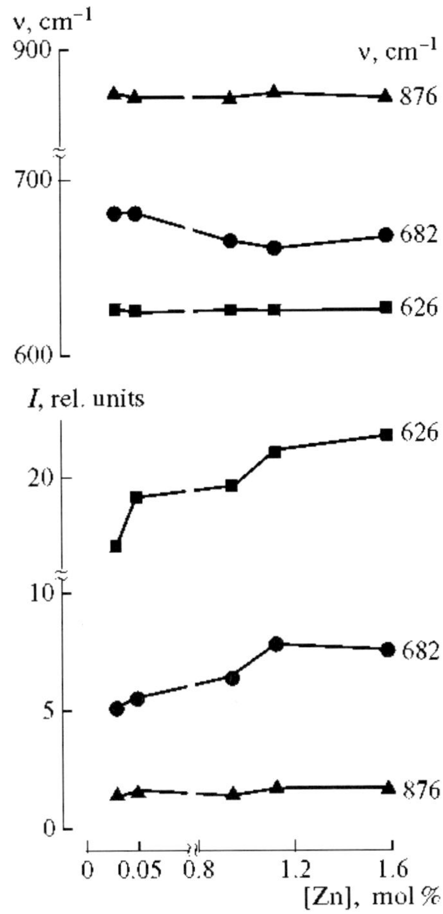

Figure 29. Dependences of the frequencies (ν, cm^{-1}) and intensities of the lines at 626 (A_1(TO)), 682, and 876 cm^{-1}, which correspond to vibrations of oxygen octahedra, in the Raman spectra of $LiNbO_3$:Zn crystals on the concentration of Zn^{2+} cations.

However, because of the presence of the photorefractive effect [44], this line is always observed in the *Y(ZX)Y* scattering geometry, with its intensity being proportional to the magnitude of the photorefractive effect; commonly, this line is used to evaluate the photorefractive properties of lithium-niobate crystals [44]. It can be seen from Figure 29 that the relative intensity of the line with a frequency of 626 cm^{-1} monotonically increases with an increase in the concentration of Zn^{2+} cations, which contradicts the data of works [44, 106, 107], in which it was shown that the electrooptical effect weakens with increasing concentration of Zn^{2+} ions. Broadening and an increase in the intensity of the line at 626 cm^{-1} can indicate that, as Zn^{2+} cations are incorporated into the lithium-niobate crystal structure, oxygen octahedra insignificantly and anisotropically expand. This is facilitated by the fact that the ionic radius of the Zn^{2+} cation is greater than those of the Li^+ and Nb^{5+} ions.

An anisotropic expansion of oxygen octahedra is also confirmed by a nonsynchronous increase in the parameters *a* and *c* of the unit cell with an increase in the content of Zn^{2+} ions [28] and by an increase in the width of the line with a frequency of 876 cm^{-1} (Figure 28), which is sensitive to clusterization of cations [123].

The presence of a region with an increased ordering of structural units of the cationic sublattice in the $LiNbO_3$:Zn crystal can be explained as follows. As is known, for a congruent lithium-niobate crystal ([Li]/[Nb] = 0.946), the basic defects are Nb_{Li}, i.e., Nb^{5+} cations that occupy positions of Li^+ cations in the ideal structure of the stoichiometric composition [44, 95, 104]. On the grounds of electrical neutrality, the formation of a Nb_{Li} defect gives rise to the appearance of four defects in the form of vacant oxygen octahedra. Incorporation of impurity Zn^{2+} cations into the structure of the congruent crystal leads, first of all, to the displacement of Nb_{Li} defects by Zn^{2+} cations. This is favorable energetically [44].

Small amounts of Zn^{2+} cations occupy lithium-oxygen octahedra (in which Nb_{Li} defects were located), order the alternation of cations and vacancies along the polar axis, and lower the defectness of the crystal with respect to Li^+ vacancies [44, 102]. Entering of a Zn^{2+} impurity cation into a vacant oxygen octahedron of an ideal structure, coupled reducing the number of Li^+ vacancies, leads to an additional increase in the defectness of the crystal structure because of the violation of the existing order of alternation of cations and vacancies along the polar axis of the crystal.

Therefore, Zn^{2+} impurity cations are incorporated into the cationic sublattice of a congruent lithium-niobate crystal by two competing mechanisms. One of them (ordering) leads to ordering of cations along the polar axis and to a decrease in the number of vacancies of cations. The other (disordering) mechanism leads to the violation of the order of sequence of cations along the polar axis caused by Zn^{2+} impurity cations themselves. At low concentrations of Zn^{2+} cations, the ordering mechanism predominates, which leads to a decrease in the widths of lines in the Raman spectrum and in the parameter *c* of the unit cell. With an increase in the concentration of Zn^{2+} impurity cations, the disordering mechanism begins to predominate and the widths of lines and the parameter *c* increase.

The data that we obtained allow us to state that, as Zn^{2+} cations enter into the structure of a congruent lithium-niobate crystal, the ordering of structural units of the cationic sublattice and the properties of the crystal vary rather smoothly, since, as the concentration of Zn^{2+} ions increases, two mutually related processes simultaneously occur; namely, Nb_{Li} defects are displaced and Zn^{2+} cations enter vacant octahedra of the ideal structure. According to the data of works [106, 107], even if the concentration of Zn^{2+} ions is as high as 3 mol %, Nb_{Li} defects are present in the crystal.

They are displaced completely only at a concentration of $[Zn^{2+}] > 8$ mol %. Therefore, Zn^{2+} cations can control the number of Nb_{Li} defects in the lithium-niobate crystal structure rather finely and efficiently, which is important for targeted creation of optical materials with tailored characteristics [44, 105].

Figure 30 presents a fragment of the Raman spectrum of $LiNbO_3$(stoich), $LiNbO_3$(congr), and $LiNbO_3$:Zn single crystals in a low-frequency range (50–140 cm^{-1}). In this range, in the spectrum of the lithium-niobate crystal, there are no lines that correspond to fundamental phonons [44, 53, 116]. In the spectrum of the $LiNbO_3$(congr) crystal recorded in the *Y(ZZ)Y* scattering geometry (in which phonons of the A_1(TO) symmetry are active), a low-intensity broad line is observed at a frequency of ~ 120 cm^{-1}, which corresponds to vibrations of quasi-particles-two-particle states of acoustical phonons the total wave vector of which is zero [44, 116]. In this case, in the Raman spectrum of highly ordered $LiNbO_3$(stoich) single crystal, this line is not observed (Figure 30) [44].

Figure 30. Fragments of the Raman spectra of crystals: (*1*) $LiNbO_3$(stoich), (*2*) $LiNbO_3$(congr), and (*3*) $LiNbO_3$:Zn ([Zn] = (*3*) 0.03, (*4*) 0.05, (*5*) 0.94, (*6*) 1.12, and (*7*) 1.59 mol %) in the range of two-particle states of acoustical phonons.

It is necessary to note that the intensity of this line is sensitive to changes in the acoustic *Q* factor of the lithium-niobate crystal [124]. The higher the intensity of this line, the lower the acoustic *Q* factor. The magnitude of the acoustic *Q* factor is the greatest in a stoichiometric single crystal, in the Raman spectrum of which the line with a frequency of 120 cm^{-1} does not manifest itself.

Processing of spectra with programs for the separation of the contours of spectral lines shows that the line at a frequency of ~ 120 cm^{-1} in the Raman spectrum of the $LiNbO_3$congr crystal has a structure and is a superposition of two lines with frequencies at ~ 104 and 117 cm^{-1} (Figure 30). The occurrence of components is also confirmed by calculations of the density of phonon states that were performed in [125]. Figure 29 shows that the intensity of the line with a frequency of ~ 120 cm^{-1} varies with a change in the concentration of the Zn^{2+} impurity in the $LiNbO_3$:Zn crystal. As the concentration of Zn^{2+} ions increases, the structure of the ~ 120 cm^{-1} line gradually vanishes (Figure 29). At a concentration of Zn^{2+} ions of 1.59 mol %, only one maximum with a frequency of 120 cm^{-1} manifests. It should be noted that, for $LiNbO_3$:Mg and $LiNbO_3$:Gd crystals, in which impurity cations enter into the lattice in a stepwise manner and more sharply compared to the $LiNbO_3$:Zn crystal [44], the broadening of lines with frequencies of 254 and 274 cm^{-1} (A_1(TO)) and the vanishing of the structure of the line with a frequency of 120 cm^{-1} and increase in its intensity occur in a narrower range of the impurity concentration [86, 87].

From our point of view, the observed effects for the line at a frequency of 120 cm^{-1} can be explained by the resonant interaction of fundamental vibrations with frequencies of 254 and 274 cm^{-1} (A_1(TO)) between each other and with two-particle states of acoustical phonons the frequency of which is in the range of 120 cm^{-1} and the total wave vector is zero. The manifestation of this interaction in the vibrational spectrum of an anharmonic crystal has been considered theoretically in [125, 126]. For a highly ordered cationic sublat-tice of a lithium-niobate crystal (in particular, for crystals that are close to the stoichiometric composition), the interaction involves a comparatively narrow range of the spectrum. For these highly ordered structures, the fundamental vibrations with the frequencies of 254 and 274 cm^{-1} (A_1(TO)) interact with two-particle states of acoustical phonons (A_1(TO)) almost independently of each other. This manifests itself in the appearance of two maxima, which are located at 104 and 117 cm^{-1} in the spectrum in the range of two-particle excitations of acoustical phonons. With an increase in the disorder in the cationic sublattice (as well as with an increase in the anharmonicity upon an increase in the temperature of the crystal), the fundamental vibrations with the frequencies of 254 and 274 cm^{-1} begin to interact with each other. In this case, the spectral range in which the fundamental vibrations and two-particle states of acoustical phonons resonantly interact broadens, which manifests itself in an increase in the intensities and widths of lines that correspond to two-particle states of acoustical phonons the total wave vector of which is zero. If this interaction is rather strong, the maxima at 104 and 117 cm^{-1} merge into a single broad maximum with a frequency of 120 cm^{-1} (Figure 30).

Therefore, the appearance of the line with the frequency at 120 cm^{-1} in the Raman spectrum can be caused by a violation of the order of sequence of cations along the polar axis in the cationic sublattice of nonstoichiometric crystals compared to the unperturbed order of sequence of cations in the ideal structure of the stoichiometric composition (Li^+, Nb^{5+}, vacant octahedron). In this case, the amount of Nb_{Li} defects increases. We can assume that the intensity of the line with the frequency at 120 cm^{-1} increases with increasing the number of defects (including Nb_{Li} defects) related to the violation of the order of sequence of Li^+ and

Nb^{5+} cations and vacant octahedra along the polar axis. Then, the splitting of the line into two components can be caused by an improvement of the selection rules in the wave vector of two-particle states of acoustical phonons of the A_1(TO) symmetry because of a decrease in the uncertainty of the wave vector of the quasi-particle upon ordering of the alternation of cations and voids along the polar axis [125, 126]. Conversely, an increase in the disorder in the cationic sublattice should lead to an increase in the uncertainty with respect to the wave vector of two-particle states of acoustical phonons of the A_1 (TO) symmetry and to an increase in the probability of two-phonon transitions [125, 126]. In this case, vibrations in an increasingly large region of the Brillouin zone will manifest themselves in the Raman spectrum. All this should cause a smearing of the structure of the line with a frequency at 120 cm^{-1}, an increase in its intensity, and a broadening of lines that correspond to the totally symmetric fundamental vibrations of cations located in octahedral voids of the structure. Ii also follows from the experimental data that we obtained that the amount of Nb_{Li} defects in the $LiNbO_3$:Zn crystal decreases, as the concentration of Zn^{2+} ions increases. However, to reliably determine the character of the dependence of the intensity of the line at the frequency of 120 cm^{-1} on the number of the Nb_{Li} defects, it is necessary to further investigate the Raman spectra of $LiNbO_3$:Zn crystals at high concentrations of Zn^{2+} ions, up to 8 mol %.

At these high concentrations of these ions, Nb_{Li} defects barely occur [107], and the intensity of the line with the frequency of 120 cm^{-1} can be close to zero.

By studying the Raman spectra of $LiNbO_3$:Zn crystals, we have revealed a region of a more ordered structure such that the order of sequence of basic ions, impurity cations, and vacancies along the polar axis of the cationic sublattice is more regular, while the oxygen octahedra are close to ideal. In this case, crystals have a higher optical quality and are more stable with respect to optical damage. A maximal ordering of the structure is observed at concentrations Zn^{2+} cations in the range ~ 0.05–0.94 mol %. In this case, the widths of lines in the Raman spectrum of the $LiNbO_3$:Zn crystal ([Zn] ~ 0.05–0.94 mol %) are smaller than those in the spectrum of the $LiNbO_3$(congr) crystal and approach the widths of lines in the Raman spectrum of the $LiNbO_3$(stoich) crystal. A region of an increased ordering of the structure can be formed because small amounts of Zn^{2+} cations displace Nb_{Li} defects, order the alternation of cations and vacancies along the polar axis, and reduce the defectness of the crystal with respect to Li^+ vacancies. Our results are important for industrial growth of optically perfect lithium-niobate crystals by doping congruent crystals with small amounts of Zn^{2+} ions. Technologically speaking, the growth regimes of these crystals almost do not differ from the growth regimes of nominally pure congruent crystals, which are well developed in industry.

7. STRUCTURAL ORDERING OF DOPED LITHIUM NIOBATE SINGLE CRYSTALS

The primary goal of doping ferroelectric crystals is to change or stabilize the properties of the matrix phase. In [18, 30, 38, 64, 72, 77, 86, 87, 94], it was discovered that an improved cation order along the polar axis, achieved because of doping, can improve the physical properties of oxygen-polyhedral ferroelectrics (such as niobates and tantalates of lithium, potassium, and others).

In particular, Raman spectra showed the following: doping cations whose ionic radii are close to the matrix cations (Li^+ and Nb^{5+}) and whose charges are intermediate between the matrix cations ($1 < Z < 5$), when doping levels are very low (a tenth or hundredth fraction of weight percent), have an ordering effect on the cation sublattice of a congruent LN crystal. Such cations can have high accommodation coefficients; in fact, they do not distort the structure but change the alternation order of structural units residing in oxygen octahedra along the polar axis. The doping cations must have no unstable variable valence (Cu^+ and Cu^{2+}, Fe^{2+} and Fe^{3+}, etc.); if they did, the photorefractive effect and optical absorption would have been dramatically increased. Probably, a similar ordering effect concerns all oxygen-polyhedral ferroelectrics that have pseudoilmenite, perovskite, layered perovskite, or other structures. The works [18, 30, 38, 64, 72, 77, 86, 87] are confined to alkali-metal niobate tantalates with a pseudoilmenite structure.

Initially, in the range of low dopant concentrations, the Raman lines are narrower than in a nominally pure congruent LN single crystal. This proves a higher structural perfection of the crystal. At higher dopant concentrations, the Raman lines broaden to exceed widths characteristic of a nominally pure crystal. Thus, the crystal lattice is ordered at low dopant concentrations, but when a threshold value is surpassed, conversely, structure is disordered. This means that small amounts of cations (boron, zinc, magnesium, gadolinium, and other), occupying lithium oxygen octahedra, decrease the structure imperfection: they reduce the concentrations of antisite defects Nb_{Li} and Li-site vacancies and, accordingly, they order cation alternation along the polar axis. In addition to the reduction of the Li-site vacancy concentration, when a doping cation is incorporated into a vacant octahedron, perturbing the alternation order of cations and vacancies along the polar axis, not only does it reduce the Li-site vacancy concentration, but it also renders the structure more imperfect. Thus, two competing mechanisms interplay when small cation amounts are incorporated into the LN or LT lattice. One (ordering) mechanism orders cations along the polar axis and decreases the Li-site vacancies; the other (disordering) mechanism perturbs the cation-alternation order along the polar axis by dopant ions. The disordering mechanism becomes dominant with an increase in dopant concentration. Interestingly, the intensity of extra lines rises along with broadening of Raman fundamentals (with increasing cation disorder) [64, 72, 77, 86, 87].

Figure 31 [48, 61] presents the 100-150 cm^{-1} fragments of the Raman spectra for LN single crystals. In this spectral range, a congruent crystal in scattering geometry *X(ZZ)Y* (where phonons A_1(TO) are active) shows a low-intensity line at 120 cm^{-1} (curve *3*) [67]. There is no consensus regarding the origin of this line. The authors of [62, 127] assign this line to E phonons, which are forbidden for scattering geometry *X(ZZ)Y* but which appear here because of internal stresses in the crystal.

The authors of [128] assign it to radiation scattering on difference optical phonons A_1 and E. In [129], slow-neutron-scattering experiments carried out in [130] were used to calculate the density function for acoustic and optical vibrations of LN; it has been shown that the two-particle states of acoustic phonons with a null overall wave vector appear in the Raman spectrum as a peak in the region of 120 cm^{-1}, Figure 32.

In the Raman spectra of high-quality congruent crystals measured in [64, 77], the line at 120 cm^{-1} is split into a pair of lines of the same polarization at 103 and 117 cm^{-1} (Figure 31).

Other authors have not observed this splitting. From curve *1* in Figure 31, it is clear that stoichiometric crystals, in which the cation sublattice is more ordered than in congruent crystals, show no lines in the range 100-150 cm^{-1}.

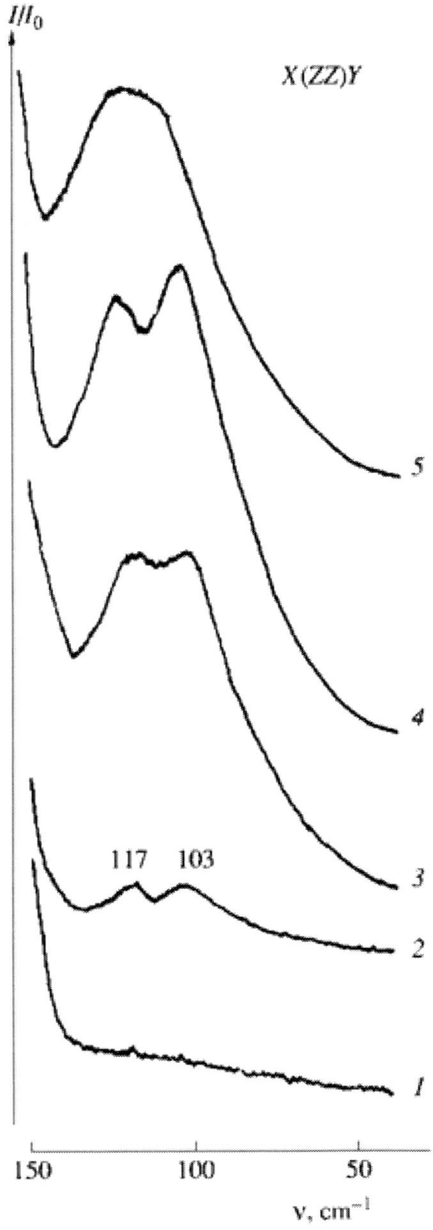

Figure 31. Fragments of Raman spectra ($T = 293$ K) for LN single crystals of various compositions in the range 100-150 cm^{-1}: (*1*) a stoichiometric composition, (*2*) a stoichiometric composition doped with Gd^{3+} (0.001 wt %), (*3*) a congruent composition, (*4*) a congruent composition doped with Mg^{2+} (0.36 wt %), and (*5*) a congruent composition doped with Gd^{3+} (0.25 wt %) and Mg^{3+} (0.75 wt %).

Lines at 103 and 117 cm^{-1} in this spectral range (Figure 31, curve *2*) appear when dopant ions having radii similar to Li^{+} or Nb^{5+} ions and charges intermediate between the Li^{+} and Nb^{5+} charges are incorporated into a stoichiometric crystal; the dopants perturb the near-ideal order of cation alternation in octahedra lying along the polar axis and induce insignificant off-stoichiometry.

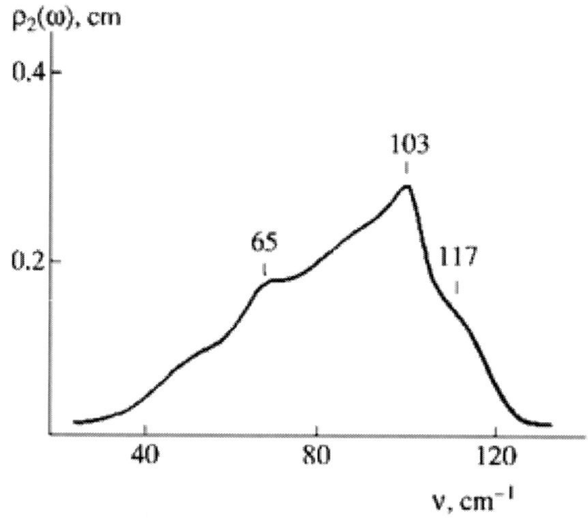

Figure 32. Density of two-phonon states for lithium niobate crystal in the range 40-150 cm^{-1}.

When these dopants are incorporated in small amounts into a congruent crystal, they first enhance splitting of the line at 120 cm^{-1} into lines at 103 and 117 cm^{-1} (Figure 31, curve *4*); then (at higher doping levels), the lines at 103 and 117 cm^{-1} broaden and merge to yield the line at 120 cm^{-1} (Figure 31, curve *5*). The structure of the two-phonon line appears synchronously with a slight decrease in the widths of fundamental lines at 254 and 274 cm^{-1} (A1 vibrations); the disappearance of the structure is also accompanied by broadening of these lines. This is unambiguous evidence that the cation sublattice of a congruent crystal is ordered and that its degree of ordering approaches the stoichiometric crystal when such impurities have certain low concentrations [38, 64, 72, 77, 86, 87, 130-134].

Thus, the Raman peak observed in an LN crystal at 100-120 cm^{-1} and associated with the two-particle states of acoustic phonons with a null overall wave vector, is sensitive to fine cation-order features. In a stoichiometric crystal, there are no Raman lines in the region of 100-150 cm^{-1} (Figure 31, curve *1*). The absence of a Raman peak can be taken as an experimental criterion to judge whether an LN crystal structurally corresponds to a high-quality stoichiometric crystal [64, 77].

The study of the photorefractive properties of lithium niobate crystals showed that the sensitivity of crystals to optical damage decreases markedly when they are doped with Mg^{2+}, B^{3+}, or Gd^{3+} to a definite level that falls within the concentration range in which the cation order is improved [18, 30, 64, 72, 77, 87, 131, 135, 136]. This means that the concentration of charged defects decreases when the structural quality of a crystal is improved.

Figure 33 [131, 136] shows fragments of Raman spectra, measured using scattering geometry *X*(*ZX*)*Y*, for doped congruent LN single crystals in the spectral region in which oxygen octahedra vibrate. In this spectral range, real crystals (either nominally pure or doped) show two intense lines at 580 cm^{-1} (*E*(TO)) and 635 cm^{-1} (A_1(TO)). The line at 635 cm^{-1} (A_1(TO)) is forbidden for this scattering geometry and appears in the spectrum as a result of the photorefraction effect.

The reduced intensity of the line at 635 cm^{-1} caused by doping proves a reduced photorefraction of the crystal and correlates with cation ordering along the polar axis found in [38, 64, 72, 77, 86, 87, 133, 134] for this range of dopant concentrations.

Two features that prove structure ordering are observed precisely in this concentration range: the line in the region of 120 cm^{-1} (due to the two-particle states of acoustic phonons with a null overall wave vector) undergoes the greatest splitting into the lines at 103 and 117 cm^{-1}, and some lines are noticeably reduced in width [38, 64, 72, 77, 86, 87, 133, 134].

In this way, photorefraction is the least in lithium niobate crystals having an improved cation order along the polar axis. Comparatively high dopant levels spoiling this order and distorting oxygen octahedra, in contrast, strengthen photorefraction and, accordingly, increase the intensities of the line at 635 cm^{-1} (Figure 11) [131, 136].

The single crystals having a more ordered cation arrangement along the polar axis, therefore, have a higher laser damage resistance.

Moreover, the mode of doping can essentially affect the optical homogeneity and laser damage resistance of single crystals. For example, in [132], lithium niobate single crystals were doped with boron conventionally, through adding boron oxide to the feed before fusing; alternatively, the dopant (boric acid) was added to the niobium stripping extract during the preparation of high-purity niobium pentaoxide. The niobium stripping extract was obtained during the extraction conversion of commercial niobium hydroxide to a high-purity product. Boric acid was added in the amount of 0.08-0.15 wt % relative to the niobium (based on niobium pentaoxide) contained in the stripping extract with allowance for the fact that part of the acid bound fluorine to HBF_4. Next, niobium hydroxide was precipitated from the stripping extract by neutralizing the extract with aqueous ammonia to reach pH 8-9.

Niobium hydroxide, washed with demineralized water, was dried and calcined to convert it to pentaoxide. The niobium pentaoxide was used to prepare the LN feedstock from which a set of single crystals was then grown.

The optical homogeneity of these single crystals containing 0.08 or 0.12 wt % was studied. From the average microdefect density visualized in a laser beam (separate defects appear in a laser beam as bright points), the crystals were found to have a high optical quality: they were completely free of microdefects (crystal quality is regarded as optical when the average microdefect density is within 10 cm^{-1}).

However, in the set of single crystals with boron levels of 0.09, 0.1, or 0.12 wt %, doped by adding boron oxide to the feedstock before fusing, a significant optical inhomogeneity was found: the microdefect density was 80-120 cm^{-3}. Both sets of single crystals were grown under identical conditions on a Kristall-2M installation equipped with an automated crystal-diameter control system and ensuring precision in growth of various crystals.

In addition, photorefraction in single crystals doped during pentaoxide preparation, which are distinguished by an improved structural order, was far lower than in nominally pure congruent single crystals. In single crystals boron-doped in the crucible before fusing, photorefraction was far higher; the forbidden line at 635 cm^{-1} in these crystals had a higher intensity than in the nominally pure congruent crystal (Figure 33d) [132].

We may state that the doping methods that yield the maximal chemical homogeneity in complex multicom-ponent systems substantially control the feasibility of growing crystals with high optical homogeneity, high structural perfection, and high laser damage resistance.

In addition, such single crystals have a larger optical transparence window and a higher photovoltaic effect; this effect repeats the shape of a nanosecond laser pulse, which is unachievable in nominally pure congruent crystals [18, 30, 38, 64, 72, 77, 86, 87, 131-136].

The above inferences are supported by the pulse photovoltaic investigations of doped LN single crystals [64, 137].

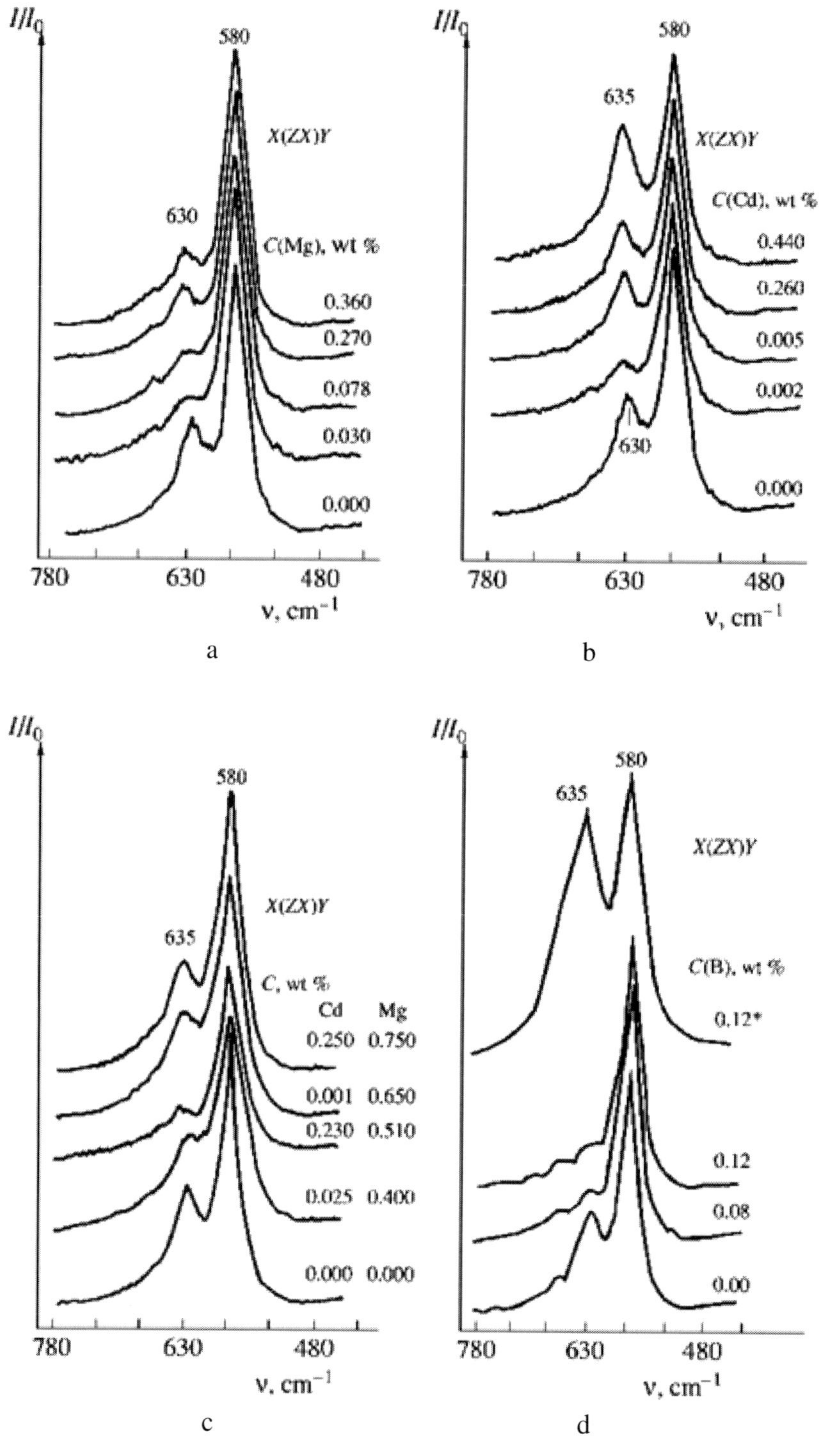

Figure 33. Fragments of Raman spectra (T = 293 K) in the region of the vibrations of oxygen octahedra for congruent LN crystals doped with (a) Mg, (b) Gd, (c) Gd + Mg, and (d) B (boron oxide was added to the feed before fusing).

It is known that a nonlinear interaction of an intense light wave with a noncentrosymmetric medium generates two effects: optical detection (OD) and photogalvanic (PG) effects. The microscopic nature of these effects lies in the asymmetry of elementary electronic processes: photo excitement, ionization, recombination, and other processes whose character depends on the amount and type of impurities and defects [137-140].

Macroscopically, these effects are manifested as follows: in a nonlinear dielectric crystal, the region exposed to laser radiation is polarized and induces a charge on capacitor plates around. Therefore, measuring the photoresponse amplitude and its kinetics, one can determine the impurity type and concentration, detect impurity complexes (i.e., evaluate the crystal quality), and use these effects for control. Measurements of the potential difference on capacitor plates in various crystallographic directions give all components of the nonlinear susceptibility tensor, which allows judging the single-domain state of the crystal.

In particular, it was shown in [64, 137] that the photoresponse in LN crystals of various chemical compositions in the time interval from 10^{-9} to 10^{-8} can be separated into two contributions: one repeats the laser pulse shape and is due to virtual transitions, and the other is a relaxation contribution associated with absorption on impurities and defects [138]. The matrix (inertialess) contribution from the optical detection effect has a characteristic time (10^{-15} s) much shorter than the laser pulse length; therefore, it repeats the laser pulse shape. The photoresponse amplitude is a function of the structure perfection of the crystal matrix.

The more perfect the crystal structure, the higher the inertialess photoresponse amplitude. The impurity contribution is due to the photoinduced alteration of the dipole moment of an impurity. Its amplitude is proportional to the impurity amount [137-139]. A method was developed for separating the components of the signal [137-140]. The behavior of each contribution was studied as a function of the type of dopant. The photoresponse kinetics was studied as a function of the intensity, wavelength, and polarization of incident radiation [64, 137].

The inertialess contribution, which characterizes the crystal matrix, in a nominally pure LN crystal changed its sign when its surface was probed with a laser beam; this meant that the crystal was not a single domain. The relaxation contribution was significant. This contribution was due to uncontrolled impurities (in particular, iron) and other structure defects and was a linear function of defect concentration. A moderate laser damage resistance was noted. In $LiNbO_3$:Gd crystals, the photo response did not change its sign during surface probing; therefore, presumably, gadolinium doping promotes the formation of a single domain state. This inference was later confirmed by electrophysical measurements [141].

The inertialess contribution first increased its amplitude (by a factor of 1.5-12) as the Gd concentration increased, and then the amplitude decreased slightly (by a factor of 1.5-3). No saturation in the amplitude was observed up to intensities of 10^8 W/cm^2. The photoresponse signal shape, in general, repeated the laser pulse shape [64, 137].

The relaxation component was saturated as the radiation intensity increased. Annealing in a hydrogen atmosphere changed only the relaxation contribution; the absorption spectrum of a $LiNbO_3$:Gd crystal became an analog of the spectrum of a $LiNbO_3$:Fe crystal doped with 0.005 wt % Fe. This means that the relaxation component of the photoresponse of the $LiNbO_3$:Gd sample is primarily governed by an uncontrolled iron impurity, which transforms from Fe^{3+} to Fe^{2+} when annealed in hydrogen.

Thus, the structure defects associated with Gd^{3+}, Mg^{2+}, or B^{3+} impurities do not form deep energy sublevels in the bandgap and do not contribute to the relaxation photoresponse;

rather, they reduce the relaxation contribution that is associated with uncontrolled impurities (e.g., Fe). Conversely, at least at low dopant levels, a doped crystal is substantially more perfect compared to nominally pure congruent crystals [64, 137].

Importantly, the maximal structure order is observed in the range of comparatively low concentrations of impurity cations (a tenth and hundredth fraction of weight percent). Such low concentrations only insignificantly change the properties of the melt; therefore, the growth schedules used for doped crystals with improved physical parameters differ little from growth schedules employed for nominally pure crystals.

8. Micro - and Nanostructures in Lithium Niobate Single Crystals Doped with Lanthanides

Active nonlinear crystals, which combine active (lasing) properties related to the presence of lanthanide impurities with the nonlinear optical properties of the matrix, are of particular interest. In such crystals it is possible for processes of lasing frequency self-conversion to be implemented when lasing at a certain frequency and the nonlinear optical conversion of this frequency occurs simultaneously in the same crystal [142, 143].

Ferroelectric crystals with regular domain structure (RDS) are promising for effective nonlinear conversion. These crystals impose no limitations on the polarization of interacting waves, and, therefore, quasi-phase-matching can be implemented in any direction relative to the crystal optical axes [142, 143].

RDS with periods from a few micrometers to several tens of micrometers in $LiNbO_3$ crystals is obtained either during crystal growth or as a result of postgrowth treatment. In the latter case, an RDS is formed in lithium niobate crystals by applying a reverse electric field [144], electron beam scanning [145], laser heating [146], or using the effect of spontaneous reverse switching [147]. Although these methods allow one to form domain structures with periods to 1-4 μm, their significant drawback is that they do not make it possible to obtain bulk (more than 0.5 mm thick) elements with a homogeneous RDS.

Samples with a larger RDS volume can be obtained based on rotational growth stripes during the Czochralski growth of $LiNbO_3$ crystals doped with rare earth and other (generally trivalent) elements [146-151].

In this paper we report the results of studying the growth of RDSs and periodic nanostructures by optical microscopy and atomic force microscopy in lithium niobate single crystals doped with lanthanides (Gd, Er). The fine features of structural ordering of lithium niobate single crystals of different compositions were investigated by Raman spectroscopy.

Lithium niobate single crystals 30-42 mm in diameter with a 60-70-mm-long cylindrical part doped with lanthanides (Gd, Er) were grown by the Czochralski method from platinum crucibles to investigate the RDSs and periodic nanostructures formed under time-dependent growth conditions.

The crystals were grown on seeds with an orientation (0001) from a charge of congruent composition (Li/Nb = 0.946) without subsequent transformation into the single-domain state. The dopant was introduced into the crucible directly before the melting in the form of the corresponding high-purity Gd_2O_3 and Er_2O_3 oxides.

In addition, nominally pure single crystals of congruent compositions were grown from a congruent melt and stoichiometric single crystals were grown from a melt containing 58.6 mol % Li_2O. The content of foreign cationic impurities in the charge did not exceed $5 \cdot 10^{-4}$ wt. %. To detect the domain structure, plates were cut from the crystals; then they were polished and subjected to selective etching in a $HNO_3 : HF = 2 : 1$ mixture at 110°C for 6 min or at room temperature for 24 h. The structures that wee revealed were investigated with a high-efficiency multipurpose Thixomet® optical image analyzer formed on the basis of the modern instrumental (Axio Observer microscope, Karl Zeiss) and program tools, as well as SMM-2000 and Nano-R atomic-force microscopes. To record Raman spectra from single crystals, samples in the form of parallelepipeds 5 × 6 × 7 mm^3 in size were cut with edges oriented along the crystallographic axes. Raman spectra were excited by an ILA-120 argon laser (λ_0 = 514.5 nm, P ~ 200 mW) and recorded on a U-1000 Ramanor spectrometer. All measurements were carried out at room temperature. The errors in measuring frequency and linewidth were ± 1.0 cm^{-1} and ± 4 cm^{-1}, respectively. The growth conditions for doped (Gd, Er) lithium niobate single crystals were chosen so as to obtain single crystals with a regular nonuniform impurity distribution and, correspondingly, with an RDS whose step was determined by the ratio of the crystal pulling and rotation rates [151].

These RDSs (Figures 34, 35; Table 7) were obtained due to the following factors:

(i) careful and long-term preparation of the melt before seeding (melt preparation with superheating and optimal exposure to exclude uncontrolled gradients of impurity concentrations in the melt);
(ii) formation of an asymmetric thermal field in the presence of a large temperature gradient above the melt (4-6 K mm^{-1}).
(iii) optimally chosen rotation and pulling rates (fairly high initial pulling rate in the cone stage and a monotonically decreasing pulling rate during cylindrical boule growth; the rotation rate was constant during cylindrical part growth);
(iv) periodic changes in the high-frequency generator power providing short-term periodic changes in temperature near the crystal-melt interface (see [151] for more details).

Controlling crystallization makes it possible to maintain the required constant change in convective flows in the melt, forming a periodic inhomogeneity of the impurity distribution due to the regular change in the growth rate at the interface.

Table 7. Statistics of linear measurements along the growth axis Z in the $LiNbO_3$: Gd single crystal ([Gd] 0.44 wt %) grown under time-dependent conditions

RDS length, μm	Average RDS period, μm	Standard deviation of the period	Confidence interval of the period	Instability of the period, %
510	7.84	0.45	0.11	1.42
513	7.86	0.5	0.19	2.49
649	7.82	0.5	0.11	1.41
254	7.93	0.066	0.023	0.28

According to Figure 34.

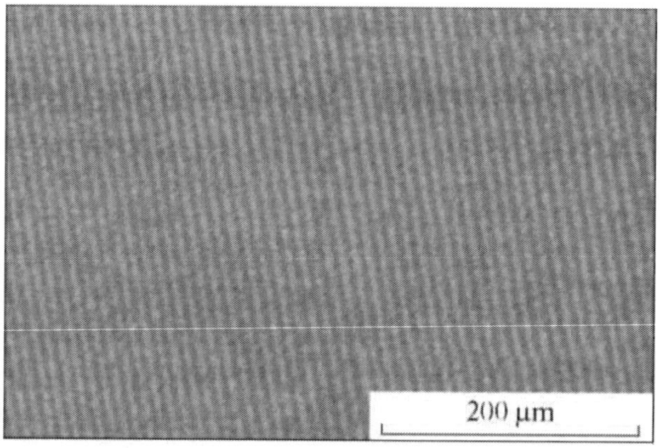

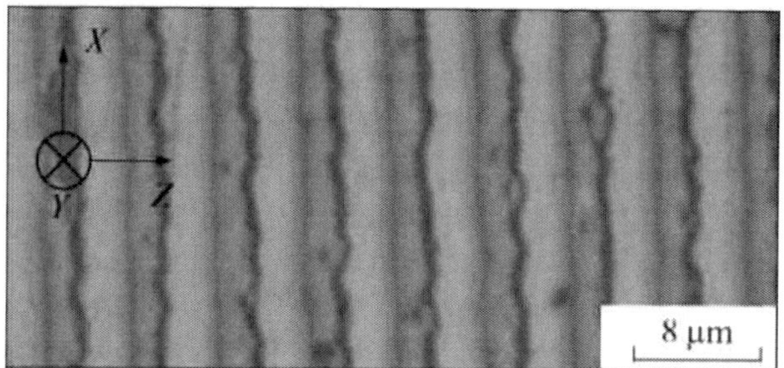

The image was obtained on a Thixomet® optical image analyzer.

Figure 34. Growth of domain structure of the $LiNbO_3$:Gd ([Gd] 0.44 wt %) single crystal.

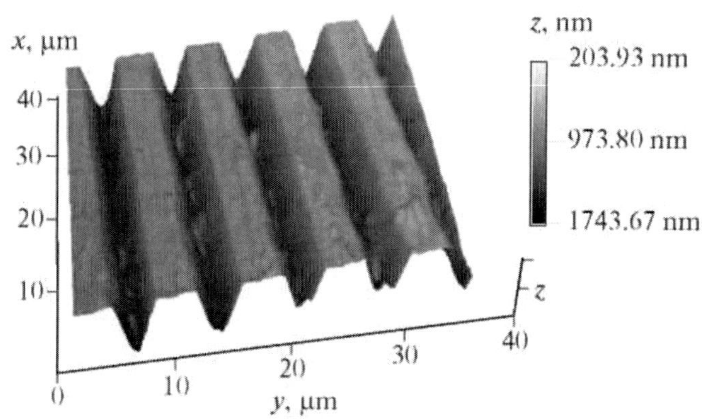

The image was obtained on a Nano-R atomic force microscope.

Figure 35. Growth of the domain structure of $LiNbO_3$:Er (2.71 wt %) single crystal.

Periodic nanostructures with a step from ~ 10 to 100 nm were recorded by atomic force microscopy in the Gd-doped lithium niobate single crystals on the negative domain wall of RDS domains after etching. The periodic partition occurs both parallel and perpendicularly to the polar crystal axis and is not likely to be limited by a region on the 10-100 nm scale (available for atomic force microscopy) (Figure 36). Obviously, the formation of such periodic nanostructures is not directly related to the growth processes as in the case of the RDS based on rotational stripes.

Apparently, the formation of such structures is due to the ordering of the clusters formed on the basis of complexes of intrinsic and impurity defects during crystallization under time-dependent thermal conditions. Such structures are obviously not domain in the ordinary sense. However, the boundaries between their individual elements are likely to be charged under nonequilibrium conditions (for example, during etching or heating).

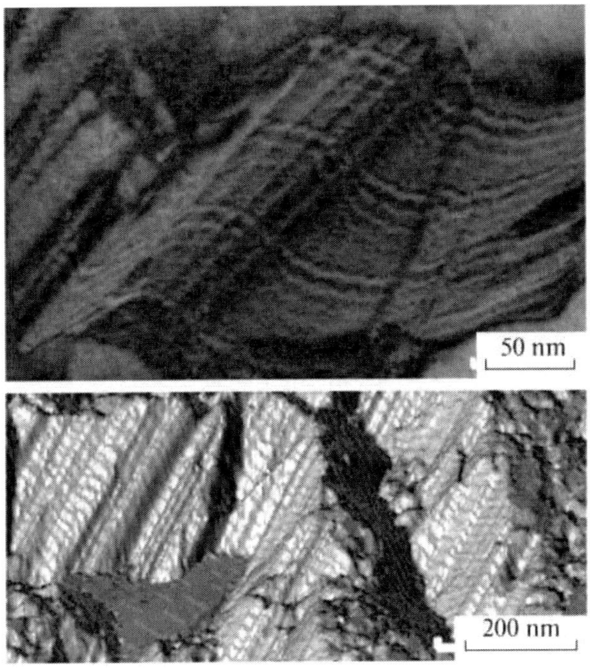

The image was obtained on a SMM-2000 atomic-force microscope.

Figure 36. Periodic nanostructures recorded on the negative domain wall of an RDS domain in a lithium niobate $LiNbO_3$:Gd ([Gd] 0.44 wt %) single crystal grown under time-dependent conditions.

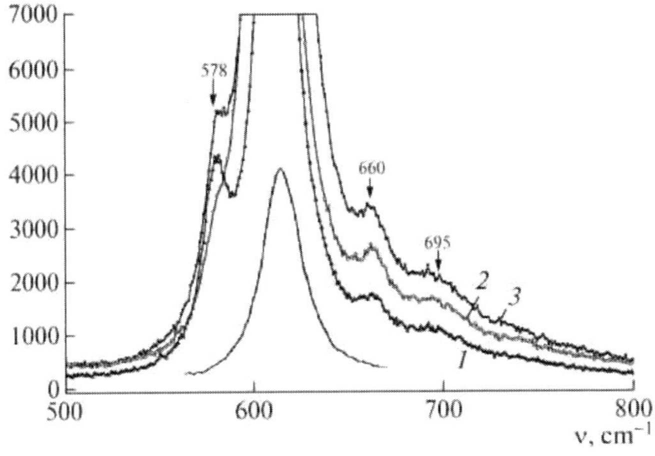

Figure 37. Fragments of the Raman spectra of lithium niobate single crystals of different compositions in the vibrational range of oxygen octahedra recorded in different scattering geometries: *(1)* $LiNbO3_{st}$ *(Y(XY)Z* geometry), *(2)* $LiNbO3_{con}$ *(Z(YY)X)*, and *(3)* $LiNbO_3$:Gd (0.44 wt %) *(Y(XX)Z)*.

Otherwise they would not be revealed by etching. The presence of a set of periodic micro- and nanostructures in a single crystal significantly changes its physical characteristics in the important temperature range (300-400 K).

This was unambiguously shown in [152], where the electrical characteristics of a $LiNbO_3$: Gd ([Gd] = 0.44 wt %) single crystal were investigated; this crystal was also used in this study to analyze the RDS and periodic nanostructures (Figures 34, 36, and Table 7).

The Raman spectra of lithium niobate single crystals of different compositions (nominally pure congruent (Li/Nb = 0.946) and stoichiometric (Li/Nb = 1) crystals and a $LiNbO_3$:Gd single crystal with periodic micro- and nanostructures) were investigated to clarify the fine features of structural ordering. The number of lines in the spectra greatly exceeds the number of lines allowed by the selection rules, with due regard to the *LO-TO* splitting. There are weak (extra) lines which are not due to the fundamental lattice vibrations in different scattering geometries (Figure 37; the weak extra lines are indicated by arrows). Weak extra lines are most sensitive to changes in the features of ordering structural units and the spatial structure of defects in lithium niobate single crystals of different compositions [44, 64, 87].

It is known that, with an increase in disordering of the crystal structure, the lines in the vibrational spectrum that are due to the fundamental lattice vibrations broaden [44]. Stoichiometric lithium niobate crystals have the most ordered structure and, correspondingly, the fundamental lines in their Raman spectra are the narrowest. The crystals of congruent compositions and, all the more, doped crystals are characterized by a much more disordered lattice than stoichiometric crystals and, accordingly, by wider fundamental lines [44, 64, 87]. Weak lines were found for the first time in the Raman spectra, the width of these lines anomalously decreases with an increase in the cation sublattice disordering when the single crystal composition changes. Table 8, along with the main parameters of the lines due to the fundamental lattice vibrations in the *Y(ZZ) Y* scattering geometry (lines ~ 631-632 cm^{-1}), contains the parameters of weak extra lines at ~ 682-693 cm^{-1}. It can clearly be seen that the width of the lines due to fundamental lattice vibrations increases with an increase in the lattice disordering from the stoichiometric to the congruent composition and then to the Gd-doped crystal, whereas the width of weak extra lines, on the contrary, decreases.

One can attribute this linewidth behavior to the existence of a superstructural sublattice of cluster defects in the crystal, which contributes its own spectrum in the form of weak extra lines, and to the ordering of this sublattice at lithium niobate lattice disordering as a whole. This suggestion is unambiguously confirmed by an increase in the width of the fundamental Raman line at ~ 631-632 cm^{-1} with a change in the crystal composition.

Table 8. Main parameters of several Raman lines of lithium niobate single crystals of different compositions in the *Y(ZZ)Y* scattering geometry

Crystal	Frequency ν, cm^{-1}	Peak line intensity I_M, rel. units	Integrated line intensity I_o, rel. units	Linewidth *S*, cm^{-1}
$LiNbO_{3st}$	632	507768	17877	20
	693	58170	611	80
$LiNbO_{3con}$	632	567642	14813	26
	686	56462	651	75
$LiNbO_3$: Gd (0.44 wt %)	631	510171	12066	30
	682	62026	671	72

The model calculations [153] show that such clusters may arise in the lithium niobate structure near intrinsic Nb_{Li} defects and form ordered sublattices with a size of several translation periods, i.e., with a step of 1-2 nm. The local symmetry of cations in octa-hedra generally changes in clusters and boundary regions; as a result, the intensity of the extra lines in the spectra may increase [44]. Thus, the lanthanide-doped lithium niobate single crystals grown under conditions far from thermodynamic equilibrium can apparently contain, along with periodic micro- and nanostructures in the scale range of ~ 10 nm-10 μm (which are recorded by optical microscopy and atomic force microscopy), ordered sublattices of cluster defects with a step of 1-2 nm.

9. Microstructural Defects and Manifestation of the Photorefractive Effect in the Ferroelectric Lithium Niobate Single Crystal

The optical characteristics of ferroelectric photorefractive single crystals are determined by fractal-type nano- and microstructures, macroheterogeneities generated in the course of nonequilibrium crystal growth, intrinsic and extrinsic defects and their clusters with trapped electrons, and laser-induced defects [39, 44, 95, 96, 99].

Formation of such structures and defects and their effect on the optical properties of materials have been inadequately studied. The task of improving structural perfection and homogeneity of single crystals, as well as the task of forming micro- and macrostructures of desired configuration in them in order to enhance or create novel physical properties, is of crucial importance in ferroelectric technology [39, 44, 95, 96, 99].

Nonlinear optical single crystal of lithium niobate $LiNbO_3$ is a phase of variable composition and is one of the most widely used materials of electronic engineering. Under certain conditions, nonequilibrium crystal growth is accompanied by formation of fractal-type macro-, micro-, and nanostructures; layers parallel and normal to the growth axis, including those tens and hundreds of nanometers thick; micron and sub-micron periodically polarized structures with plane boundaries [154-156]. Crystals with periodically polarized domains are good candidates for use as optical converters and nonlinear gain media [44, 95, 96, 99]. One of the most promising materials for these purposes is stoichiometric lithium niobate crystal (R = Li/Nb = 1) owing to the low coercive field (five or more times as low as in the congruent crystal (R = 0.946)) [[44, 95, 96, 99]. However, stoichiometric crystals exhibit a high (as compared to the congruent crystals) photorefractive effect (optical damage) significantly limiting laser beam generation and laser frequency conversion [44, 95].

The photorefractive effect in the lithium niobate crystal has been well studied, both experimentally and theoretically [44, 95, 97]. The photorefractive effect consists in that, in the place where a laser beam passes and in some adjacent region being as large as a few millimeters, there is a noticeable change in the refractive index and distortion of the crystal structure persisting for a long time after the laser beam has been removed.

Despite serious publications (see review in [44, 95, 97]), fine features of this distortion have been inadequately studied.

In particular, the effect of the macro-, micro-, and nanostructure of a crystal on the onset of the photorefractive effect still remains quite unclear.

Meanwhile, in a crystal illuminated with a laser beam, the refractive index can change differently in macro- and microscopic regions differing by their physical properties. Thus, the heterogeneity of a crystal can influence not only the optical characteristics of materials but also the generation and time dynamics of the photorefractive effect.

In this work, we have demonstrated for the first time that exposure of a lithium niobate crystal to visible laser light initially leads to formation of fluctuating micro- and nanostructural defects with physical parameters other than those of the single crystal in the absence of the photorefractive effect. As the laser beam intensity or illumination duration increases, the content of such microstructural defects builds up and they transform into static micro- and macroscopic objects, which are subsequently transformed into a continuous laser track.

It has been shown that stoichiometric lithium niobate crystals have a sufficiently high photorefractive effect to be used as materials for data recording and storage.

Single crystals were grown by the Czochralski method under different conditions, including those far from the thermodynamic equilibrium [44, 95]. Specimens for study—parallelepipeds 5 × 6 × 7 mm in size with the edges coinciding with the crystallographic axes *X*, *Y*, and *Z* (*Z* is the polar axis of the crystal) or plates of different orientation-were cut out of a single crystal boule.

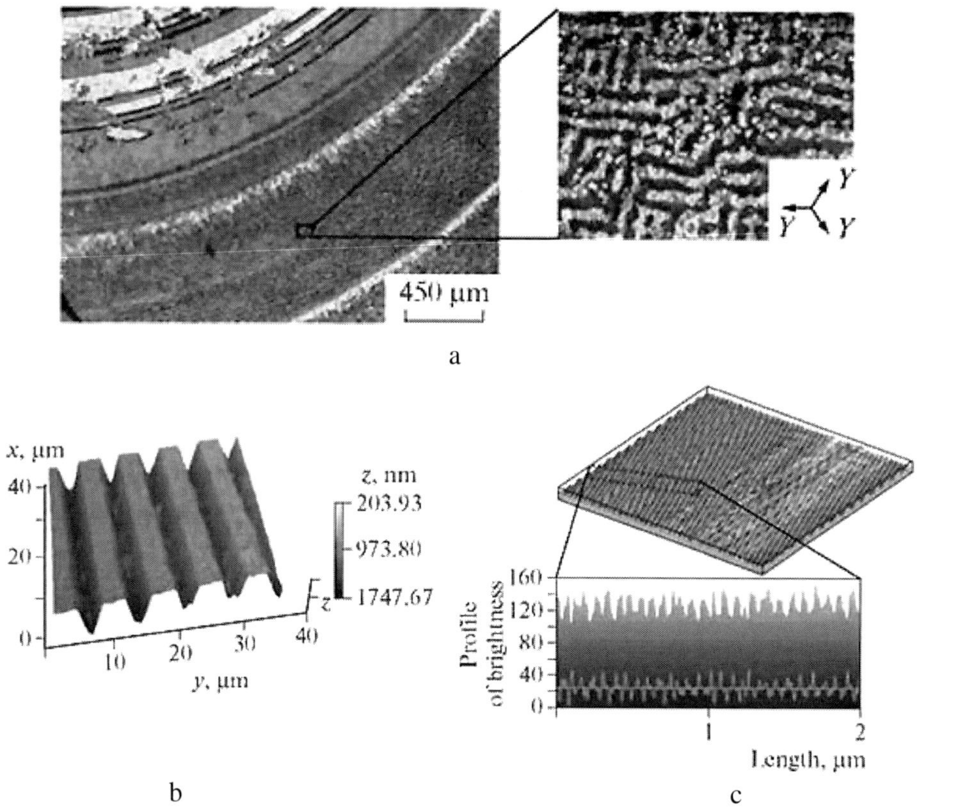

Figure 38. Typical growth domain structures (GDS) on doped lithium niobate crystals grown under nonstationary conditions. (a) Self-similar fractal domain microstructures of the $LiNbO_3$:Tm crystal ([Tm] = 1.8 wt %); (b) regular domain structure of the $LiNbO_3$:Er crystal ([Er] = 2.71 wt %); (c) submicron regular domain structure of the $LiNbO_3$:Gd crystal ([Gd] = 0.44 wt %).

The parallelepiped and plate faces were thoroughly polished. The micro- and nanostructures in crystals were visualized using chemical etching in a mixture of mineral acids. A parallelepiped specimen was illuminated with a laser beam at a wavelength of 514.5 nm (a Spectra Physics argon laser). All images were shot on a digital camera.

Typical growth of spatially heterogeneous structures in the lithium niobate crystal is shown in Figure 38. When a laser beam passes through such heterogeneous crystals, which in addition have a large number of charged defects, rather nonequilibrium conditions can be created in the illuminated region [157]. The crystal structure can acquire the ability for self-organization, since there is a necessary prerequisite for this in the form of an energy flow supplied by an external source and dissipated by the structure.

Due to this flow, the nonequilibrium system becomes active so that, in addition to the structures (including those inside structures) formed upon nonequilibrium crystallization, there can appear laser-induced macro-, micro-, and nanostructures [157].

The type and dimension of micro- and nanostructures can dramatically influence the photorefractive effect and its time dynamics and the physical characteristics of single crystal optical materials [158].

Thus, in the regions where the laser beam passes through a crystal under severely nonstationary conditions, some instability can develop and structures on different scales with clear signs of self-organization can form. These structures are self-similar at different scale levels and can be identified as fractals.

In the course of our experiments on the propagation of linearly polarized laser light in photorefractive lithium niobate crystals of different composition, we revealed for the first time that, at short illumination times or low pump-beam powers, a laser track does not instantaneously formed in the place where the laser beam passes; first, the laser beam induces the formation of local micro- and macrostructures with a refractive index differing from that of the single crystal in the absence of the photorefractive effect. At the early stage of illumination, the microstructures are fluctuating.

These micro- and macrostructures are clearly observed in the experiment (Figure 29a). With an increase in the illumination duration or laser beam power, the number of such structures progressively increases (Figures 39b, 39c), and they gradually transform into a continuous track (Figure 40). This track can persist in the crystal for a long time caused by the Maxwell relaxation time [44] (up to a year in the dark). The existence of the track is evidence that this material can be used for optical information recording.

It is worth noting that the detected stepwise appearance of the laser track and its evolution in time in the crystal (fluctuating microstructures–static microstructures–continuous laser track) correlates well with the development of a three-layer speckle pattern of the photorefractive light scattering (PRLS) in the lithium niobate crystal [159].

PRLS is an interfering factor for data recording.

We observed for the first time the periodic pattern of the laser beam propagating along the polar axis Z in the stoichiometric lithium niobate crystal grown from a melt with 58.6 mol % Li_2O (Figure 39d). The propagation period (m) was ~ 0.33 mm. Such a periodic pattern was absent at the first moment of irradiation.

No such effect was observed when the laser beam propagated along the crystallographic axes X and Y. Analogous studies were performed for nominally pure and doped lithium niobate single crystals grown from a congruent melt and for nominally pure stoichiometric single crystals grown from a congruent melt with adding K_2O.

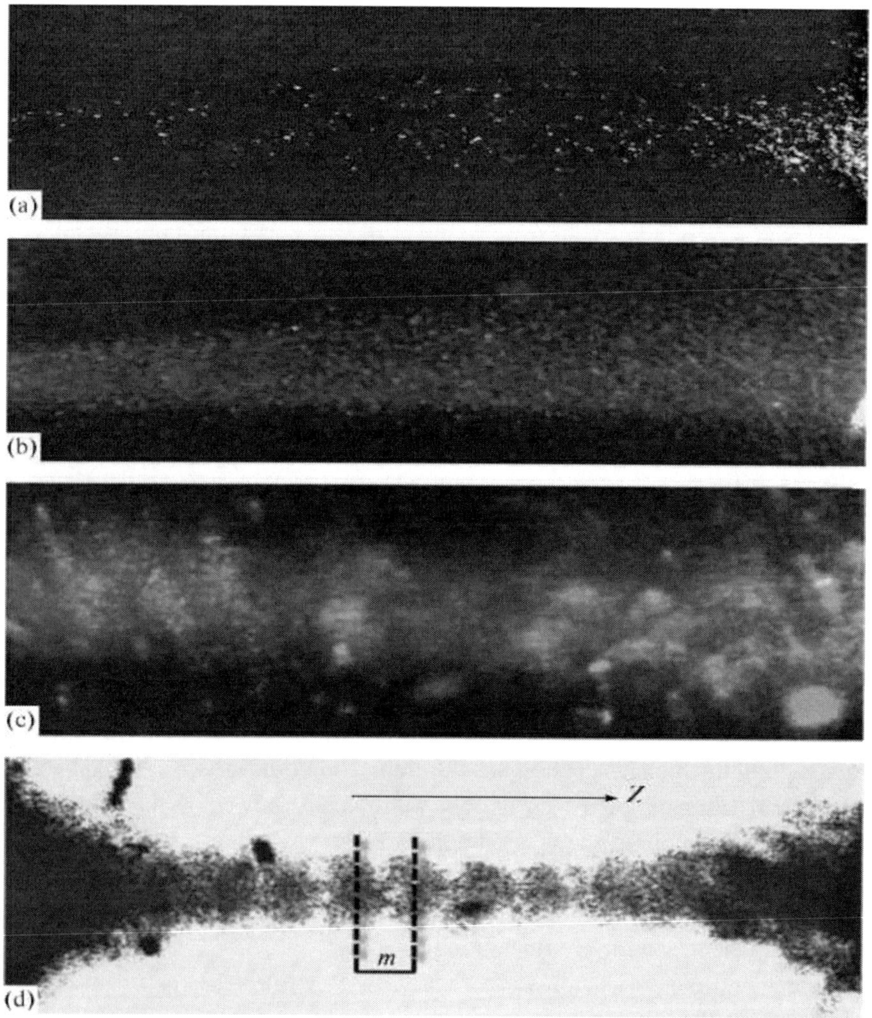

Figure 39. Laser beam (a-c) propagation and (d) periodic pattern in the stoichiometric lithium niobate single crystal grown from the melt with 58.6 mol % Li_2O.

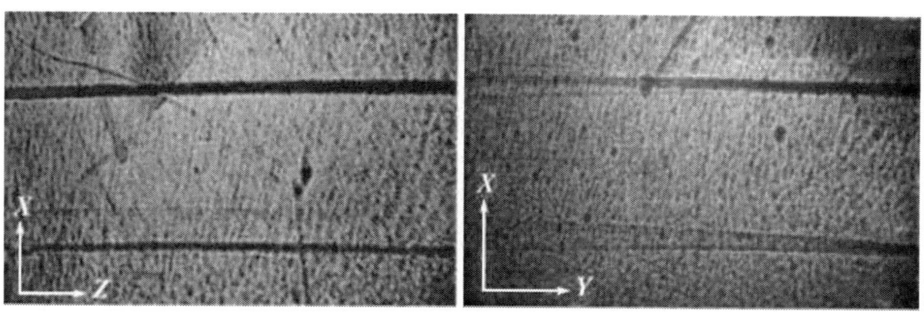

Figure 40. Image of the laser track in the stoichiometric lithium niobate single crystal. The laser beam vector (*E*) is aligned with the polar axis.

In these crystals, no periodic pattern of the laser beam was observed irrespective of its propagation direction. The periodic pattern of the laser beam can be caused by gyrotropy in the stoichiometric lithium niobate single crystal grown from the melt with 58.6 mol % Li_2O.

The emergence of the periodic pattern of the laser beam in the stoichiometric lithium niobate single crystal grown from the melt with 58.6 mol % Li_2O can be associated with specifics of the crystal growth process [44]. This growth technique results in single crystals with significant heterogeneity of composition along the growth axis. At the same time, our studies did not reveal the periodic pattern of the laser beam in lithium niobate single crystals (of nearly stoichiomet-ric composition) grown from the congruent melt with adding K_2O, characterized by the higher homogeneity of the refractive index along the growth axis.

It is also worth noting that the procedure of growing lithium niobate single crystals from a melt with K_2O affords crystals of nearly stoichiometric rather than exactly stoichiometric composition [20, 122, 148].

Thus, we can assume that the presence of the periodic pattern of the laser beam propagating along the polar axis is evidence of the stoichiometric composition of a lithium niobate single crystal. This fact, along with the absence of the band at 120 cm^{-1} in the Raman spectrum excited by visible light [157], can be used for assessing the degree of stoichiometry of a crystal.

Finally, it is worth noting that the laser-induced fluctuating and static microstructures and the speckle pattern of PRLS are characteristic of lithium niobate crystals of different composition, both nominally pure and doped. However, such defects and light scattering on them have their individual fine features, and studying the latter can provide information on the structure and micro- and macroheterogeneity of crystals.

Further studies of specific features of the laser beam propagation in lithium niobate crystals of different composition grown by different methods and under different conditions are of evident interest for design of materials with desired optical properties.

10. OPTICAL PROPERTIES OF $LiNbO_3$:MG(5.21 MOL %) AND $LiNbO_3$:FE(0.009 MOL %):MG(5.04 MOL %) CRYSTALS

Ferroelectric lithium-niobate ($LiNbO_3$) single crystals grown from a congruent melt and doped with photovoltaically inactive ("nonphotorefractive") cations (Mg^{2+}, Zn^{2+}, Gd^{3+}, B^{3+}, etc.)[2] are characterized by a weak photorefractive effect and are promising for use as nonlinear optical materials for transformation and generation of laser radiation [44, 95, 106, 122, 160]. Commonly, the photorefractive effect is suppressed most at a high level of doping (for Mg^{2+}, above 5 mol %) [122]. However, photoinduced (photorefractive) light scattering[3], which is caused by spatial microdefects with a static or fluctuating refractive index that are induced by laser radiation, gives rise to a strong destruction of the laser beam in the crystal, and is a factor that impedes the generation and transformation of the radiation [44, 97]. Therefore, obtaining $LiNbO_3$ single crystals with a low level of photoinduced light scattering is a problem important for practice. These crystals are also characterized by an increased optical strength.

---

[2] Under the action of light, photorefractive (multiply charged) cations in the crystal change their charge and enhance the photorefractive effect. Nonphotorefractive cations possess a permanent charge and are capable of lowering the photorefractive effect in the crystal under certain conditions.

[3] In the Russian-language literature, the terms "photoinduced light scattering" and "photorefractive light scattering" are used interchangeably. In the English-language literature, this phenomenon is widely known as the "photorefractive beam fanning effect." Below, in this work, we will use the term "photoinduced light scattering."

In this work, using methods of laser conoscopy, photoinduced light scattering, electronic spectroscopy, and Raman light-scattering spectroscopy, we study the optical homogeneity, electronic-absorption and transmission spectra, and photorefractive properties of single crystals $LiNbO_3$:Mg(5.21 mol %) and $LiNbO_3$:Fe(0.009 mol %):Mg(5.04 mol %) that were grown from congruent melts. Conoscopic patterns and Raman spectra of $LiNbO_3$:Mg crystals at low doping levels (up to 1 mol %) have been studied in [86, 87, 122, 161]. To our knowledge, conoscopic patterns and Raman spectra of $LiNbO_3$:Mg(5.21 mol %) and $LiNbO_3$:Fe(0.009 mol %):Mg(5.04 mol %) crystals have not been previously investigated.

Crystals were grown in air atmosphere by the Czochralski method. We used a lithium-niobate batch that was synthesized from solid Nb_2O_5:Mg and Nb_2O_5:Fe,Mg precursors that were obtained by homogeneous doping of a reextract with magnesium at the stage of extraction of Nb_2O_5. This method of synthesis ensures a significant improvement in the chemical homogeneity of the lithium-niobate batch, a decrease in the number of defects with localized electrons, and an improvement in the optical homogeneity of crystals, as well as in their resistance to optical damage.

The technique of crystal growth and preparation of the batch with the use of methods of homogeneous doping of pentoxides Nb_2O_5:Mg and Nb_2O_5:Fe,Mg was described in greater detail in [163]. The chemical composition of plates cut from the top and bottom parts of boules of grown $LiNbO_3$:Mg (5.21 mol %) and $LiNbO_3$:Fe(0.009 mol %):Mg (5.04 mol %) single crystals was determined by the spectral-analysis method. Table 9 shows contents of cationic trace impurities in the $LiNbO_3$:Mg(5.21 mol %) crystal. It can be seen from this table that the crystal is characterized by a high homogeneity with respect to the content of impurities. Similar results were also obtained for the $LiNbO_3$:Fe(0.009 mol %):Mg(5.04 mol %) crystal.

Specimens for research were cut in the shape of parallelepipeds with a dimension of 5 × 6 × 7 mm and with their edges being parallel to the crystallographic axes. Faces of parallelepipeds were thoroughly polished. The optical-absorption spectra of crystals were registered using an SF-256 UVI spectrophotometer. In order to examine the optical homogeneity of a single crystal specimen by the laser-conoscopy method, it was arranged on a two-dimensional optical table, the table was installed between a crossed polarizer and analyzer, and the specimen to be examined was exposed to a divergent beam of laser radiation such that the transmission axis of the polarizer made an angle of 45° with the vertical. These experiments were performed using an MLL-100 Y:Al garnet laser, $\lambda_0 = 532$ nm. The axis of the laser beam coincided with the polar axis of the crystal and was perpendicular to its input face. The conoscopic pattern was observed on a semitransparent screen and was registered with a digital camera. Since the crystals under study are photorefractive, the details of the observed conoscopic pattern should depend on the power of the laser radiation; for this reason, this power was varied between 1 and 90 mW. The setup for laser-conoscopy studies is described in more detail in [164, 165]. In experiments on photoinduced light scattering, we also used the MLL-100 Y:Al garnet laser, $\lambda_0 = 532$ nm. A single crystal under study was placed in the path of the laser beam such that the wave vector of the beam was directed along the *Y* axis normally to the input face of the crystal, while the vector of the electric field of the laser radiation was parallel to the polar *Z* axis. With this geometry, the photoinduced light scattering is most intense [122]. The radiation scattered by the crystal was incident on a semitransparent screen, which was installed behind the crystal and registered with a digital camera. In more detail, the experimental technique was described in [99, 161].

Table 9. Results of spectral analysis of trace impurities in plates cut from top and bottom parts of a $LiNbO_3$: Mg (5.21 mol. %) crystal

Impurity	Impurity content, mas %	
	top part	bottom part
Zr	$<10^{-3}$	$<10^{-3}$
Mo	$<10^{-3}$	$<10^{-3}$
Ca	$<5\cdot10^{-4}$	$<5\cdot10^{-4}$
Fe	$<5\cdot10^{-4}$	$<5\cdot10^{-4}$
Ti	$<10^{-3}$	$<10^{-3}$
Si	$<5\cdot10^{-4}$	$<5\cdot10^{-4}$
Pb, Ni, Cr, Co	$<5\cdot10^{-4}$	$<5\cdot10^{-4}$
Al	$<10^{-3}$	$<10^{-3}$
Cu	$<10^{-3}$	$<10^{-3}$
Mn, V, Sn	$<5\cdot10^{-4}$	$<5\cdot10^{-4}$

Raman spectra were excited by radiation at 514.5 nm of an argon laser (Spectra Physics) and were registered with a T64000 spectrograph (Horiba Jobin Yvon), which was equipped with a confocal microscope. The power of the laser radiation that emerged from the microscope to excite the specimen did not exceed 3 mW. Since photorefractive crystals exposed to laser radiation can experience temporal changes [122], their Raman spectra were registered roughly 1 h after the beginning of the laser irradiation of the specimen, when its structure was stabilized, and almost no changes were observed. All the spectra were recorded at room temperature with a resolution of 1.0 cm^{-1}. Processing of contours of complex spectral lines and determination of their basic parameters (frequencies, widths, intensities) were performed using the programs LabSpec 5.0, Origin 8.0, and Bomem Grames/386 (Version 2.03), Table 10. The determination error of the line frequency (ν) was ± 1.0 cm^{-1}, the error for the linewidth (S) was ± 2.0 cm^{-1}, and that for the intensity (I) was 5%.

Figure 41 presents optical absorption and transmission spectra of single crystals $LiNbO_3$:Mg(5.21 mol %) and $LiNbO_3$:Fe(0.009 mol %):Mg(5.04 mol %). The $LiNbO_3$:Fe(0.009 mol %):Mg(5.04 mol %) crystals were slightly reddish-brown, whereas the $LiNbO_3$:Mg(5.21 mol %) crystals were absolutely colorless. It can be seen from the optical absorption and transmission spectra of the $LiNbO_3$:Mg(5.21 mol %) and $LiNbO_3$:Fe(0.009 mol %):Mg(5.04 mol %) crystals, which were obtained based on the Nb_2O_5:Mg and Nb_2O_5:Fe,Mg precursors, that the optical characteristics of these compounds are significantly different.

By extrapolating rectilinear parts of the spectra, we found that the absorption edges of the crystals correspond to $\lambda_{LiNbO3:Mg;\ Fe}$ = 363.3 nm and $\lambda_{LiNbO3:Mg}$ = 308.8 nm. That is, the fundamental absorption edge of the $LiNbO_3$:Fe(0.009 mol %):Mg(5.04 mol %) crystal is sharply shifted (by 54.5 nm) toward the long wavelength range, which testifies that a large amount of charged defects, as well as structural inhomogeneities, are formed in the crystal structure. The optical-absorption spectrum of the crystal exhibits weakly pronounced absorption bands in the range of ~ 400 – 600 nm. This spectral range was processed with the program Origin, and the wavelengths of the absorption maxima were determined to be at ~ 485.2 and 497.1 nm. According to the data of [166], the former maximum corresponds to the intracenter transition Fe^{3+}[Nb]–Li^{+}[V] of the Fe^{3+} ion, while the latter maximum reflects the photoionization of Fe^{2+} ions, which occupy positions of Li^{+} cations in the structure.

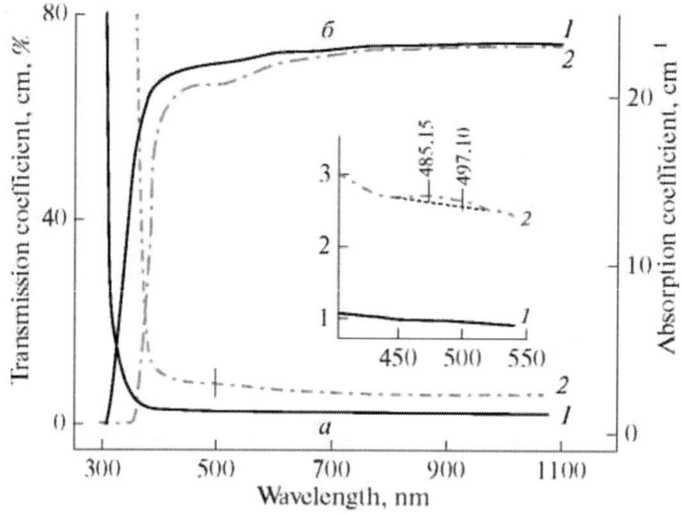

Figure 41. Spectra of (a) optical absorption and (b) transmission of *(1)* $LiNbO_3$:Mg (5.21 mol %) and *(2)* $LiNbO_3$:Fe(0.009 mol %): Mg(5.04 mol %) crystals.

The presence of Fe dopants in the lithium-niobate crystal should also manifest itself in photoinduced light scattering and in Raman spectra, whereas structural inhomogeneities should manifest themselves in conoscopic patterns.

Photoinduced light scattering in a ferroelectric crystal is a consequence of the photorefractive effect and is caused by static and dynamic (fluctuating) microinhomogeneities of the structure that are induced by the laser radiation [44, 97, 167]. The shape and dimensions of the indicatrix of the speckle structure of photoinduced light scattering in a ferroelectric crystal are very sensitive to the magnitude and particular features of the manifestation of the photorefractive effect in it [99, 167, 168].

The speckle structure of the lithium-niobate crystal is three-layer. Its peculiarities are determined by subtle features of the crystal structure the occurrence of multiply charged "photorefractive" cations and defects with localized electrons [168].

Our data on photoinduced light scattering (Figure 42) show that the occurrence of photorefractive iron cations in the $LiNbO_3$:Fe(0.009 mol %): Mg(5.04 mol %) crystal does not cause a significant increase of the photorefractive effect.

From Figure 42, it is also seen that, even at a comparatively high power of the laser radiation (170 mW), the pattern of the photo-induced light scattering in the $LiNbO_3$:Mg(5.21 mol %) and $LiNbO_3$:Fe(0.009 mol %):Mg(5.04 mol %) crystals is not developed and only an insignificant circular scattering is observed, which indicates that the photo-refractive effect is weak. These data are supported by conoscopic patterns and data on Raman spectra.

Figure 43 presents conoscopic patterns of investigated single crystals that were obtained at laser radiation powers of 1 and 90 mW. Upon scanning of the plane of the input face by the low-power radiation (1 mW), conoscopic patterns correspond to those of an uniaxial crystal (Figures 43a, 43c). In this case, if the optical axis coincides with the normal and is orthogonal to the input face of the crystal, isochromatic curves (lines of the same phase shift) form a series of concentric circles the center of which is located at the exit point of the optical axis. Against the background of the circular isochromatic curves, a black "Maltese cross" retains a minimal intensity in the range from the center to the periphery of the field of view.

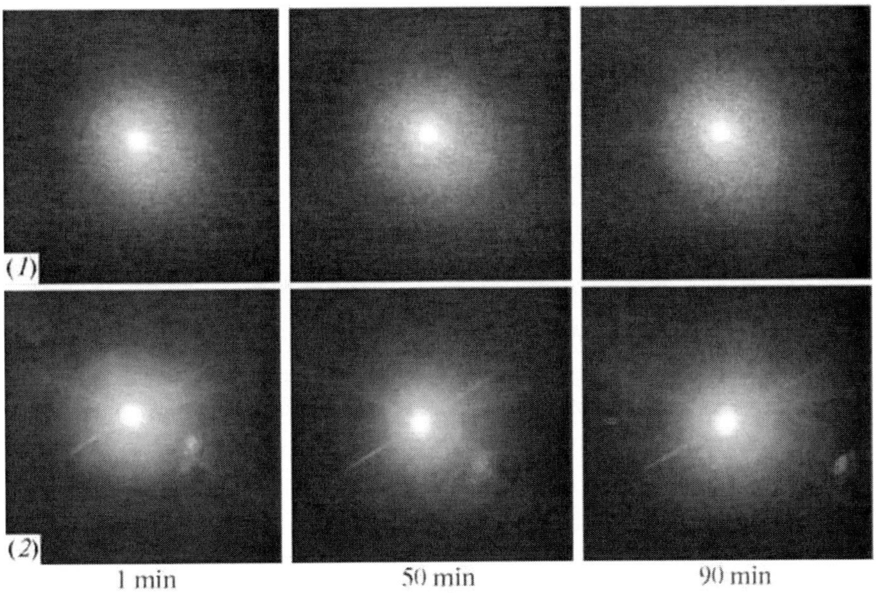

Figure 42. Photoinduced light scattering in (*1*) $LiNbO_3$:Mg (5.21 mol %) and (*2*) $LiNbO_3$:Fe(0.009 mol %):Mg(5.04 mol %) crystals.

However, it should be noted that, even at a low power, the cono-scopic patterns reveal the manifestation of the photo-refractive effect; thus, the contrast of images is somewhat lowered and, at a certain diffuseness and absence of clear-cut contours of interference fringes, the speckle structure is more pronounced (Figures 43a and 43c).

As the power of the laser radiation is increased to 90 mW (Figures 43b and 43d), the conoscopic patterns of the $LiNbO_3$:Mg(5.21 mol %) and $LiNbO_3$:Fe(0.009 mol %): Mg(5.04 mol %) crystals become more pronounced and contrasting and have an appearance that, on the whole, corresponds to those of uniaxial crystals.

However, they reveal weak but well-pronounced interference anomalies, which testify to the appearance of a weak optical biaxiality. Thus, whereas in the lower half-plane of the conoscopic pattern of the $LiNbO_3$:Fe(0.009 mol %):Mg(5.04 mol %) crystal (Figure 43b) the branches of the Maltese cross have the shape that is standard for uniaxial crystals, in the upper half-plane, in the area of the left branch of the cross, a discontinuity is observed, as well as displacements and pairwise joining of isochromatic curves at the border between adjacent quadrants (Figure 43b). These anomalies are most distinguishable in the interval from the fourth to the tenth isochromatic curves counting from the center of the conoscopic pattern.

In the area of a minimal intensity of the left branch of the Maltese cross, the third and the fourth, the fifth and the sixth, the seventh and the eighth, and the ninth and the tenth isochromatic curves join pairwise with the fourth, sixth, eighth, and tenth isochromatic curves, respectively, in the adjacent vertical quadrant. The third, fifth, seventh, and ninth isochromatic curves of the vertical quadrant have a discontinuity in the area of the left branch of the Maltese cross and their intensities differ from minimal intensities, characteristic of strictly uniaxial crystals. In the area between the second and seventh isochromatic curves, the left upper branch of the Maltese cross (Figure 43b) contains an additional system of vertical interference fringes against the background of the main conoscopic pattern of the crystal.

Similar anomalies in the shape of additional interference structures have also been observed for $LiNbO_3$:Mg (5.0 and 5.5 mol %) crystals [87].

In the lower half-plane of the conoscopic pattern, the right branch of the Maltese cross retains its standard form, whereas, in the area of the left branch of the Maltese cross, beginning from the second isochromatic curve counting from the center, the following distortions are observed: (i) a pair joining of the second and third isochromatic curves of the lower quadrant with the second isochromatic curve of the left adjacent quadrant; (ii) a joining of the third and fourth and of the fifth and sixth isochromatic curves of the right quadrant with the fourth and the sixth isochromatic curves of the lower quadrant, respectively; (iii) a displacement of the fourth, sixth, and seventh isochromatic curves on the borders of the lower quadrant by one order; and (iv) a discontinuity of the fifth isochromatic curve of the lower quadrant and of the fourth isochromatic curve of the left quadrant on the adjacent border. In the area of a minimal intensity of the right upper branch of the Maltese cross, we observed (i) a discontinuity of the third isochromatic curve, (ii) a pair joining of the third and fourth isochromatic curves of the upper quadrant with the fourth isochromatic curve of the right quadrant, and (iii) a joining of the fifth and sixth isochromatic curves of the right quadrant with the fifth isochromatic curve of the upper quadrant.

At all powers of the used radiation, conoscopic patterns of $LiNbO_3$:Mg(5.21 mol %) crystals that were grown from a batch without adding Fe have the appearance that, on the whole, correspond to those of uniaxial crystals (Figures 43c, 43d). An increase in the power of the laser radiation that was used in experiments from 1 (Figure 43c) to 90 mW (Figure 43d) leads to a certain increase in the contrast and image sharpness; however, in this case, no appreciable interference anomalies occur in conoscopic patterns.

Laser-conoscopy and photoinduced light-scattering methods do not yield any information on particular features of the internal structure of crystals and defects, which determine their photorefractive properties. Raman spectroscopy is an informative method of investigation of subtle features of the crystal structure, the state of its defectness, and changes in the structure that are caused by doping and photorefractive effect.

Raman spectra are highly sensitive to changes in interactions between structural units of the crystal, as well as to defects both intrinsic and induced by the laser radiation [44].

At present, Raman spectroscopy is the sole method for simultaneous investigation of the photorefractive effect and changes in the crystal structure caused by it.

In the Raman spectrum of a lithium-niobate crystal, the photorefractive effect (as well as the photoinduced light scattering) manifests itself maximally if it is induced by the laser radiation that is polarized along the polar *Z* axis, i.e., in the polarization scattering geometries *(ZX), (ZY),* and *(ZZ)* [1]. In this case, in the crystal, the energy of the exciting laser radiation is transferred to scattered light and, since the refractive index changes predominantly along the *Z* axis, the crystal strongly defocuses the laser beam and the pattern of the photoinduced light scattering becomes developed [44, 99, 161]. Since the exciting radiation is defocused, in particular, the geometry *X/Z(Z/XX)Y* (*E*(TO) phonons are active) is transformed into the geometry *X/Z(Z/XX)Y,* and lines that correspond to A_1(TO) phonons, which are forbidden by selection rules in the geometry *X(ZX)Y but* are allowed in the geometry *Z(XX)Y* [1], appear in the spectrum. By measuring the intensity of lines in the spectrum that correspond to phonons forbidden in the given scattering geometry, one can estimate the magnitude of the photorefractive effect. Theoretically, it makes no difference which group of lines in the Raman spectrum is selected for the estimation of the relative intensity [44].

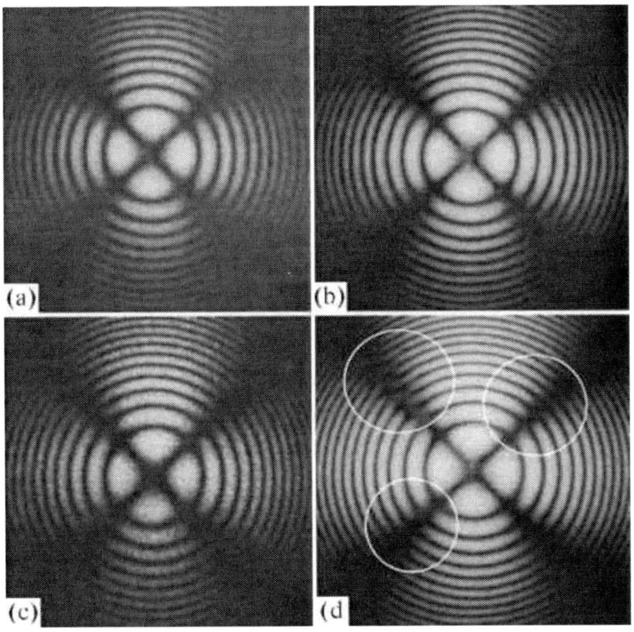

Figure 43. Conoscopic patterns of $LiNbO_3$:Mg (5.21 mol %) crystal: (a) laser-radiation power, 1 m W; (b) laser-radiation power, 90 mW; and $LiNbO_3$:Fe(0.009 mol %):Mg (5.04 mol %) crystal: (c) laser-radiation power, 1 mW; (d) laser-radiation power, 90 mW.

It is only necessary that the energy-transfer mechanism are of the same type; in this case (for the *Y(ZX)Y* scattering geometry), it is the *E(*TO*)* — > *A* $_1$(TO) phonon transition. In the Raman spectrum of the lithium-niobate crystal, the line with a frequency of 578 cm^{-1}, which corresponds to doubly degenerate vibrations of oxygen octahedra and belongs to the *E*(TO) symmetry species [44], is the most convenient analytical line for estimation of the magnitude of the photorefractive effect.

Figure 44 presents the Raman spectra of the $LiNbO_3$:Mg(5.21 mol %) and $LiNbO_3$:Fe (0.009 mol %): Mg(5.04 mol %) crystals that were measured in the *Y(ZX)Y* scattering geometry. According to the selection rules [44], in the *Y(ZX)Y* scattering geometry, in the range of 500-650 cm^{-1}, only one line should be observed in the absence of the photorefractive effect, which belongs to the *E*(TO) symmetry and is located at a frequency of 574 cm^{-1}. The line at a frequency of 628 cm^{-1}, which is reliably observed in the spectra of the $LiNbO_3$: Mg (5.21 mol %) and $LiNbO_3$:Fe (0.009 mol %):Mg(5.04 mol %) crystals in Figure 44, is forbidden by the selection rules in the *Y(ZX)Y* scattering geometry and is observed in this geometry due to the photorefractive effect.

In this case, to estimate the magnitude of the photorefractive effect in the lithium-niobate crystal, based on the dispersion dependence of frequencies [44], it is convenient to use not the absolute intensity of the line, but, rather, relative intensity I_{rel}, which is determined by the formula $I_{rel} = (I_{630}/I_{580}) \times 100$. The relative intensities of the "forbidden" lines that were determined in this way were $I_{rel\ LiNbO_3:Mg} = 22\%$ and $I_{rel\ LiNbO:Mg:Fe} = 30\%$. Therefore, the relative intensity of the forbidden line in the spectrum of the $LiNbO_3$:Fe(0.009 mol %): Mg (5.04 mol %) crystal is higher than in the spectrum of the $LiNbO_3$:Mg (5.21 mol %) crystal, which testifies to a larger magnitude of the photorefractive effect in this compound. In our opinion, this is caused by the presence of the photorefractive iron dopant.

It is known that, in congruent lithium-niobate single crystals, iron is present in two valence forms, Fe^{2+} and Fe^{3+}, the ratio between which depends on the thermochemical history of the growth of the crystal [44, 154].

Under the action of light, Fe^{2+} and Fe^{3+} exchange electrons in accordance with the reaction

$$Fe^{3+} + h\nu \rightarrow Fe^{2+} + \text{hole}.$$

In this reaction, illuminated and dark regions act as electron donors and acceptors, respectively [44]. The photoionization of Fe^{3+} in the $LiNbO_3$:Fe(0.009 mol %): Mg(5.04 mol %) crystal finds its confirmation in the electronic-absorption spectra (Figure 41).

The width of the forbidden line with a frequency of 628 cm^{-1} in the spectrum of the $LiNbO_3$:Fe (0.009 mol %):Mg(5.04 mol %) crystal is considerably greater (by 15 cm^{-1}) than that in the spectrum of the $LiNbO_3$:Mg(5.21 mol %) (Table 10) crystal, which, seemingly, is related to an insignificant deformation of oxygen octahedra, which is stronger in the $LiNbO_3$: Fe(0.009 mol %):Mg(5.04 mol %) crystal than in the $LiNbO_3$:Mg(5.21 mol %) crystal. According to the data of [169-171], the threshold concentration of Mg^{2+} cations in a $LiNbO_3$: Mg congruent crystal at which it becomes stable with respect to optical damage is 4.6 mol %.

It is assumed in this case that, in the crystal of this composition, Mg^{2+} cations predominantly occupy positions of Li^{+} cations in the ideal structure, displacing Nb^{5+} cations from these positions [44, 169, 170].

However, the question of the localization of Mg^{2+} cations in the structure of lithium niobate remains debatable. Thus, it was substantiated in [44, 171] that, upon the addition of Mg^{2+} cations to the crystal, they displace Nb^{5+} and Fe cations from lithium positions of the ideal structure. In this case, depending on the concentration of the Fe dopant and on the Li/Nb ratio, it is assumed that the replacement process can proceed by one of the two mechanisms. If the concentration of Mg^{2+} cations is lower than the threshold concentration, the dominating process is that in which Mg^{2+} cations replace Nb^{5+} cations in lithium positions in the ideal structure, whereas Fe_{Li} antisite defects remain intact. When the majority of Nb^{5+} cations are displaced from lithium positions, Mg^{2+} cations begin to displace Fe_{Li} defects.

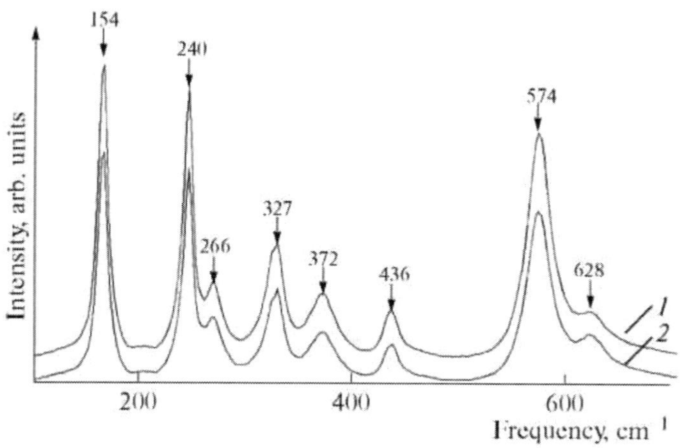

Figure 44. Raman spectra of (*1*) $LiNbO_3$:Mg(5.21 mol %) and (*2*) $LiNbO_3$:Fe(0.009 mol %):Mg(5.04 mol %) single crystals measured in the *Y*(*ZX*)*Y* scattering geometry.

In turn, iron begins to incorporate itself into the positions of Nb^{5+} cations. It should be noted that correct investigation of the distribution of main and doping cations over positions in the structure of the lithium-niobate crystal is a complicated problem that requires performing experiments by the method of full-profile X-ray diffraction analysis in combination with calculations with the use of vacancy split-models [106, 107, 172].

In the case of the $LiNbO_3$:Fe(0.009 mol %):Mg(5.04 mol %) crystal that we investigated, it is likely that incorporation of Mg^{2+} cations occurs by the second mechanism, since the concentration of Mg^{2+} cations in the crystal exceeds the threshold concentration (4.6 mol %) by 0.2 - 0.4 mol %. The ionic radius of iron is 0.8 Å, whereas those of lithium and niobium are 0.67 Å each [44].

Therefore, the displacement of Nb^{5+} cations from oxygen octahedra by Fe cations should lead to a distortion of the whole oxygen framework and, as a consequence, to a broadening of the line with a frequency of 628 cm^{-1}, which corresponds to totally symmetric vibrations of oxygen octahedra.

Using the complex of methods (electronic spectroscopy, laser conoscopy, photoinduced light scattering, and Raman spectroscopy), we have investigated the optical homogeneity, optical transmission, and photorefractive properties of $LiNbO_3$:Mg(5.21 mol %) and $LiNbO_3$: Fe(0.009 mol %):Mg(5.04 mol %) single crystals that were grown from congruent melts with the use of solid precursors Nb_2O_5:Mg and Nb_2O_5:Fe,Mg obtained by homogeneous doping with magnesium of reextracts at the stage of extraction of Nb_2O_5.

We have ascertained that doping with Mg^{2+} cations results in the suppression of the photorefractive effect in the lithium-niobate crystal. In this case, upon double doping (Fe:Mg) at concentrations of Mg^{2+} cations above threshold concentrations, when the photorefractive effect is almost zero, photorefractive Fe cations do not affect so significantly the photorefractive effect, as in the case of normally pure congruent crystals doped with Fe.

However, an appreciable deformation of oxygen octahedra caused by Fe cations is observed. For investigated crystals, the indicatrix of photoinduced light scattering is not developed even at comparatively high powers of the exciting laser radiation (170 mW) and only insignificant circular scattering is present, which indicates that the photorefractive effect is weak. For the use of single crystals in holography to be successful, it is necessary to suppress photo-induced light scattering in them but to retain their photorefractive properties.

Table 10. Basic parameters of lines that are observed in Raman spectra of $LiNbO_3$: Mg (5.21 mol %) and LiNbO3 : Fe (0.009 mol %) : Mg (5.04 mol %) single crystals in the Y(ZX)Y scattering geometry

$LiNbO_3$: Mg : Fe			$LiNbO_3$: Mg		
ν, cm^{-1}	*s*, cm^{-1}	*I*, arb. units	ν, cm^{-1}	*s*, cm^{-1}	*I*, arb. units
153	11	23171	154	11	29816
240	12	20618	240	11	26226
266	17	6394	266	15	7557
328	18	9088	327	19	11268
372	26	5022	372	26	6524
436	16	3842	436	16	5064
574	27	16601	574	26	21924
626	47	4974	628	62	4882

11. Laser Conoscopic Research Technique for Single Crystals $LiNbO_3$: Mg

Initially, the conoscopic patterns obtained with a polarizing microscope were used in mineralogy in order to identify minerals based on the data on crystal symmetry and orientation [173]. Conoscopic pattern informativity provides for the possibility to determine orientation and nature of optical indicatrix, measure an angle between the optical axes of a biaxial crystal, determine an optical sign of the crystal, detect optical axes dispersion, identify qualitative and quantitative changes in the optical indicatrix in response to external action, etc. [173-183]. Conoscopic method is one way to analyze the properties of optical crystals, which allows to determine their functionality, and which has long and successfully used in scientific research and a variety of optical devices.

In this article, it is proposed to obtain conoscopic patterns using an optical system where diverging laser radiation is let pass through an anisotropic crystal placed between the polarizer and analyzer, rather than using a polarizing microscope. The pattern on the screen is recorded by a digital camera and displayed on a computer.

Where the point symmetry group of the crystal is already known, the practical importance of such conoscopic studies lies in detection and analysis of various distortions of optical elements of actual crystals [182, 183]. Modern industrial technology for growing single crystals of lithium niobate doped with different dopants influencing the composition of the crystals and physical properties, allowing them to adjust to a wide range. One of the main criteria is the quality of produced crystals of optical homogeneity. Use as a dopant cations Mg^{2+} provides lower interfering effect photorefraction in lithium niobate single crystals, however, can complicate the structure is strong enough crystal and, as a consequence, lead to the optical inhomogeneity. The possibility of observing conoscopic patterns of large-scale appears when you use the laser system in which divergent wide-beam radiation is obtained through the diffuser placed in front of the front face of the crystal [177].

The significant size of the image allows you to perform a detailed analysis of subtle features of the structural distortions in the crystal, as in the center of the field of view, and in the peripheral region of the conoscopic patterns.

The development of laser conoscopic method also relevant to studies of thin structural distortions, arising in photorefractive crystals, for the detection and investigation of subtle features of structural distortions, as well as micro-and nanostructures, inevitably present in doped single-crystal materials [122].

In this paper, a laser conoscopic method investigated the fine features of structural distortions in a series of single crystals of lithium niobate ($LiNbO_3$) congruent ($R = Li/Nb = 0.946$), doped with Mg^{2+}, characterized by low effect photorefraction (optical damage), promising as materials for electronics [44, 122]. Used as a relatively lightly doped crystal $LiNbO_3$:Mg[0.01 - 1.5 mol·%], and crystals with a high concentration of Mg^{2+} ($LiNbO_3$:Mg [3.0 - 5.5 mol·%]), photorefractive effect in which is almost equal to zero [44].

The test samples were cut from a single crystal boule grown in the direction of the Z (the polar axis of the crystal). In order to evaluate the optical homogeneity of single crystal boules grown samples were cut from different parts of the boule. The cylindrical portion of the boule was cut into transverse disks from which the samples were cut into parallelepipeds ~ 8×6×4.7 mm^3 edges parallel crystalophysical axes X, Y, Z, respectively.

Faces of the parallelepiped and plates carefully polished. Methods of crystal growth and preparation of samples for research are described in more detail in [122]. When conducting an experiment to observe conoscopic patterns of optical crystals with optical system (Figure 45), consisting of a source of radiation, polarizer, diffuser, crystal, analyzer and the screen, which allows you to receive conoscopic pattern of considerable size (0.5 meters or more). Investigated crystal plate is located on the two-coordinate optical mobile stand that allows you to scan the entire plane with a laser beam entrance face and get a series of conoscopic patterns. In the experiments, radiation He-Ne laser (λ = 632.8 nm) power not exceeding 1 mW in order to minimize the possible impact of the photorefractive effect on conoscopic pattern. To investigate the defect micro and macrostructure single-crystal $LiNbO_3$:Mg was applied high-performance and flexible image analyzer Thixomet®, based on modern hardware (microscope of *Carl Zeiss* - Axio Observer) and software [184].

Conoscopic picture of perfect uniaxial crystals obtained with linearly polarized radiation is well known, explained and described in the literature [173, 182, 183].

This picture of the propagation of a diverging beam of light along the optical axis is composed of concentric rings centered at the output of the optical axis. Rings superimposed on the characteristic intensity distribution -black "maltese cross" In this case, each ring is the same line of the phase shift and the cone of rays with the same angle of incidence at the coincidence of the axis of the conical radiation beam with the optical axis of the crystal.

Izohrom form depends on the orientation of the optical axis with respect to the input face of the crystal. With some of the angle between the optical axis and the normal to the entrance face of the ring transformed into ellipses. Samples with significant corners form izohrom approaching hyperbole.

Branch of the "maltese cross", consisting of two isogyre minimum intensity, intersect in the center of the visual field, perpendicular to each other and coincide with the axes of transmission of the polarizer and the analyzer.

The characteristic feature of arising anomalous optical biaxiality in which there is a deformation of the optical indicatrix of the crystal is the rupture of black “maltese cross” in two parts with the enlightenment in the center of the visual field.

In our experiments, for samples $LiNbO_3$:Mg[0.01-1.5 mol %] were observed conoscopic pattern of the standard form, in which the black “maltese cross” preserves the integrity of the center of the field of view, and isochromes have the form of concentric circles.

For samples with the same thickness in the direction of the optical axis, but with a different concentration of dopant Mg, for example, $LiNbO_3$:Mg [0.5 mol %] and the $LiNbO_3$: Mg [1.0 mol %] general view of conoscopic patterns has coincided Figure 46 with preservation of diameter ring-izohrom. Conoscopic pattern crystal $LiNbO_3$:Mg [0.01-1.5 mol %] and $LiNbO_3$:Mg [3.0-5.5 mol %] are very different.

When scanning the plane entrance face with a rather high concentration of the impurities $LiNbO_3$:Mg [3.0 mol %], in addition to standard patterns and similar in appearance (Figure 47(a)) were observed and the distorted conoscopic pattern (Figures 47(b)-(e)).

On conoscopic pattern (Figure 47(b)) black “maltese cross” cut in half with the enlightenment in the center field of view. The azimuthal direction of displacement on parts of “maltese cross” amounts to the angle of ~ 10 - 13 clockwise from the vertical. Isochromen keep integrity, but some extend in the direction of displacement of the fragments of the cross and take the form of ellipses with the attitude of the minor and major axes of ~ 0.9:1.

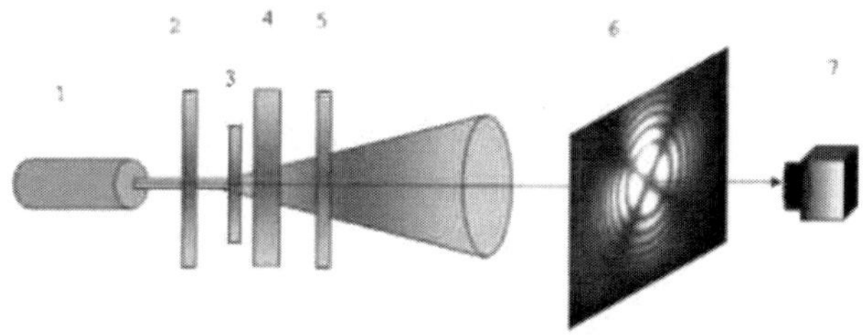

Figure 45. Diagram of the optical sign identifying system: 1-He-Ne laser; 2-polarizer; 3-diffuser; 4-investigated crystal plate; 5-analyzer crossed with polarizer; 6-screen; 7-camera.

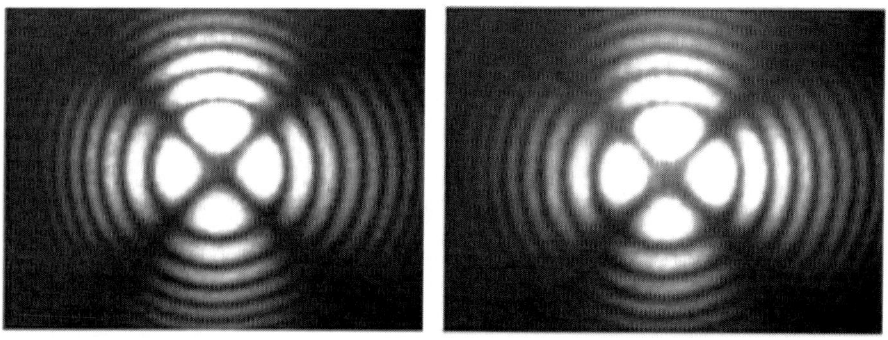

a b

Figure 46. Conoscopic pattern of single crystals of $LiNbO_3$: Mg:(a)-[0.5 mol %]; (b)-[1.0 mol %].

On conoscopic pattern (Figures 47(c), (d), (e)) the black "maltese cross" in the center of the field of view, on the contrary, is an integer, and retain the form isochromen rings. However, in the periphery of the field of view at a considerable angular distance from the center of the picture, starting with a 5 - 6th isochromen, in only one branch of the "maltese cross" is observed by imposing additional interference structure. While the remaining three branches "maltese cross" retain their usual form.

All observed by scanning the plane entrance face conoscopic pattern $LiNbO_3$:Mg [5,0 mol. %] Characteristic of uniaxial crystals, as indicated by the black "maltese cross" on the background of the rings-izohrom (Figures 48(a)-(e)).

However, on some conoscopic patterns on a small angular distance from the center of one of the four branches of the "maltese cross" there is the imposition of additional distinct interference fringes (Figures 48(b)-(e)).

Conoscopic pattern samples with the highest concentration of the dopant $LiNbO_3$:Mg[5.5 mol. %]. Characteristic of uniaxial crystals (Figures 5(a)-(e)), but at some point the input face light up with some pictures of conoscopic anomalies. One type of anomaly is a superposition of additional interference pattern at an angular distance from the center, corresponding to a 3 - 4th isochromen, inone branch of the "maltese cross" (Figures 5(b)-(c)).

Another kind of anomaly is manifested as additional interference pattern, but in the center of the field of view of a conoscopic pattern on the background of black crossing branches "maltese cross" It should be noted that the conoscopic patterns of each of the three samples $LiNbO_3$:Mg[3.0 - 5.5 mol- %] with circular polarizer and the analyzer, which allows you to remove beclouding "maltese cross" have a standard form of rings and show no noticeable distortion (Figures 47(f), 48(f) and 49(f)).

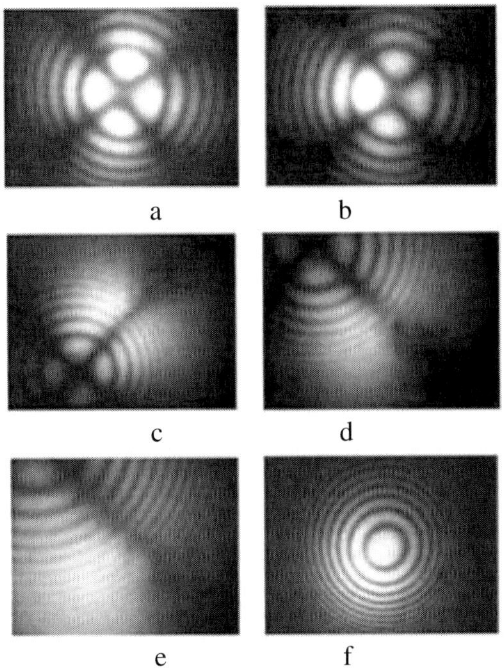

Figure 47. Conoscopic pattern of single crystal of $LiNbO_3$: Mg:[3.0 mol %].

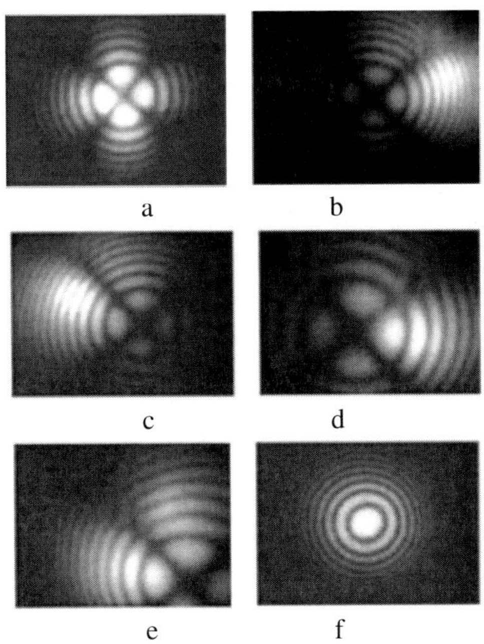

Figure 48. Conoscopic pattern of single crystals of $LiNbO_3$: Mg[5.0 mol %].

Results conoscopic method study of crystals $LiNbO_3$, doped Mg cations to varying concentrations of interest, grown under different conditions, show that lightly doped lithium niobate samples containing Mg [0.003-1.0 mol·%] have a high optical homogeneity.

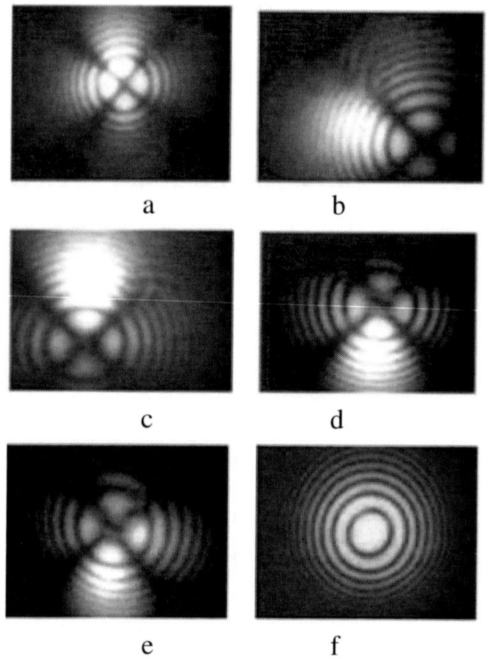

Figure 49. Conoscopic pattern of single crystals of $LiNbO_3$: Mg [5.5 mol %].

Analysis of the effect of the dopant Mg on the form the conoscopic pattern $LiNbO_3$:Mg showed that when the concentration of Mg dopant in the samples with the same geometric parameters of the scale of the conoscopic pattern, intensity distribution, shape and size of the "maltese cross" and izohrom saved.

Conoscopic technique to study samples of lithium niobate with the content of Mg [3.0 - 5.5 mol - %] suggests that a stronger doping Mg cations while maintaining overall uniaxial crystal leads to the appearance of local birefringent inclusions, which are recorded in the form of additional interference pattern on the background main conoscopic pattern in the center of the field of view, and in its peripheral region.

Small anomalous biaxiality in a bounded domain is registered for a sample $LiNbO_3$: Mg [3.0 mol - %], which is confirmed by the break and enlightenment "maltese cross" in the center of the conoscopic pattern of the crystal.

The differences in conoscopic patterns of single crystals of $LiNbO_3$: Mg [0.01 - 1.5 mol- %] and $LiNbO_3$:Mg [3.0 - 5- 5 mol - %] can be explained as follows.

Feature of lithium niobate single crystals doped with cations Mg^2 + at relatively high (> 3 mol • %) dopant concentration is uneven of impurity [44, 184] and, therefore, the appearance of growth bands associated with gradients of dopant concentration, as in plane perpendicular to and in the plane parallel to the growth axis (Figures 50, 51).

Banding is accompanied by the growth of microde-fects in the form of dislocations, microdomains of domain walls and block structure, especially in the high impurity concentration gradients at the boundaries of growth bands (Figure 52).

Growth bands, the gradient of the impurity concentration, the concentration of microdefects lead to a local change of the elastic characteristics of the crystal and appearance of mechanical stress [184], locally distorting the optical indicatrix of optically uniaxial crystal.

This leads to a distortion of the conoscopic patterns (Figures 47-49).

Moreover, the maximum distortion is observed for the conoscopic patterns on the borders growth bands, where the concentration of structural defects and the dopant concentration gradients are maximized.

In the series of crystals investigated by us striation of samples, in general, decreases with increasing impurity concentration from 3.0 to 5.5 mol % (Figures 50, 51). In the same row is somewhat reduced degree of distortion conoscopic patterns (Figures 48, 49).

Thus, the deficiency of the crystal associated with the inhomogeneity of admixture disposition, passes through a maximum at a certain concentration of ~ 3 mol % Mg^{2+}. The latter may be due to a change in the mechanism of admixture disposition when changing dopant concentration [44, 185].

In particular, the research methods of microanalysis found a reduction ratio R = Li/Nb (0.94) at a concentration in the crystal Mg^{2+} ~ 3% [16].

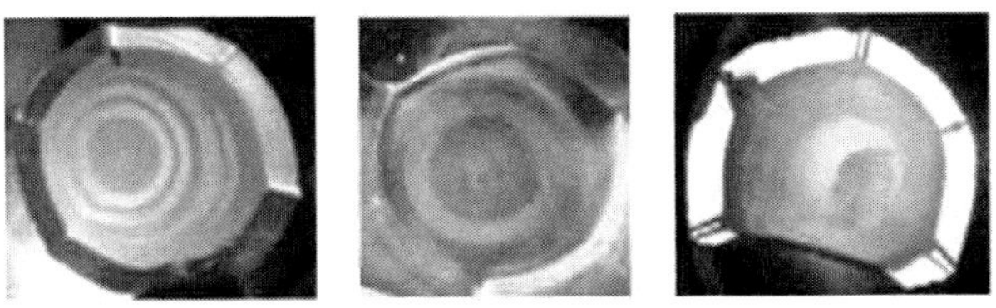

a b c

Figure 50. Bands of crystal growth LiNbO3: Mg in the plane perpendicular to the growth: (a) - [3,0 mol %]; b - [5,0 mol %]; c - [5. Mol %].

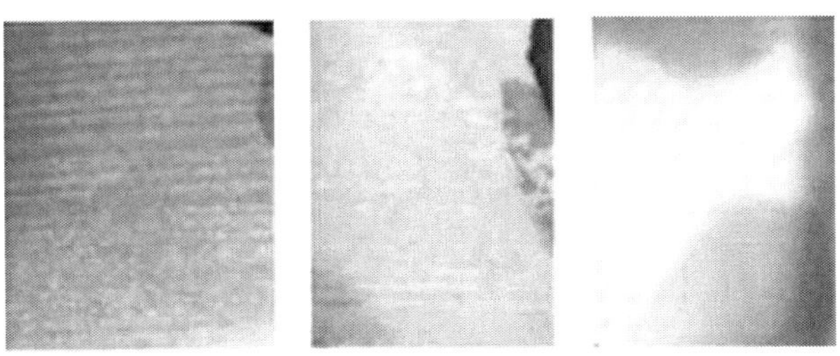

a b c

Figure 51. Bands of crystal growth LiNbO3: Mg in the plane parallel to the growth:(a) - [3.0 mol %];(b) - [5.0 mol %];(c) - [5.5 mol %].

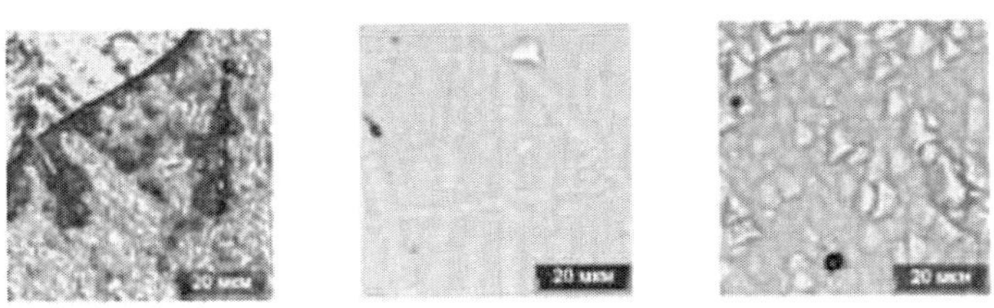

a b c

Figure 52. Microdefects at the boundaries of crystal growth bands $LiNbO_3$: Mg in the plane perpendicular to the axis of growth: (a) - 3.0 mol %];(b) - [5.0 mol %];(c) - 5.5 mol %].

With such a concentration of Mg^{2+} defects Nb_{Li} (cation Nb^{5+}, are in the positions of lithium of the ideal structure of the stoichiomet-ric composition) completely forced out in the cation sublattice [44].

At a concentration of $Mg^{2+} > 3\%$ are replaced by base cations Li^{+}, accompanied by an increase of defects V_{Li} (vacant oxygen octahedrons, which in an ideal structure stoichiometric composition should be located the cations Li^{+}) [44, 186].

When approaching the value R = Li/Nb to the value of 0.84 [$Mg^{2+} \geq 8\%$], corresponding to the stability boundary $LiNbO_3$ phase in the phase diagram [13], the cations Mg^{2+} are included in both (lithium and niobium) position of the ideal structure of stoichiometric composition with a simultaneous decrease concentration of compensating defects V_{Li} [44, 186].

Results conoscopic method study of lithium niobate crystals doped with cations Mg^{2+}, show that samples containing Mg^{2+} (0.01 - 1.5 mol %), even cut from different crystals, grown under different conditions, have high optical homogeneity. And the scale of the conoscopic pattern, intensity distribution, shape and size of the “maltese cross” and izohrom are fully maintained for samples with the same geometric parameters.

Stronger doping cations Mg^{2+} (3.0 - 5.5 mol %), while maintaining overall uniaxial crystal leads to the appearance of local birefringent inclusions, which are recorded as additional interference pattern on the background main conoscopic pattern in the center of the field of view and in its peripheral region.

A small anomalous biaxiality in a limited region is registered for a sample containing Mg^{2+} 3.0 mol %, which confirmed the break and enlightenment “maltese cross”in the center of the conoscopic pattern of the crystal.

Defect structure of the crystals, associated with irregular occurrence impurities leads to a local change of the elastic characteristics of the crystal and the appearance of mechanical stresses that cause distortion of the conoscopic patterns. And the imperfection of the crystal samples and the degree of distortion conoscopic patterns, generally decrease with increasing impurity concentration from 3.0 to 5.5 mol %.

Thus, laser conoscopic research technique on the proposed scheme in this paper to identify subtle changes in the optical properties of the crystal at its doping. Conoscopic same study samples under a polarizing microscope, in which the light source is used as a incandescent lamp, make it possible to observe a picture of the crystal as a whole, without detailing the fine features of the structure in the form of local distortions.

12. Determination of Photoelectric Fields in a Lithium Niobate Crystal by Parameters of Indicatrix of Photoinduced Scattered Radiation

One of the most effective photorefractive monocrystalline material is a ferroelectric single crystal of lithium niobate possessing unique combination of optical, electro-optical, piezoelectric, pyroelectric, photovoltaic characteristics. It allows using this material in electronic devices. Lithium Niobate is a phase of a variable composition.

A wide range of homogeneity (44.5-50.5 mol. % Li_2O at 1460 K [1]) allows to grow (by method of Chokhralski) nominally pure and doped with a wide range of “photorefractive” and "non-photorefractive" cations[4] single crystals with different value R=Li/Nb from the melts of different composition [1]. The peculiarities of the crystal structure, its physical characteristics in particular the electrooptic effect and therefore the effect of photorefraction can vary within very wide limits [187].

Passing of laser radiation through the photorefractive crystal is accompanied by effect of photoinduced (photorefractive) light scattering (PLS) which has a complicated structure is dynamic and occurs at the induced by laser radiation micro and macro defects with fluctuating or static physical parameters (refractive index, conductivity etc.) [99, 166].

As a result of photo excitation the spatial charges transfer (drift or diffusion) and their subsequent capture by deep levels with the formation of the space charge field takes place. The appearance of this field leads to the change of refractive index.

In the noncentrosymmetric Lithium Niobate crystal where the primary mechanism of photorefraction is photovoltaic mechanism at the expense of linear electro-optic effect [99, 187], the value of which determines the value of the angle opening of PLS, which occurs mainly along the polar axis of the crystal [99, 168]. The magnitude and speed of the angle opening determine the sensitivity and speed of the photorefractive holographic information recording, electro-optical modulators and valves [187].

Using the PLS, an attempt to perform a comparative evaluation of photovoltaic fields in Lithium Niobate crystals of different composition (nominally pure and alloyed), grown by the Chohralski method in different ways is made. Working-out of methods for the experimental evaluation of photovoltaic fields in the crystal is an important task for the creation of new materials based on single crystal of Lithium Niobate with given structural and photorefractive properties for the generation and conversion of laser radiation, recording and storage of information and for testing of industrial technology of single crystal growth.

As the research objects nominally pure single crystals of Lithium Niobate of stoichiometric composition (R=Li/Nb = 1) grown from the melt of 58.6 mol. % Li_2O ($LiNbO_3$ stoich.) and also nominally pure single crystals of congruent composition alloyed by Cu^{2+}, Zn^{2+}, Gd^{3+} were used. All the single crystals are grown by the Chokhralsky method. The content of foreign cationic impurities was not less than 10^{-3} wt.%. The technique of the single crystal charge growing is set out in detail in work [122].

In Figure 53 a scheme of experimental setup for determination of intensity and angle of the scattered radiation is shown. As the source of radiation He-Ne laser with power of P = 60 mW (k_o = 0.6328 μm) was used. Diameter (d) of the light beam was 3 mm. The laser beam is directed through the hole (d = 1 mm) in the blackout chamber 2 on the researched sample 3. Crystal is oriented in the blackout chamber so that the polar axis of the crystal (Z) was placed in the horizontal plane, and a laser beam propagated along the X-axis. The tension vector of electric field of the light wave is oriented along the Z axis to investigate the interaction of its type [99]. A photodiode 4 is situated inside the chamber. It can move in horizontal plane in the range of -51° to +51°. Accuracy of moving is 0.5°. The photodiode is connected to a multimeter 5, which transmits data to the computer 6.

The samples of single crystals had been irradiated by the laser for 60 min to achieve a steady state of the indicatrix PLS.

[4] "Photorefractive" cations (cations with variable valence) change their charge in the crystal under the action of light and increase the effect of photorefraction. "Non-photorefractive" cations under the action of light change their charge in the crystal and under certain conditions change the crystal's structure so that the effect of photorefraction decreases.

Firstly the photodiode was installed in the central area of the scattering pattern, i.e. in the area of the laser beam. Within 60 min the changes in the intensity of crystal's radiation had been fixed. Then according to the scattering angle in the horizontal plane in each 3° the intensity of radiation in the steady state of indicatrix PLS was measured.

As it's known, when a crystal is irradiated with laser radiation during the expansion of the indicatrix PLS the intensity of the central beam decreases and the energy transfer from the central beam into the scattered radiation is observed [161, 188, 189]. Under the conditions of our experiment PLS of *ee*-type is observed [4]. Indicatrix PLS of *ee*-type has the form of an asymmetric "eight" elongated along the polar axis of the crystal [102, 190], that allows according to the indicatrix form determine the direction of the polar axis. PLS intensity measurement results for different single crystals in the central beam are given in Table 1. It can be seen that the most efficient energy transfer occurs in the crystal LiNbO3:Cu + Gd (0.57 + 0.07 wt.%), and the least efficient occurs in stoichiometric $LiNbO_3$ crystal.

The dependence of the relative intensityI/I_0 from the scattering angle forthe studied crystals is shown in Figure 54. Maximum scattering angle after 60 min of radiation is observed for stoichiometric crystal ($LiNbO_3$ stoich). It reaches 48° with a relative decrease in the intensity 2.522 x 10^{-5}. From Figure 54 it can be also seen that the scattered radiation has asymmetric form because to the same value of the relative intensity different angles of PLS correspond.

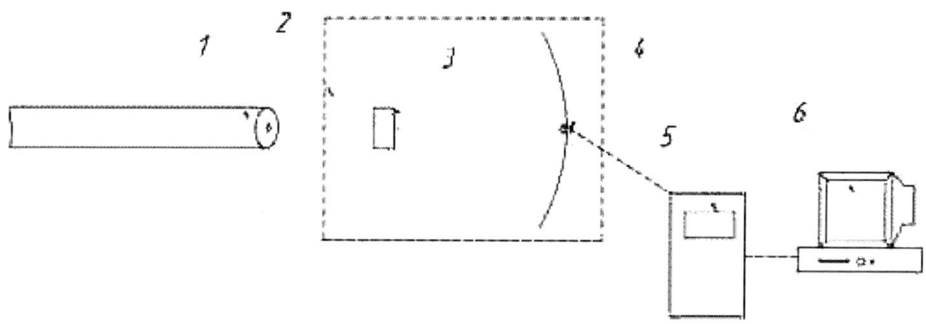

Figure 53. Scheme of experimental setup for determining the intensity and angle of the scattered radiation.

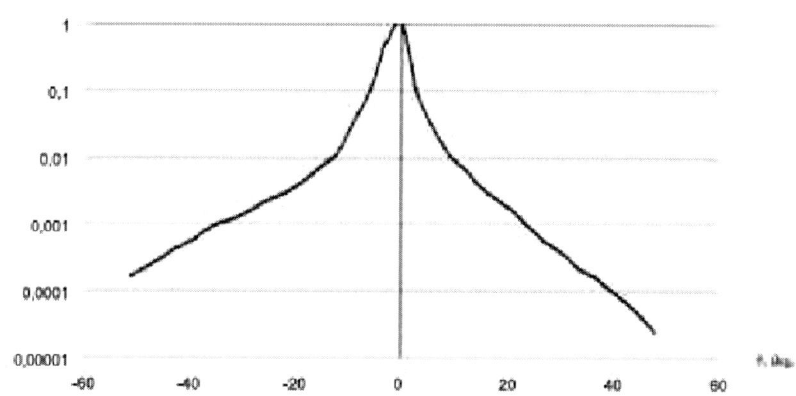

Figure 54. Dependence of I/I_0 from the angle of scattered radiation in the crystal LiNbO3 stoich. after 60 min radiation.

As it is shown in the work [102, 109] the biggest scattering angle is observed in the positive direction of the polar axis of the crystal.

According to the parameters of PLS indicatrix it is possible to estimate the total value of the diffusion and the photovoltaic fields [191]:

$$E_{\rho v} = \frac{\lambda(\Gamma_{-c}+\Gamma_{+c})}{2\pi n_e^3 r_{33}\sqrt{cos\theta_s^{in}\cos\left(\frac{\theta_s^{in}}{2}\right)}n_e n_o^2 r_{51}\sqrt{tan\theta_s^{in} sin\theta_s^{in}\sin(\frac{\theta_s^{in}}{2})}}, \quad (1)$$

$$E_D = \frac{\lambda(\Gamma_{-c}+\Gamma_{+c})}{2\pi n_e^3 r_{33}\sqrt{cos\theta_s^{in}\cos\left(\frac{\theta_s^{in}}{2}\right)}n_e n_o^2 r_{51}\sqrt{tan\theta_s^{in} sin\theta_s^{in}\sin(\frac{\theta_s^{in}}{2})}}, \quad (2)$$

where E_{pv} - photovoltaic field, E_D - diffusion field, k - length of the wave, θ^{in} - angle of scattered radiation, Γ_{-c} and Γ_{+c} - gain coefficients (indices ⟨⟨–⟩⟩ and ⟨⟨+⟩⟩ indicate the direction of the scattered radiation against and along the direction of the polar axis of the crystal respectively) n_e and n_o - refractive indices of extraordinary and ordinary beam respectively, r_{33} and r_{51} - electro-optical coefficients for $LiNbO_3$.

Depending on the angle of the scattered radiation the gain coefficient is calculated:

$$(\theta_s^{in}) = \frac{1}{l_{eff}(\theta_s^{in})} ln \frac{I_s(\theta_s^{in})}{I_{SO}^{\Omega}(\theta_s^{in})}, \quad (3)$$

where I_S - intensity of the scattered radiation, I_{SO} - intensity of primary scattering (of the incident beam), l_{eff} - effective interval of interaction is calculated in dependence on scattering angle according to the following formulae [10]:

$$l_{eff} = \frac{d}{cos\theta_s^{in}} \text{ for} \theta_s^{in} < arctan(\frac{w_p}{2d}), \quad (4)$$

$$l_{eff} = \frac{d}{2sin\theta_s^{in}} \text{ for } \theta_s^{in} \geq \arctan(\frac{w_p}{2d}), \quad (5)$$

where d - thickness of the crystal, w_p - diameter of the laser beam. Thus, from the PLS experiments' conditions according to the formulae (4) or (5) it is possible to calculate l_{eff}. Using l_{eff} and the data of Table 1 it is possible to calculate the gain coefficient r according to the formula (3).

Then using the data received by the formulae (1) and (2) it is possible to calculate the values of the photoelectric fields E_{pv} and ED and their dependence on the PLS angle. The results of calculations for the studied single crystals are represented in Figure 55.

Knowing the value of photoelectric field one can calculate the value of crystal birefringence $\Delta n = 0{,}5(n_e^3 \cdot r_{33} - n_e^3 \cdot r_{13}) \cdot E_{\rho v}$ [95, 99], where: n - value of crystal birefringence, n_e - refraction index of the extraordinary beam, n_o - refraction index of the ordinary beam. r_{13} and r_{33} - electro-optical coefficients for $LiNbO_3$. The results of calculations are summarized in Table 2.

From the received calculated data it is seen that the highest value of the photovoltaic field and diffusion field is observed for the crystal of stoichiometric composition. Similar results were obtained by the authors of works [192-194].

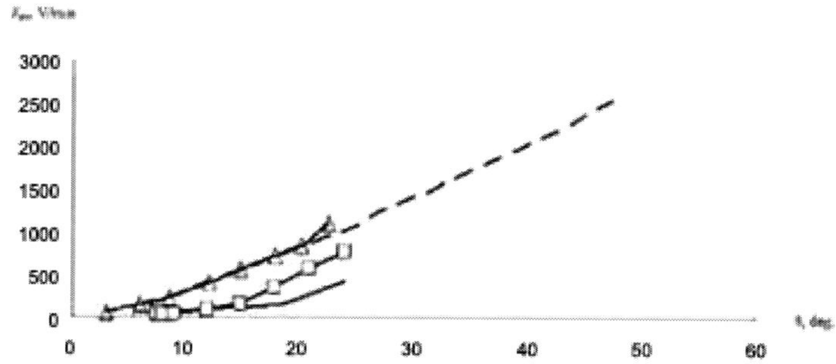

a

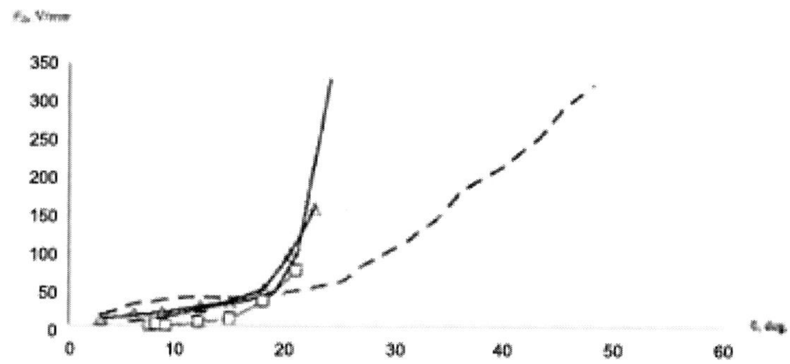

b

Figure 55. Dependence of photoelectric fields on PLS angle in the crystals of Lithium Niobate. (a) Photovoltaic field; (b) diffusive field.

Table 11. Energy transfer from central area into scattered radiation

No *n/n*	Crystal	Intensity I_0, mA, at t=0min.	Intensity I_{60}, mA, at t=60min.	Intensity decrease, I_0- I_{60}/ I_0, %
1	$LiNbO_3$:Cu + Gd (0.57 + 0.07 wt.%)	0.432	0.201	53.47
2	$LiNbO_3$:Zn (0.03 wt.%)	0.456	0.254	44.3
3	$LiNbO_3$:Cu	1.032	0.994	3.68
4	$LiNbO_3$ stoich.	0.405	0.396	2.22

I_0 - intensity of radiation in the initial period of time, I60 - intensity of radiation after 60 min of radiation.

Table 12. Photoelectric fields

No *n/n*	Crystal	Maximum value of diffusive field E_D, V/mm	Maximum value of photovoltaic field E_{pv}, V/mm	Δ_n, x10^{-5}
1	$LiNbO_3$: Cu + Gd (0.57 + 0.07 wt.%)	327	428	4828
2	$LiNbO_3$: Zn (0.03 wt.%)	56	772	8705
3	$LiNbO_3$: Cu	155	1106	12,476
4	$LiNbO_3$ stoich.	303	2592	29,243

The value of the photoelectric field in congruent Lithium Niobate crystal reaches 2500 V/cm, and in the stoichiometric 7000 V/cm with a power density 10W/cm^2 and a wavelength of 488 nm [192, 193].

Taking into account the conditions of our experiment the values of photovoltaic fields are comparable to the results of other authors [192-194].

The value Δn in dependence on correlation [Li]/[Nb] is in good accordance with the work's results [195] thus, for the congruent crystal it is ~ $(4\text{-}10)\times10^{-5}$.

According to PLS characteristics obtained experimentally, quantification of photoelectric fields in photorefractive single crystals of Lithium Niobate of different composition is done. The obtained results are well comparable with the data of other authors [192-195]. When excite PLS by He-Ne laser radiation (P = 60 mW, λ_o = 632.8 nm) the highest value of the photovoltaic field, and hence one has stoichiometric $LiNbO_3$ crystal, that proves prospects of stoichiometric single crystal as a photorefractive material for optoelectronics. Thus, the obtained data suggest that even low-power radiation of helium-neon laser quite actively excites photo processes in Lithium Niobate crystals of stoichiometric composition which is in agreement with the model of the photorefractive effect in stoichiometric crystals [187]. There is no photorefractive effect and therefore PLS in the crystals of congruent composition under similar experiment conditions [99]. It is observed because the stoichiometric Lithium Niobate crystals grown from the melt with 58.6 mol.% Li_2O have a lot of shallow electrons traps in the band gap [95]. These electrons as our experiments show can be excited when irradiated by the laser beam with a low energy, for example He-Ne laser radiation.

Finally, it is necessary to note the following: simplicity of the experimental setup and the conditions of correct execution of the experiment prove the acceptability of the method and open good opportunities to study the electro-optic characteristics of photorefractive crystals by PLS method.

13. Optical Characteristics of Doped Lithium Niobate Single Crystals

At present, special attention is given to the development of new functional materials and optimisation of their characteristics. In this case, the dielectrics, in particular ferroelectric crystals, have been used in the development of a large number of new directions in electronics, acoustics and optoelectronics, integrated optics, laser technology, communication and automation systems, optical memory media, and the technology of treatment of materials and medical instruments.

The most important dielectric materials, used widely in these applications, include the materials based on oxide compounds of niobium and tantalum, with the most important ones being the ferroelectric single crystals of lithium niobate and tantalate ($LiNbO_3$ and $LiTaO_3$), with the structure of pseudo-ilmenite, characterised by the efficient combination of the electrooptical, pyroelectric, piezoelectric and non-linear optical characteristics.

This determines the mass application of these materials.

The important special feature of the crystals of lithium niobate and tantalate (like of many crystals with the oxygen-octahedral structure) is the presence of a wide region of homogeneity on the phase diagram.

The composition of congruent melting of the crystals does not coincide with the stoichiometric composition. The structures are usually characterised by the extensive spatial heterogeneity and the complicated spectrum of the point and extended defects, forming a complicated difficult-to-model structural disorder [1-4]. The physical characteristics of the materials, produced from crystals of this type, especially the optical characteristics, are greatly controlled by the special features of the formation of defects in various sublattices of the structure, already formed in the stage of preparation of the charge and the stage of growth of the single crystals.

In this connection, the development of technological regimes of the synthesis of the charge and the doping methods, the methods of the growth of single crystals based on niobium compounds with the oxygen-octahedral structure, the development of a complex of effective methods of controlling the quality of the material and also the efficient examination of the properties of the disordered crystal phases, the processes of transition from the ordered to disordered, are of considerable interest for the development of materials with the given physical parameters. This investigations have the direct applied value because in particular the imperfections of the crystal structure (intrinsic and impurity defects) has a controlling effect on the quality of the physical characteristics of materials.

In the currently used approach to the development of ferroelectric materials there are two main directions: the synthesis of new structures and the modification of the existing structures in order to produce materials of high-quality, characterised by high electro-optical and nonlinear-optical coefficients. The first direction is usually associated with the production of complicated multicomponent compounds with different degrees of ordering of the structural units. The second approach, based on taking into account the special features of structural ordering, is especially important because, in this case, the materials with the qualitatively new characteristics can be obtained in the bases of the already available technologies.

We carried out a detailed analysis of that structural special features of the cation sub lattice of the real crystals of lithium niobate of different chemical composition (nominally pure and doped) in the examination of the Raman scattering spectra. We justify the assumption according to which the cation sublattice of the crystals with the composition differing from the stoichiometric composition is characterised by the formation of an ordered sub lattice of cluster-like intrinsic and impurity defects which generates its vibration will Raman scattering spectrum in the form of low intensity ('surplus') lines, differing from the spectrum of fundamental vibrations (Figure 56). The experimental results show that this sublattice of the defects is not present in the highly ordered crystals of the stoichiometric composition [44, 53, 77, 86, 87, 115, 134, 196].

It has also been established that the maximum in the spectrum of Raman scattering of the crystal of lithium niobate in the range 100-120 cm^{-1}, corresponding to the two-frequency states of the acoustic phonons with the total wave vector equal to zero, is sensitive to the fine special features of the structural ordering of the cation sublattice. The experimental results show that the spectrum of the crystal with the kinetic composition with a high degree of perfection contains almost no lines of Raman scattering in the range 100-150 cm^{-1} (Figure 31, curve *1*). The absence of the maximum in the Raman scattering spectrum may be accepted as the experimental criterion of the correspondence of the crystal of lithium niobate to the stoichiometric composition with the high degree of structural perfection.

The introduction of a small amount of the impurity ions into the structure of the crystal with the stoichiometric composition disrupts the ideal nature of alternation of the cations, results in a small the deviation of the composition of the crystal from the stoichiomet-ric composition, and determines the formation in this range of the spectrum of the lines 103 and 117 cm^{-1} (Figure 31, curves 2).

The introduction of small amounts of cations (B^{3+}, Mg^{2+}, Zn^{2+}, Gd^{3+}, etc.) to the structure of the crystal with the congruent composition initially increases the splitting of the 120 cm^{-1} line into two components (103 and 117 cm^{-1}) in comparison with the splitting detected in the nominally pure crystal with the congruent composition, and a further increase of the concentration of these ions results in the broadening and merger of the lines 103 and 107 cm^{-1} to the 120 cm^{-1} line (Figure 31, curves 3-5).

This fact shows unambiguously the ordering of the cation sublattice of the crystal with the congruent composition and the approach, as regards the degree of ordering, to the sublattice of the crystal stoichiometric composition in the presence of specific concentrations of impurities.

The main aim of doping the ferroelectric crystals is the directional variation of stabilisation of the properties of the main phase. We have shown that the physical parameters of the oxygen-polished neutral ferroelectrics (such as the crystals of lithium niobate and calculate, potassium, etc.) can be improved by increasing the degree of structural ordering of the cation sublattice along the polar axes by doping.

On the basis of the examination of the Raman scattering spectra it was established that the impurity cations with the ion radii, close to the radii of the main cations (Li^{+} and Nb^{5+}) and the charges, intermediate between the charges of the main cations ($1<Z<5$) in the region of small concentrations, have an ordering effect on the cation sublattice of the congruent crystal of lithium niobate [44, 53, 77, 86, 87, 115, 134, 196].

In addition to this, the alloying elements should not have the unstable variable valency (Cu^{+}-Cu^{2+}, Fe^{2+}-Fe^{3+}) because in this case the photorefractive effect and optical absorption greatly increase. Evidently, this relates to all oxygen-polyhedral ferroelectrics, characterised by the structures of pseudo-ilmenite, layered perovskite, etc.

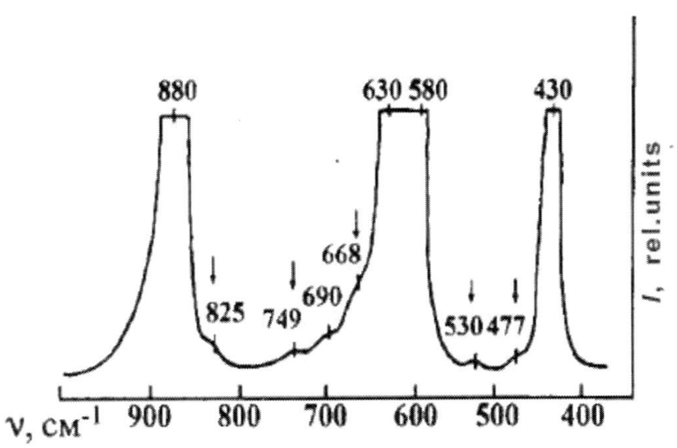

Figure 56. Fragment of the spectrum of Raman scattering of the single crystal of lithium niobate with congruent composition, subjected to heating for 6 hours at 1200 K, $T = 77$ K. "Excess lines" are indicated by arrows.

Thus, on the example of the single crystal of lithium niobate it has been shown unambiguously that in doping with the cations, characterised by the previously mentioned characteristics, in the specific range of concentration, the degree of ordering of the cation sublattice of the crystal greatly increases [44, 87, 196]. This is accompanied by a large increase of the resistance of the crystal to the damage by laser radiation. Figure 33 shows the fragments of the spectra of the Raman scattering of doped single crystals of lithium niobate of congruent composition in the range of oscillations of the oxygen octahedrons. In the research, the spectrum of the real crystals of lithium niobate in the scattering geometry X(ZX)Y contains two high-intensity lines, 580 cm^{-1} E(TO) and 635 cm^{-1} A_1(TO). The line 635 cm^{-1} A_1(TO) is forbidden for the given geometry of scattering and is reflected in the spectrum as a result of photorefraction. The effects of the decrease of the intensity of the line with a frequency of 635 cm^{-1} indicate the decrease of the intensity of photorefraction in the alloying of the crystal and this is in good correlation with the detected ordering of the cation sublattice along the polar axes for this range of the concentration of the doping additions. In this range of the concentrations, there is extensive splitting into two components (the lines 103 and 117 cm^{-1}) of the line in the range 120 cm^{-1}, determined by the scattering of light on the two-frequency acoustic phonons, and the large reduction in the width of certain lines, indicating the ordering of the structure.

Photorefraction is minimum in the crystals of lithium niobate, differing by the increased structural ordering of the cations along the polar axis. On the one hand, this fact may indicate a decrease in the number of charged structural defects with an increase in the degree of structural perfection of the crystals. On the other hand, photorefraction becomes stronger when an increase in the concentration of the introduced impurity not only increases the disordering of the cation sublattice but also results in the deformation of the oxygen frame of the crystal which leads correspondingly to an increase of the intensity of 635 cm^{-1} line (Figure 33). In this case, the number of charged structural defects evidently increases. Thus, in the ordered crystals, characterised by the reduced number of defects, a small amount of electrons may transfer from the forbidden band into the conduction band with further capture in deep traps. Correspondingly, the strength of the non-compensated electrical field, affecting the refractive index and the photon refractive properties of the crystal, becomes lower.

On the other hand, it was established in [44, 64, 87, 131] that the non-photorefractive impurities in lithium niobate may form fine electron traps, for example, the 'complex Mg^{+}' which is represented by the Mg^{+} ion in the area of Li^{+} with the electron de-localised on a number of surrounding ions. This is accompanied by a large decrease in the intensity of the photorefractive effect as a result of the efficiency of emission recombination of the photon-excited carriers without capture on the deep levels. The efficiency of this the recombination greatly determines the intensity of luminescence in these doped crystals.

The application of cathode excitation has made it possible to increase the intensity of the glow of the lithium niobate to a greater extent than in for example excitation with ultraviolet light and is facilitating the obtaining of more specific data on the effect of the composition of the specimen and the intensity of luminescence.

Figure 57 shows the spectra distribution of the cathodoluminescence of the crystals of LiNbO3:Gd. The spectral curves show a peak with a maximum at the wavelength of 430-460 nm (similar to that detected for the crystals of $LiNbO_3$:Mg and $LiNbO_3$ in [44]). The highest intensity of luminescence was recorded for the specimens with a gadolinium concentration of ~ 0.05 wt%.

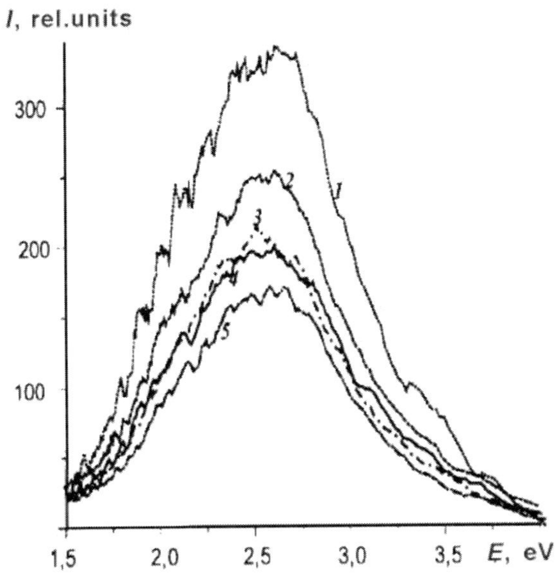

Figure 57. Cathodic luminescence spectra of crystals of lithium niobate doped with Gd^{3+} (in wt.%): 1) 0.05; 2) 0.4; 3) 0.65; 4) 0.002; 5) 0.45.

The intensity of the forbidden line with a frequency of 635 cm^{-1} in the Raman spectrum of this specimen was minimum [44] and, consequently, the photorefraction was also minimum.

Thus, the single crystals, characterised by the more ordered distribution of the cations along the polar axis are characterised by the maximum intensity of luminescence and increase the resistance to optical damage. Consequently, there is a significant relationship between the ordering of structural units and the condition of the electron subsystem of the crystal. This relationship requires further examination.

Examination of the macroscopic optical homogeneity of a series of doped single crystals with a high degree of structural ordering in respect of the mean density of the microdefects, visualised in the laser beam (individual defects have the form of the glowing spots in the laser beam) established that they have very high optical quality: the microdefects in the single crystals were almost completely absent (it is assumed that the quality of crystals determine the bases of this criterion corresponds to the optical quality of the mean density of the microdefects is not higher than 10 cm^{-2}). At the same time, in the series of the disordered single crystals with a higher content of the impurity, examination showed a considerable optical heterogeneity, the number of defects was 15-120 cm^{-3}. Both batches of the single crystals were grown in the same conditions in equipment fitted with the automatic system of controlling the diameter of the crystal, ensuring the reproducibility of the results in the growth of different single crystals.

Thus, it was established for the first time that the high macroscopic optical homogeneity of the single crystals (lithium niobate) is obtained in doping in the concentration range of non-photorefractive impurities in which the doped crystals are characterised by the maximum degree of structural ordering and increased resistance to optical damage.

In addition to this, the optical homogeneity in laser strength of the single crystals may be greatly affected by the method of doping.

For example, the doping of the single crystals of lithium niobate with boron was carried out by both the conventional method, by the addition of boron oxide to the charge prior to melting of the crucible, and by the addition of the alloying impurity into the re-extract with the formation of special purity niobium pentoxide.

In the latter case, the boron-containing reagent (boric acid) was introduced directly into the niobium re-extract produced in the process of extraction processing of the commercial niobium hydroxide to high purity substance.

The boric acid was added on the basis of the calculated amount of 0.08-0.15 wt% in relation to niobium (in calculation for the pentoxide), present in the re-extract, and taking into account the fact that part of the acid bonded the fluorine in HBF_4.

Subsequently, the niobium hydroxide was precipitated from the re-extract, and was neutralised with ammonia water to pH 8-9. The niobium hydroxide, rinsed in de-mineralised water, was dried and subsequently baked to produce the pentoxide.

A charge of lithium niobate was synthesised from the niobium pentoxide and used for growing a batch of single crystals.

Examination of the optical homogeneity of the single crystals with a boron content of 0.08 and 0.12 wt% in respect of the mean density of the microdefects, visualised in the laser beam, showed that they are characterised by very high optical quality: no defects were found in the single crystals.

At the same time, in a batch of single crystals with a boron content of 0.09, 0.1 and 0.12 wt%, doped by addition of the boron oxide into the charge prior to melt in the crucible, examination showed significant optical heterogeneity, the number of microdefects was 80-120 cm^{-3}.

In addition to this, the photorefraction in the single crystals, doped in the stage of production of the pentoxide, characterised by increased structural ordering, was considerably smaller in comparison with the nominally pure single crystals of the congruent composition (grown in the same conditions), and in the single crystals, alloyed by the additional boron oxide to the charge prior to melting it was considerably higher.

In the latter case, the intensity of the 'forbidden' line of 635 cm^{-1} was considerably higher than in the nominally pure single crystal of the congruent composition (Figure 33).

The alloying methods which make it possible to ensure the maximum degree of chemical homogeneity of the complex multicomponent systems greatly determine the possibility of producing single crystals of lithium niobate of high optical homogeneity in structural perfection characterised by increased resistance to laser radiation damage.

In addition to this, in the single crystals the window of optical transparency is greater, the value of the photovoltaic effect is higher and this effect also repeats the form of the nanosecond laser pulse, which is not possible for the nominally pure crystals of lithium niobate of congruent composition [44, 122, 196].

It is important to note that the maximum degree of ordering of the structural units, detected in the range of the relatively low concentrations of the impurity cations (fractions of mass percent).

The very low concentrations change the properties of the melt only slightly and, consequently, the technological conditions off the growth of the adult crystals with improved physical characteristics this only slightly from the conditions of growing nominally pure crystals.

14. Radiation Hardness of Lithium Niobate Nonlinear Optical Crystals Doped with Y, Gd and Mg

Intense defect formation in crystals can be caused not only by nonequilibrium growth conditions but also by ionizing radiation. A number of applications of optical devices based on lithium niobate crystals involve exposure to ionizing radiation.

In this connection, there is currently great practical interest in assessing the radiation hardness of lithium niobate crystals doped with rare-earth and alkaline-earth metals and finding out whether radiation-induced changes in the optical characteristics of such crystals can be used for ionizing radiation dosimetry.

The atomic and electronic structural imperfections produced by ionizing radiation in a crystal lattice determine in many respects the properties of the crystal. The generation of color centers, which contribute to optical absorption, is one of the most prominent examples of defect formation in crystals. Color centers result from carrier capture at native lattice defects or impurities [44, 197].

Influences different in nature, such as annealing in a reducing atmosphere and gamma irradiation, have seemingly the same effect on lithium niobate crystals, producing coloration, increasing their photorefractive sensitivity, and changing their optical absorption [44, 198-200]. Their optical absorption increases in a broad spectral range, from = 380 to 700 nm. This effect is difficult to interpret with certainty because, in this spectral range, the optical spectra of lithium niobate show a number of broad, overlapping absorption bands [201]. Previous work [202] addressed the formation mechanisms of electronic and point defects and allowed a model of ionizing-radiation-induced processes to be proposed.

In particular, defects in the oxygen sublattice (single charged oxygen ions in interstitial and octahedral sites) were shown to play an important role in radiation-induced coloration and bleaching. Doping may significantly change the optical properties of crystals, e.g., their sensitivity to ionizing radiation damage [202, 203].

In this paper, we report the growth of nominally undoped, rare earth-doped, and alkaline earth-doped lithium niobate crystals: $LiNbO_3$, $LiNbO_3$:Y (0.46 wt %), $LiNbO_3$:Y,Mg (0.32, 0.24 wt %), $LiNbO_3$:Mg (0.27 wt %), and $LiNbO_3$:Gd (0.004, 0.04, 0.26, 0.43 wt %). We examine the effect of crystal composition and ionizing radiation dose on their spectroscopic characteristics (transmission spectra).

Samples for characterization had the form of rectangular parallelepipeds measuring 5x7x9 mm in dimensions, with their edges parallel to the crystallographic axes. The crystals were irradiated in a ^{60}Co gamma source in an MRKh-γ-20 system to gamma doses of ~ 1 Gy to $5 \cdot 10^4$ kGy at a dose rate of ~ 0.5 $Gy \cdot s^{-1}$. Their transmission spectra were taken on Specord.

A comparative study of the optical characteristics (optical transmission spectra) of unirradiated and gamma-irradiated lithium niobate crystal of various compositions (nominally undoped, rare earth-doped, and alkaline earth-doped in a wide concentration range) showed that the effect of ionizing radiation on the optical transmission of the crystals depended on both the nature of the dopants and their concentration (Figures 58, 59). The response of the optical characteristics of the doped crystals to ionizing radiation can be both substantially stronger and substantially weaker than that of the nominally undoped crystals (Figure 58).

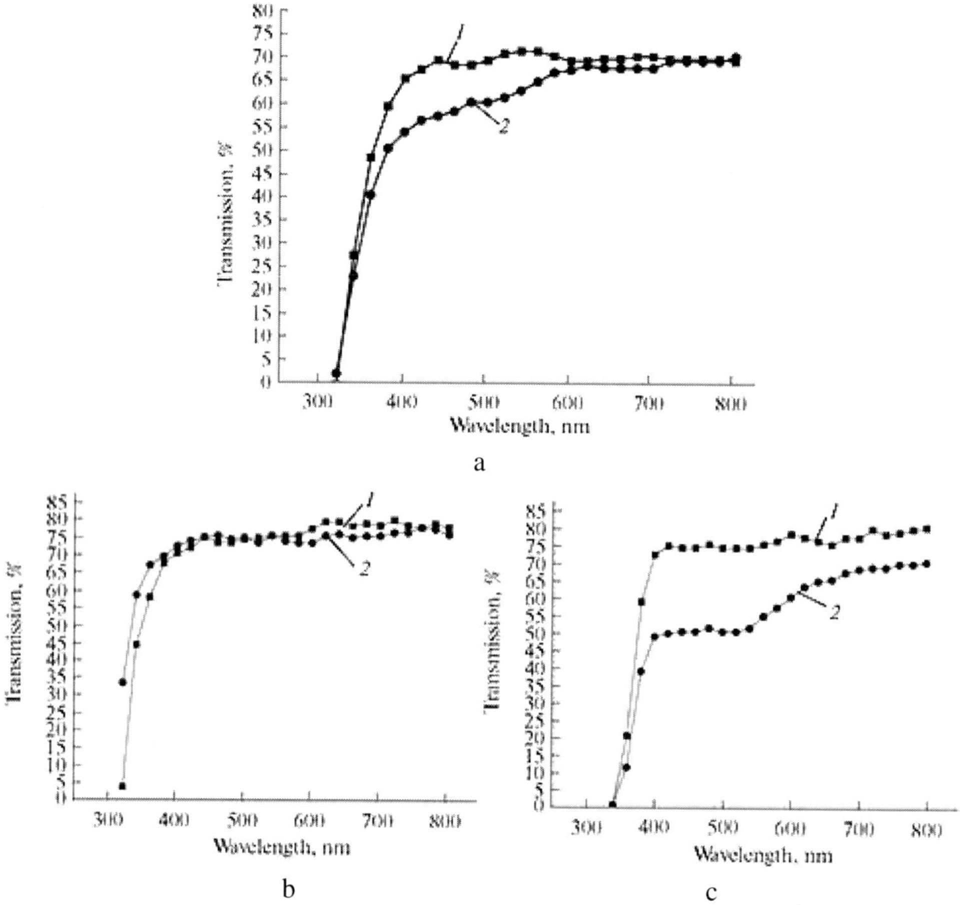

Figure 58. Optical transmission spectra of (a) a nominally undoped $LiNbO_3$ single crystal, (b) $LiNbO_3$: Mg (0.27 wt %), and (c) $LiNbO_3$:Gd (0.004 wt %) *(1)* before and *(2)* after γ-irradiation to a dose of ~ $5{\cdot}10^4$ kGy.

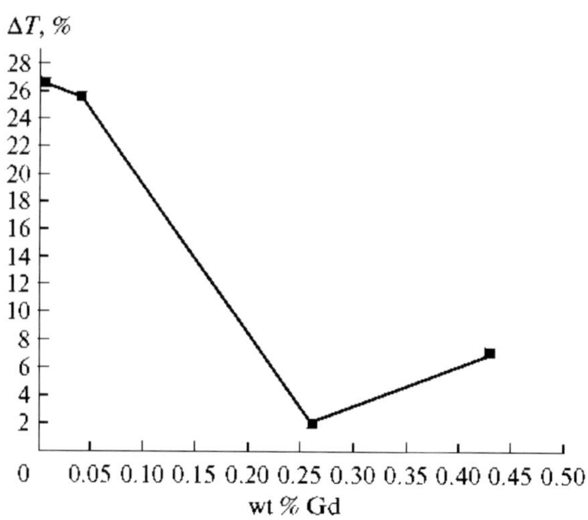

Figure 59. Change in optical transmission induced with ionizing radiation as a function of Gd concentration for $LiNbO_3$:Gd single crystals after γ-irradiation to a dose of ~ $5{\cdot}10^4$ kGy.

The optical transmission of the $LiNbO_3$:Gd crystals experienced the largest changes under γ-irradiation. The response had a maximum (up to = 35%) at comparatively low Gd concentrations (0.004 and 0.04 wt %) and was lowest at 0.26 wt % Gd (=3%) (Figures 58c, 59, 60, a). The highest resistance to ionizing radiation damage was offered by the $LiNbO_3$:Gd (0.26 wt %), $LiNbO_3$:Gd (0.43 wt %), and $LiNbO_3$:Mg (0.27 wt %) crystals: γ-irradiation had little effect on their optical transmission (Figures 58b, 60b, 60c).

The codoped crystal, $LiNbO_3$:Y,Mg (0.32, 0.24 wt %), showed an intermediate ionizing-radiation-induced change in its optical transmission: it was smaller than that in the $LiNbO_3$: Gd (0.004 and 0.04 wt %) crystals and greater than that in the nominally undoped and Mg- and Gd-doped (0.26 and 0.43 wt %) crystals (Figures 58, 60).

It is worth pointing out that, in the case of the $LiNbO_3$:Gd crystals with relatively low Gd concentrations (0.004 and 0.04 wt %), which had the largest ion-izing-radiation-induced change in optical transmission, we observed a significant shift of the fundamental absorption edge to longer wavelengths relative to both the nominally undoped and doped crystals (Figure 61). This indicates the formation of charged defects in the oxygen sublattice of the crystals.

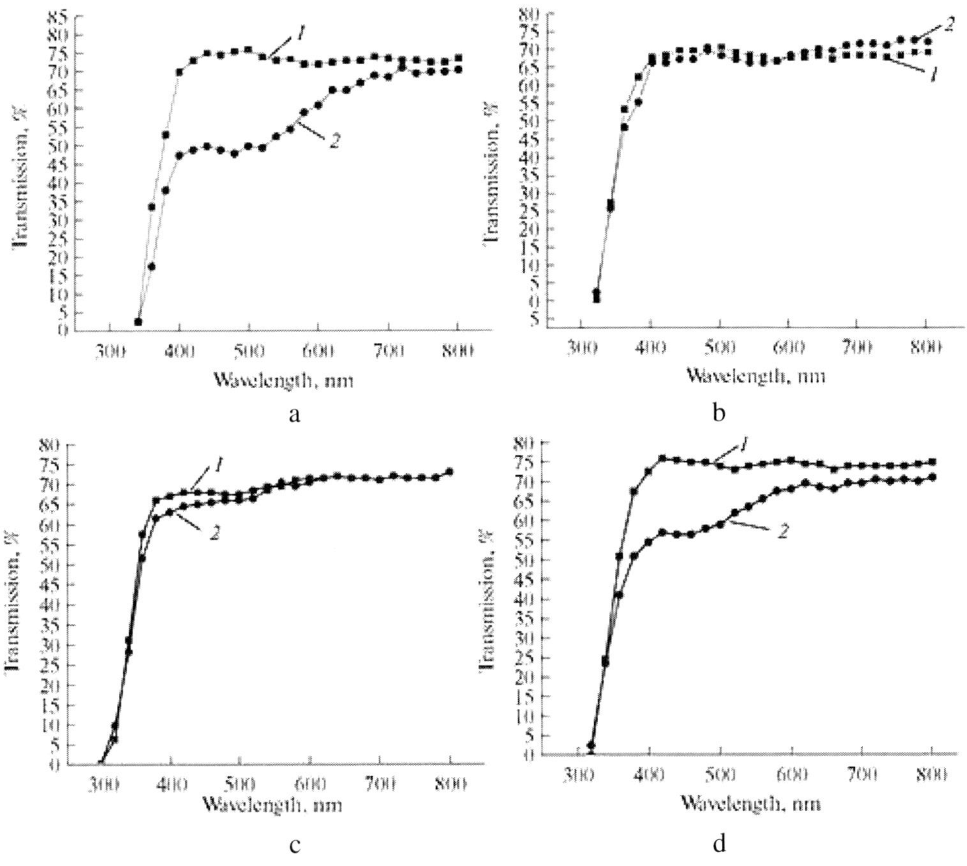

Figure 60. Optical transmission spectra of (a) $LiNbO_3$:Gd (0.04 wt %), (b) $LiNbO_3$:Gd (0.26 wt %), (c) $LiNbO_3$:Gd (0.43 wt %), and (d) $LiNbO_3$:Y (0.46 wt %),Mg (0.32, 0.24 wt %) single crystals (*1*) before and (*2*) after γ-irradiation to a dose of ~ $5 \cdot 10^4$ kGy.

Such defects seem to be responsible for the high sensitivity of the optical characteristics of those crystals to gamma irradiation.

The present results point to significant changes in the optical transmission of the $LiNbO_3$:Gd crystals with low Gd concentrations (0.004 and 0.04 wt %) at comparatively low ionizing radiation doses (1 – 160 Gy), where the optical transmission is an almost linear function of dose (Figure 62). At high doses, we observe saturation of the radiation-induced coloration, which may be due to radiation-induced defect annealing.

In particular, gamma doses of ~ 160 Gy and $5 \cdot 10^4$ Gy have almost identical effects on the optical transmission of the $LiNbO_3$:Gd (0.04 wt %) crystal (cf. Figures 59, 60, and 62).

The effect of gamma irradiation on the optical transmission of the $LiNbO_3$:Gd (0.004 and 0.04 wt %) single crystals can be used for ionizing radiation dosimetry. The purpose of dosimetry is to quantify the effect of ionizing radiation on a particular object. The magnitude of the effect is uniquely determined by the absorbed energy and is a measure of this energy.

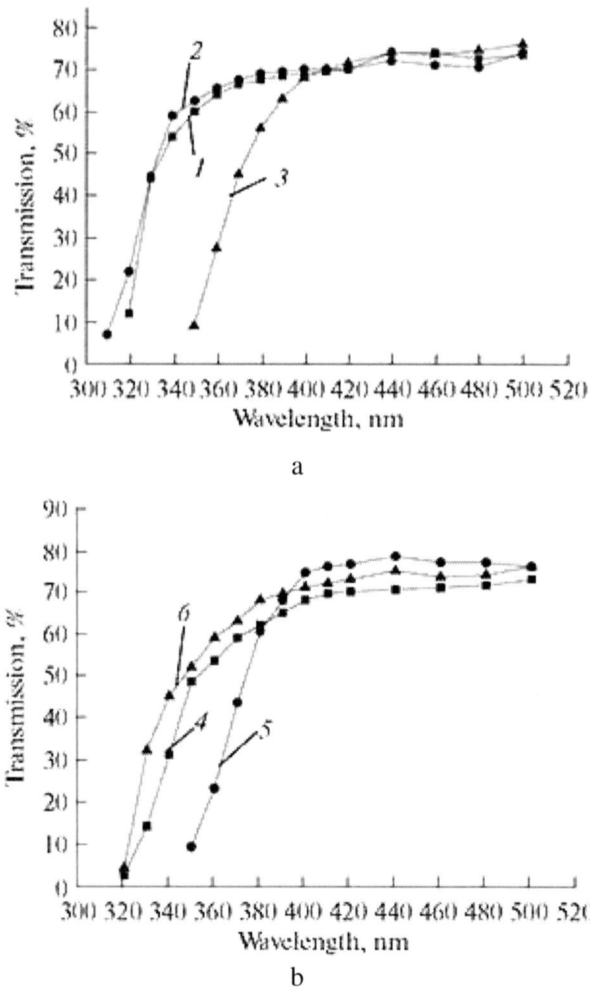

Figure 61. Optical transmission spectra of unirradiated lithium niobate single crystals: (a) (*1*) $LiNbO_3$, (*2*) $LiNbO_3$:Gd (0.43 wt %), (*3*) $LiNbO_3$:Gd (0.04 wt %), (b) (*4*) $LiNbO_3$:Gd (0.26 wt %), (*5*) $LiNbO_3$:Gd (0.004 wt %), and (*6*) $LiNbO_3$:Mg (0.27 wt %).

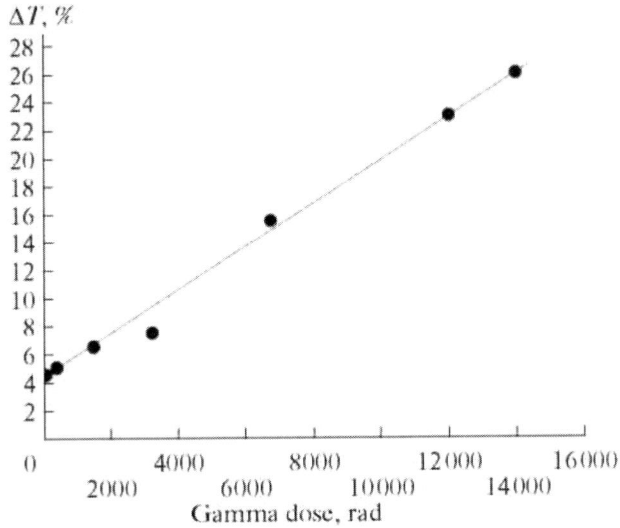

Figure 62. Gamma_induced change in 440 nm optical transmission as a function of gamma dose for a $LiNbO_3$:Gd (0.04 wt %) crystal.

To select a "working medium" of a dosimeter, one must create a material with a property that varies as widely as possible in a preset range of ionizing radiation doses in order to ensure the highest sensitivity. Other requirements include ease of handling, a simple recording procedure, good reproducibility, and reliable information storage. These requirements are fully met by $LiNbO_3$:Gd (0.004 and 0.04 wt %) crystals. Comparatively low gamma doses provide sizeable changes in their absorption (up to =30% for $LiNbO_3$:Gd (0.004 wt %) crystals. Changes in optical absorption at a fixed wavelength can be recorded using the simplest photocalorimeters. During storage at room temperature in the dark, radiation-induced changes in the optical absorption of a $LiNbO_3$ crystal persist for a long time (years).

Crystals can be used as dosimeters many times. Annealing at 180°C for 40 min eliminates radiation-induced changes in optical absorption, bringing the optical transmission of the crystal to the level of lithium niobate that has never been irradiated, so that the crystal can be again used as a dosimeter. Single crystals can also be bleached by high-intensity illumination at ~ 480 nm [202].

We have grown nominally undoped, rare earth-doped, and alkaline earth-doped lithium niobate crystals: $LiNbO_3$, $LiNbO_3$:Y (0.46 wt. %), $LiNbO_3$:Y, $LiNbO_3$:Mg (0.27 wt. %), and $LiNbO_3$:Gd (0.004, 0.04, 0.26, 0.43 wt. %). Studies of the optical characteristics (optical transmission spectra) of the crystals before and after gamma irradiation to various doses allowed us to assess the effects of the nature of dopants, their concentration, and ionizing radiation dose on the optical transmission of lithium niobate and to find out whether radiation-induced changes in the optical transmission of such crystals can be used for gamma radiation dosimetry. The ionizing-radiation-induced change in the oxygen transmission of the crystals depends on both the nature of the dopants and their concentration. The response of the optical characteristics of the doped crystals to ionizing radiation can be both substantially stronger and substantially weaker than that of nominally undoped crystals.

Among the crystals studied, the largest gamma-induced change in optical transmission was observed for the $LiNbO_3$:Gd (0.004–0.04 wt. %) crystals.

In addition, the fundamental absorption edge of those crystals was found to be shifted to longer wavelengths, which was tentatively attributed to the formation of charged defects in the oxygen sublattice of the crystals. At high γ-irradiation doses, we observed saturation of the radiation-induced coloration, which was due to radiation-induced defect annealing.

In particular, gamma doses of ~ 1 ÷ 160 Gy had almost identical effects on the optical transmission of the $LiNbO_3$:Gd (0.04 wt. %) crystal. The present findings suggest that changes in the optical transmission of $LiNbO_3$:Gd (0.004-0.04 wt. %) crystals can be used for gamma dosimetry in the dose range ~ 1 ÷ 160 Gy.

15. Curie Point Versus Concentration in Lithium Niobate and Lithium Tantalate and the Positions of Dopant Cations in the Structure

In the pseudoilmenite structure (LN and LT have this structure), octahedra are rather rigid and are distorted only insignificantly. The octahedra are centered with matrix (Li^+, Nb^{5+}, Ta^{5+}) or impurity (B) metal cations (Figure 1) [9]. A lithium ion is in fact accommodated loosely in an octahedron, whereas Nb^{5+}, Ta^{5+}, and most of the impurity cations are fixed by a comparatively rigid covalent bond [1-9]. For this reason, there are two metal–oxygen distances (B-O-B) in the octahedra occupied by Nb^{5+}, Ta^{5+}, and most of impurity cations; this is responsible for the high anisotropy of the crystal field in the direction of the polar axis [1-3]. The type of cations, their position in octahedra, and their distribution over octahedra substantially control the ferroelectric properties of alkali metal niobate tantalates.

Recall that in the pseudo-ilmenite structure as little as about two-thirds of the octahedra are occupied by matrix or impurity cations. The others remain vacant [1-3].

Many impurity ions comparatively easily incorporate from a melt into a growing LN or LT crystal because of the high mobility of lithium ions in a crystal at temperatures near the melting point, where the ionic vibra-tional amplitude is high. For example, in the LN paraphase, a Li^+ ion easily transfers from octahedron to octahedron [1-3]. The Li^+ position in the crystal lattice in this case can be regarded as the statistical average of all possible positions, and vacant octahedra at near-melting temperatures are as if absent. Favorable conditions appear both for the substitution of an impurity ion (or a Nb^{5+} ion) for a Li^+ ion during crystal growth and for impurity incorporation into a vacant octahedron [1].

The ionic radius of the impurity must fit the size of the octahedron. For crystals doped with ions whose ionic radii are close to the Li^+ or Nb^{5+} ionic radius, the arrangement of impurity and matrix cations in octahe-dra and the distribution of octahedra along the polar axis will be dictated primarily by the character of bonding between the ions and the oxygen sublattice, by cation charges, and probably by the cation mass [11, 44, 78, 204, 205].

Currently, no straightforward reliable experimental methods exist for studying the arrangement of matrix and impurity cations in oxygen-octahedral ferroelectrics or for studying random cation distribution over octahedra. Impurity cation distribution is studied, as a rule, qualitatively: the symmetry of an impurity center is determined experimentally, mainly by means of EPR or NMR spectroscopy, and the type of defect structure is simulated by testing various vacancy models (the Li-site vacancy model and the Nb-site vacancy model are used most widely [44]).

However, recent results [78, 79, 205] give serious evidence in favor of the Li-site vacancy model. Hereafter, all considerations are in terms of the Li-site vacancy model.

Lithium niobate and lithium tantalate crystals have been studied most extensively [1–3, 44]. It is commonly supposed that impurity Ta^{5+} ions in LN crystals are iso-structural to Nb^{5+} ions and prefer to occupy octahedra designated for Nb^{5+} ions in the ideal structure (Ta^{5+} and Nb^{5+} ionic radii are both equal to 0.66 Å) [1, 44, 206, 207]. Mg^{2+}, Zn^{2+}, and other divalent cations, at low concentrations, prefer to occupy the Li^+ sites of the ideal structure (the Li^+, Mg^{2+}, and Zn^{2+} ionic radii are 0.68, 0.74, and 0.83 Å, respectively) [1, 44, 208, 209]. As the Mg^{2+} concentration in a congruent crystal increases, however, Mg^{2+} ions first substitute for the Nb^{5+} antisite ions occupying the Li^+ sites of the ideal structure, then for the Li^+ ions residing in their major sites of the ideal structure. Opinions differ as regards the positions occupied by tervalent impurity ions, in particular, Er^{3+}, Gd^{3+}, Y^{3+}, and others, in the oxygen octahedra of LN [210-213]. A tervalent dopant, most likely, will substitute lithium and niobium to produce charge-balanced pairs in neighboring Li and Nb sites [16, 44, 214]. For Eu^{3+} ions, ion backscattering showed that about 64% of Eu^{3+} substitute for niobium ions and 36% for lithium ions [215]. Interstitial Eu^{3+} ions have not been detected in notable amounts. Logically, a similar situation must occur for Gd^{3+}, another rare-earth ion. This is supported by the EPR study of $LiNbO_3$:Gd single crystals [44, 196, 216]. The signal intensity per unit mass in a set of samples was proportional to the Gd concentration [44, 196, 216]. Results obtained in [44, 212, 216] imply definitely that Gd^{3+} ions in the LN lattice occupy two physically nonequivalent sites differing not strongly in their magnetoresonance parameters. Most likely, these two states are identical to the ones described in [212] and denoted as PI and PII. If we take the intensity of the observed absorption line as a measure of the amount of this type of paramagnetic center in the sample, then the amount of PII centers contained in the crystals studied is 1.5 times the amount of PI centers [44, 216]. According to [212], a PII defect has an orthorhombic symmetry; PI has a lower symmetry. When we refer to cationic sites occupied by lithium and niobium ions as the most probable positions for Cd^{3+} ions, the dopant ions must have a trigonal symmetry. The dominance of axial parameters in the crystal field [212] confirms the fact that Gd^{3+} ions, indeed, lie on the threefold axis; i.e., they substitute for Li^+ and Nb^{5+} ions.

Orthorhombic distortion in a PII defect in [212] is attributed to the near-lying charge compensators. From the consideration of the LN crystal lattice [1, 44], it follows that a cationic site occupied by a niobium ion has a higher symmetry of oxygen surroundings compared to a lithium site. From this, the inference that PII centers are Gd^{3+} ions in niobiumion sites and that PI centers are Gd^{3+} ions in lithium-ion sites is physically consistent.

The defect structure of $LiNbO_3$:Gd with balanced charged reads fied model. It ignores other possible defects, such as niobium vacancies, gadolinium incorporation into vacant oxygen octahedra, precipitation of impurity phases, and dopant concentration gradients in the bulk of a crystal.

The arrangement of cations residing in the vacant octahedra of oxygen-octahedral ferroelectrics is, without question, of interest in order for understanding the origin of ferroelectricity and in the context of the preparation of materials with novel or improved properties.

The basic physical parameters of LN vary monoton-ically over the homogeneity range (in particular, the Curie point T_c increases with increasing Li/Nb ratio [1-3, 10]); that is, in the framework of the existing models of the defect structure, formally, the parameters are controlled by the amount of cationic vacancies.

The authors of [206] claim that the trends of property versus concentration diagrams for doped LN crystals are dictated by the valence of the dopant, the positions it occupies in the lattice, and the amount and positions of cationic vacancies. Therefore, the trend of such a diagram, e.g., for the Curie point, can serve as an indicator of the dopant position in the lattice.

In terms of the Li-site vacancy model, the amount of lithium vacancies is controlled by the Li/Nb ratio or, in accordance with formula $[Li_{1-5x}Nb_x(Li_v)_{4x}][Nb]O_3$, by the x value. The structural parameters of a crystal vary systematically with its composition. The Nassau and Lines model [44, 217] suggests that the unit cell parameters of LN increase as the Li concentration decreases; the c parameter of the hexagonal unit cell increases the greatest because, given a niobium excess, the strongest effect is expected from an increase in the Nb–Nb distance along the polar axis [44, 217]. These inferences agree with results obtained by Lerner et al. [10]: when the Li_2O concentration changes from 50 to 47.5 mol %, Δc is + 0.014 Å, whereas the a parameter increases by as little as 0.003 Å. The unit cell volume increases accordingly. The molecular weight and density also increase because of an increase in the atomic fraction of the heavy niobium ions. The Curie point T_c shows an inverse tendency: T_c increases with lithium concentration.

It is commonly supposed that the Curie point of LN in the solid-solution region is dictated by the character and concentration of lattice defects arising from doping or changing the Li/Nb ratio. It is stated in [206] that cationic vacancies (in the defect-structure model at hand, lithium vacancies) have the greatest effect on T_c. Indeed, a one-to-one correlation between the cationic vacancies and T_c is in some cases clear cut. For example, an increase in Li_2O concentration in nominally pure lithium niodate is accompanied by a reduction in lithium vacancy concentration (to reach zero for Li/Nb = 1 and for an ideal structure) and an increase in T_c [216, 217]. The situation is the same when LN is doped with Mg^{2+} cations, which are located in lithium sites: as the mag nesium concentration increases, the Li-site vacancies decrease and the Curie point increases [27, 218, 219].

In the set of $LiNbO_3$:Zn crystals studied in [27, 218, 219], the Curie point also monotonically elevated from 1145 to 1173°C when the zinc content increased from 0.003 to 0.88 wt %. According to the defect-structure formula $Li_{1-5x}Nb_{1+x-0.6y}Gd_y(V_{Li})_{4x-0.4y}O_3$, the Li-site vacancy concentration of $LiNbO_3$:Gd crystals also drops, but together with a decrease in the Curie point [44, 216] (Figure 63). In LN doped with Er^{3+}, the Curie point also drops with a rise in the dopant concentration (Figure 63) [44, 216].

Likely, a more intricate correlation exists between the Curie point and the type of defect structure; factors other than concentration and positions of cationic vacancies also count. In general, the type of defect structure in doped single crystals is a function of many factors: the dopant valence, the positions occupied by dopants, cationic vacancy positions, antisite substitutions of matrix cations, and the type of cluster (density inhomogeneity) created by an irregular alternation (compared to the ideal structure) of matrix cations and by dopant cations.

The authors of [216, 219] suppose that the position of impurity cations is the key factor in the trend of T_c as a function of concentration. Divalent and tervalent impurity cations in a certain concentration range can order the cation sublattice of lithium niobate and, through this, can render the crystals more perfect [18, 30, 38, 64, 72, 77, 86, 87, 131-136]. The structure transition point in a crystal is known to drop as the structure perfection of the crystal is deteriorated. In nominally pure LN ceramics, the Curie point drops as R = [Li]/[Nb] decreases [216, 219]. The occupancy of lithium sites Time must decrease, and the cation sublattice is markedly disordered [18, 30, 38].

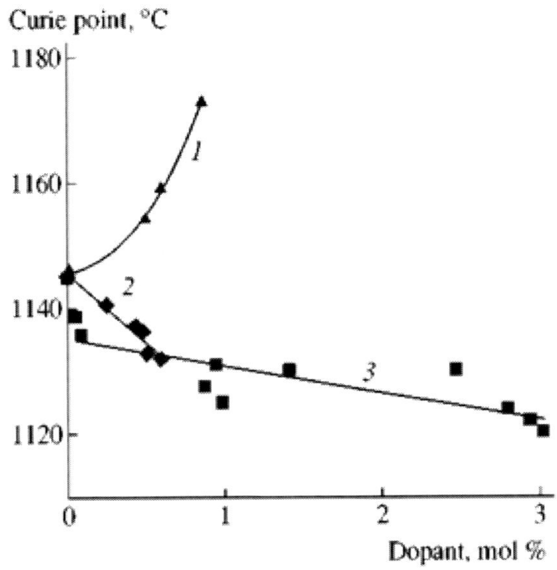

Figure 63. Plot of the Curie point vs. dopant concentration for single crystals (*1*) $LiNbO_3$:Zn, (*2*) $LiNbO_3$:Gd, and (*3*) $LiNbO_3$:Er.

Therefore, according to the above speculations, the positions occupied by impurity cations, controlling the perfection of the cation sublattice, are the key factor that governs the T_c versus concentration trend in doped LN crystals. If dopant cations having $1 < Z < 5$ prefer to occupy the Li sites of the ideal structure (i.e., they force out excess niobium ions (antisite defects Nb_{Li}) from these sites), this crystal is ordered and its Curie point rises.

Conversely, T_c drops when the dopant disorders the structure through substituting for niobium ions. When several dopants are introduced, the ordering and disordering mechanisms can cancel each other out; that is, the dopant action on T_c is additive. The Curie point of such a crystal will not change within some range of dopant concentrations.

For example, if dopant cations occupy only niobium sites (as Ta^{5+} cations do) and if Mg^{2+} cations occupy only lithium sites, the $LiNbO_3$:Ta,Mg crystal with a certain Ta-to-Mg ratio has T_c equal to the Curie point of a nominally pure crystal [216, 219].

In nominally pure LN, the rising Li/Nb ratio increases the occupancy of lithium sites; in $LiNbO_3$:Zn and $LiNbO_3$:Mg, Zn^{2+} and Mg^{2+} also occupy lithium sites; in $LiNbO_3$:Gd, Gd^{3+} cations are located in both lithium and niobium sites. The Gd^{3+} cations in niobium sites are about 1.5 times more abundant (EPR results; [216, 219]).

This circumstance is likely responsible for the drop in T_c with increasing Gd^{3+} concentration, despite the decrease in the total concentration of Li-site vacancies.

The increasing Gd^{3+} concentration within the concentration range studied changes the lattice parameters as follows: a monotonically increases, and c remains constant within the experimental error [216]. In the $LiNbO_3$:Ta system, the increasing tantalum concentration increases a and dramatically decreases c with a concurrent drop in T_c [220, 221].

In this, $LiNbO_3$:Gd and $LiNbO_3$:Ta crystals are different from a nominally pure lithium niobate crystal (in which both *a* and *c* decrease as the Li_2O concentration rises) and from a $LiNbO_3$:Mg crystal (in which both parameters increase as the Mg^{2+} concentration increases) [27, 218, 219].

Therefore, the concentrational behavior of the unit cell parameters does not seem to be an indicator of the T_c versus concentration trend. It is supposed in [216, 219] that the dopant position is a more important factor: if the dopants occupy Li sites (as Li^+ cations do in nominally pure LN when the Li_2O concentration increases, or Mg^{2+} and Zn^{2+} cation do), the Curie point rises with an increase in cation concentration.

In cases where lithium and niobium sites are both occupied but where the niobium sites are preferred, the Curie point drops. When dopants occupy niobium sites exclusively (as Cu^{2+} ions do [222] or as occurs in $LiNbO_3$:Ta, where Ta^{5+} cations do not change the cation vacancy concentration [206]), the Curie point drops more abruptly with an increase in dopant concentration. In a set of $LiNbO_3$:Cu crystals, the Curie point decreased monotonically from 1145 to 1136°C as [Cu] increased from 0 to 0.05 wt % [216]. The $LiNbO_3$:Ta system also experiences a dramatic drop in T_c caused by the increasing Ta concentration [220, 221].

Thus, the Curie point versus concentration dependence is a comparatively simple circumstantial indicator of the dopant position in a lithium niobate structure.

T_c determination can serve as a method for finding Li/Nb(Ta) both in the feed and in LN and LT single crystals. The Curie point is known to rise almost linearly within the homogeneity range of nominally pure $LiNbO_3$ and $LiTaO_3$ phases of variable composition were discussed and compared in [1, 34].

This correlation was used in [111, 112] to develop an original procedure in order to determine Li/Nb(Ta) starting from the Curie point; T_c versus Li_2O concentration curves were plotted for nominally pure lithium niobate and lithium tantalite.

Figure 64 plots heating curves for an LN sample of the congruent melting composition (48.65 mol % Li_2O [18, 30]). An endotherm is due to melting (Figure 64a). The ferroelectric transition region in the DTA curve is marked.

Given the standard sensitivity of the instrument, T_c is not detected. Heating curves, measured with higher resolution using a procedure developed in [216, 219] and illustrated by Figure 64b, show a feature associated with a phase transition (close to a second-order transition) and give T_c. The uncertainty in T_c determined using computer processing is ± 0.5°C [216, 219]. Figure 65 shows a T_c versus composition plot for nominally pure LN together with data points. The plot is fitted by polynomials

$$T_c = -442.77 + 32.617C \text{ and}$$

$$T_c = -11328 + 447.77C - 4.551C^2,$$

where C is the Li_2O concentration, mol %.

For nominally pure lithium tantalite, the T_c versus composition plot together with data points is shown in Figure 66. The plot is fitted by a first-order polynomial

$$T_c = -1065.0 + 34.755C,$$

where C is the Li_2O concentration, mol % [111].

The above method for determining the composition of single crystals [216, 219] was tested during lithium niobate and lithium tantalate crystal growth at the Institute of Chemistry, Kola Research Center, Russian Academy of Sciences.

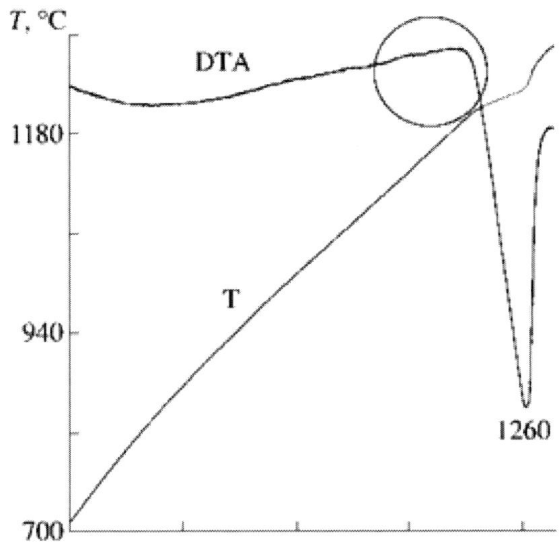

a

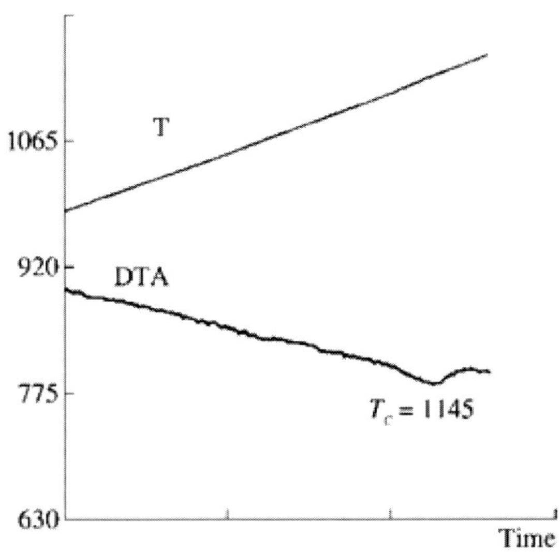

b

Figure 64. Panel (a): DTA curves for lithium niobate containing 48.65 mol % Li_2O. The melting point: 1260°C. The range of a ferroelectric transition is separated out [111]. Panel (b): a fragment of the DTA curve for LN containing 48.65 mol % Li_2O measured with a higher resolution. A ferroelectric transition is detected on the DTA curve. The Curie point: 1145°C.

This procedure, along with electron probe microanalysis, powder X-ray diffraction, Raman spectroscopy, electronic absorption spectroscopy, electrophysical measurements, and optical homogeneity control, has been successfully employed to control the feed quality and the chemical homogeneity of crystals.

The high heating rates are suitable for rapid measurements [216, 219].

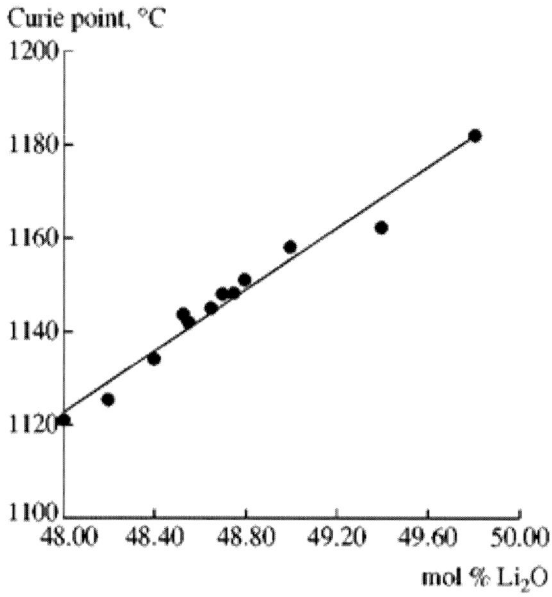

Figure 65. Plot of the Curie point vs. chemical composition for lithium niobate.

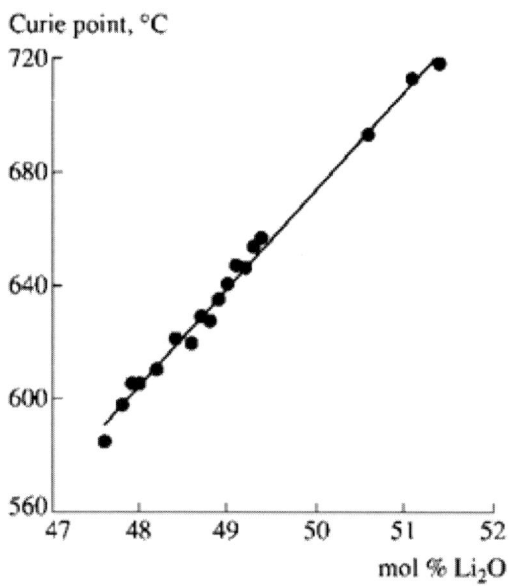

Figure 66. Plot of the Curie point vs. chemical composition for lithium tantalate.

16. Electrophysical and Spectroscopic Characteristics of Lithium Niobate Single Crystals

Anomalies of the physical properties of lithium niobate are observed in the temperature range of application (300 – 400 K).

Their origin is ambiguous. Since these anomalies can directly affect the performance of devices based on lithium niobate, they have been extensively studied.

Reported were the anomalous temperature dependences of optical, dielectric, and pyroelectric properties and of conductivity, as well as the characteristic temperature evolutions of polarized-light images, in nominally pure and doped LN crystals at 300-400 K [1, 223-231]. The anomalous temperature behavior of physical parameters in nominally pure and, more often, in doped LN crystals was observed by many researchers.

However, most results are irreproducible; they are essentially affected by the thermal and field history and the real structure of crystals.

In [141, 219, 232], in order to gain experimental information on the nature of anomalies observed in LN crystals doped with B^{2+}, Zn^{2+}, or Gd^{3+}, the domain structure, static and dynamic piezoelectric and dielectric properties, and conductivity were studied in the specified temperature range and over a wide frequency range. This was the first attempt was to match the observed anomalies with structure ordering and electronic processes in a crystal.

Figure 67 [219, 232] plots the real part of the dielectric constant measured as a function of temperature ε'_{33} (T) at various fixed frequencies for a Gd-doped LN crystal. The ε'_{33} (T) curve displays a significant anomaly in the region 330-380 K; the anomaly decreases with an increase in frequency f and practically disappears when $f \geq 10$ kHz. The conductivity versus temperature curve shows an anomaly, appearing as a jump, in the same temperature range (Figure 68) [219, 232]. These curves were measured under temperature elevation; importantly, both anomalies decrease by more than an order of magnitude during subsequent thermocycling. Note that the ε'_{33} (T) curve measured at 10 kHz (Figure 67) for a doped crystal fully coincides with curves measured for a *z*-cut sample of a nominally pure crystal.

The relaxation character of the observed anomalies can be inferred from the above results. The dielectric dispersion results are illustrated in Figure 69 [219, 232] by Cole-Cole diagrams for a $LiNbO_3$:Gd crystal measured at various temperatures (frequencies are expressed in Hz; next to the Cole-Cole curves, the exposure time (in hours) at T = 344 K is specified). From the diagrams, the dielectric dispersion of $LiNbO_3$:Gd in the frequency range from 1 Hz to 1 MHz is due to a single Debye relaxation process with a characteristic relaxation time at room temperature $i \sim 2.5\ 10^{-2}$. Neither the dispersion depth nor the dielectric constant changes until samples are heated to temperatures above ~ 340 K, but the dielectric relaxation time shows a temperature dependence satisfying the Arrhenius law:

$$\sigma(T) = \sigma_0 \exp(U_a/kT),$$

where the activation energy and the frequency factor are U_a = 0.23 eV and $T \sim 2.0 \cdot 10^{-6}$ s, respectively [219, 232].

An increase in temperature $(T > 340$ K) abruptly decreases the dispersion depth and increases the relaxation time (Figures 69, 70). These changes in dielectric properties occur in jumps; in the region of $T \sim T_0 \sim$ 340-350 K, they develop with time, while the dynamic dielectric constant eL remains constant. Exposure at $T_0 \sim$ 340-350 K for 4 h fully eliminates

Debye dispersion. In this case, both the frequency and the temperature dependences of the dielectric constant become analogous to the known properties [21] of nominally pure lithium niobate crystals [219, 232].

The temperature dependence of bulk static conductance $\boldsymbol{c_s}$ derived from complex admittance diagrams (Figure 70), like the T(T) curve, shows a thermoactivation character at $T < T_0$ and has an Arrhenius plot as $T = A_0 exp(-H_a/kT)$ with the enthalpy of activation H_a = 0.22 eV and A_0 ~ 0.19 K/(Q m).

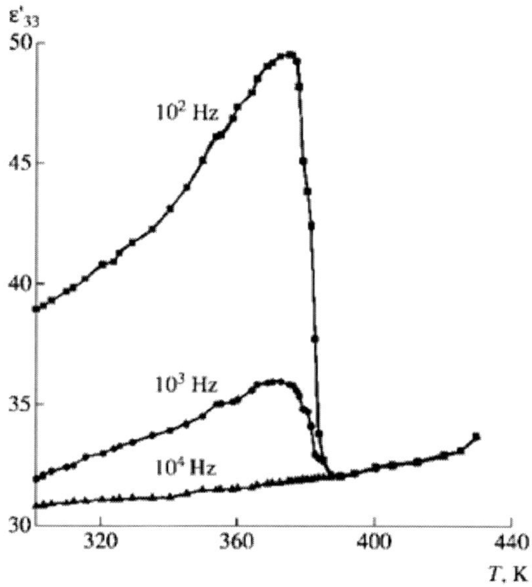

Figure 67. Plot of the dielectric constant vs. temperature for a $LiNbO_3$:Gd crystal (0.44 wt %, *z* direction) measured at fixed frequencies.

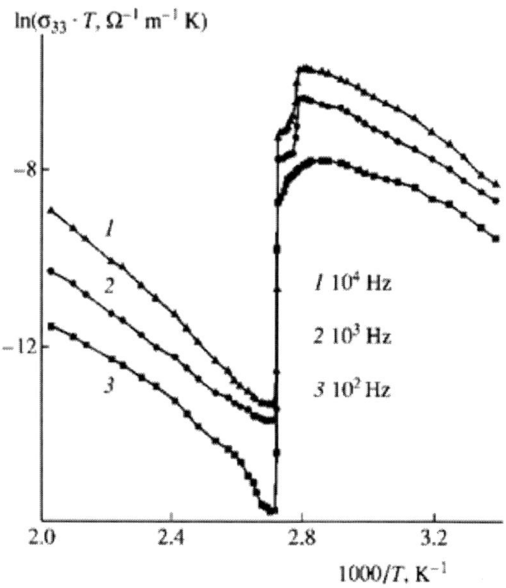

Figure 68. Plot of the specific conductivity vs. temperature for a $LiNbO_3$:Gd crystal (0.44 wt %, *z* direction) measured at fixed frequencies.

In the vicinity of T_c, static conductance drops (by more than two orders of magnitude) to values characteristic of nominally pure LN crystals at the same temperature [141, 219, 232].

Dielectric relaxation experiments in displacing electrical fields (Figure 71) show that an increase in the displacing field strength E_{dis} from 0 to 20 kV/cm even at room temperature appreciably decreases the dispersion depth but does not change the Debye character of dispersion. After switching on the displacing field, the dispersion depth recovers its initial value only in a long period of time [2, 32, 141].

The examination of etching figures on $LiNbO_3$:Gd [219, 232] implied a regular domain structure based on rotational growth stria, like the one observed in LN crystals doped with Y^{3+}, Dy^{3+}, or Nd^{3+} [147].

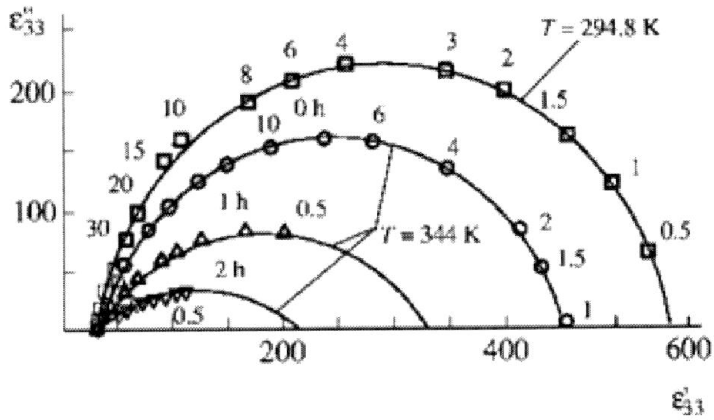

Figure 69. Cole-Cole diagrams for a $LiNbO_3$:Gd crystal (0.44 wt %, z direction) at various temperatures.

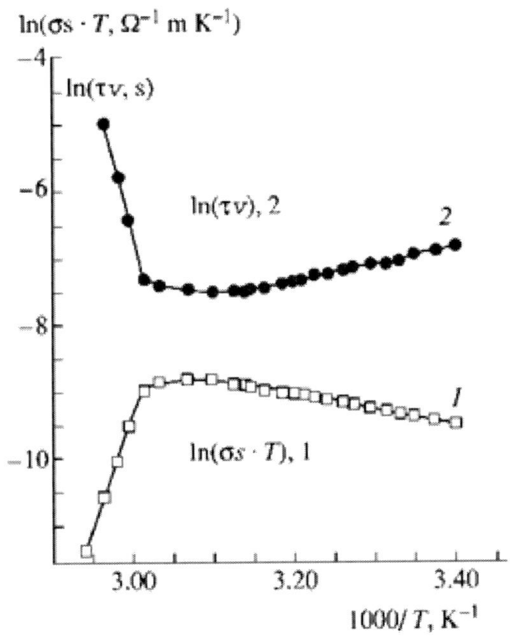

Figure 70. Plots of the dielectric relaxation time and static specific conductivity for a $LiNbO_3$:Gd crystal (0.44 wt %, z direction).

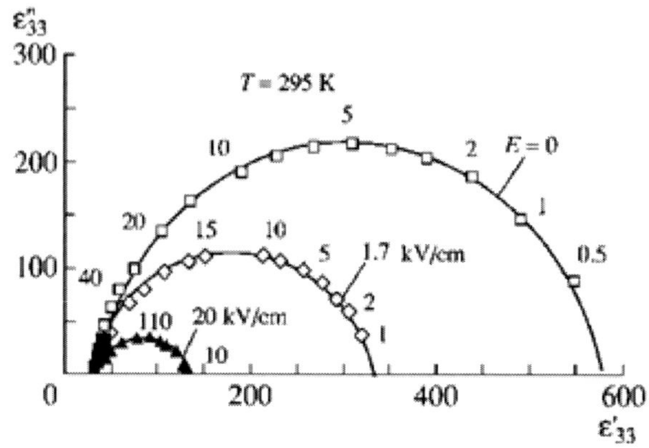

Figure 71. Cole-Cole diagrams for a $LiNbO_3$:Gd crystal (0.44 wt %, *z* cut) in displacing electric fields (T = 295 K).

Gadolinium (like yttrium, dysprosium, and neodymium) forms a regular domain structure, because it has an unbalanced charge, a large ionic radius (Gd^{3+}, 0.94; Y^{3+}, 0.97; Dy^{3+}, 0.88; Nd^{3+}, 0.99 Å), and an effective segregation coefficient $K_{eff} < 1$.

In [233], Y^{3+} distribution over rotational growth stria in $LiNbO_3$:Y single crystals was studied. Dopant distribution was measured along a normal to domain boundaries. Domains were found to form when the yttrium concentration is near a maximum or minimum. The same must be observed for Gd^{3+}, because yttrium and gadolinium belong to one family in the periodic system and because they have identical charges and similar atomic and ionic radii. When the crystal experiences a ferroelectric transition, the Gd^{3+} charge is not fully shielded.

Therefore, an inhomogeneous dopant distribution is equivalent to an inhomogeneous charge distribution and, accordingly, to an inhomogeneous internal field and the formation of domains with opposite polarization.

From the anomalous temperature behavior, the specific slow evolution of dielectric properties, the character of dielectric dispersion, the values of relaxation times and the activation energy (at least, in the temperature range in which the Arrhenius law holds for the conductance and relaxation time), and the type of domain structure, one may suggest that the observed low-temperature dielectric dispersion is due to the relaxation of point defects (associated with dopant Gd^{3+}) that interact with domain boundaries in the initially polydomain crystal. However, the increase in relaxation time with temperature elevation is atypical; this can result from the rearrangement of the domain structure accompanied by a significant increase in domain size and, accordingly, a change in the interaction character of point defects with domains [141, 219, 232].

In order to prove the existence of a labile domain structure in $LiNbO_3$:Gd, the static and dynamic piezoelectric effects were studied as a function of temperature [232]. Ignoring the possibility of weak natural uni-polarity, one must believe that the macroscopic piezoelectric effect in polydomain samples is absent. In a single-domain state far from the Curie point, the macroscopic piezoelectric module has the highest possible values [21].

These values, determined experimentally, can serve as a measure of unipolarity for particular crystal samples. These speculations provided the basis for direct measurements of the static macroscopic piezoelectric module d_{33} [141, 232]; the measured temperature dependence is plotted in Figure 72.

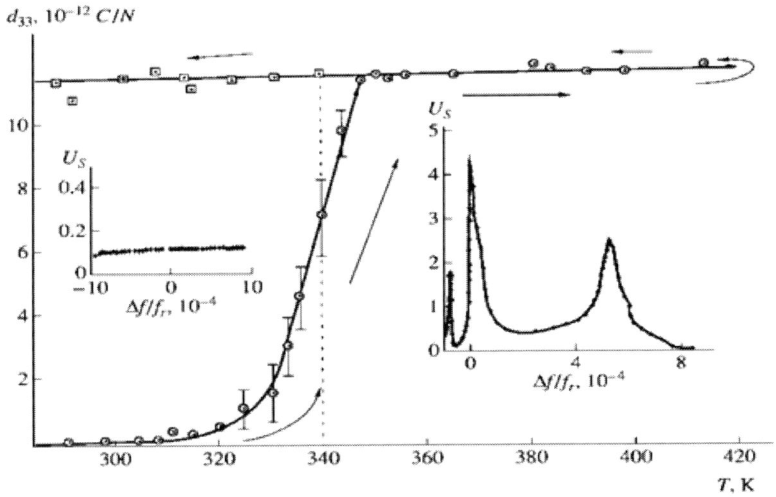

Figure 72. Plot of the static piezoelectric module d_{33} vs. temperature for a $LiNbO_3$:Gd crystal (0.44 wt %, z direction); in the insets, piezoelectric resonance signal vs. frequency curves at $T > T_0$ and $T > T_0$.

The scatter in the transition temperature range shown in Figure 72 means that the measured values experienced a temporal drift at T = const (toward increasing d_{33} upon heating). The results of these experiments indicate that, when a sample is heated in the range of $T < 340$ K, d_{33} has low values, likely controlled by weak natural unipolarity; in the temperature range corresponding to the discovered anomalies of dielectric properties and conductance, conversely, the piezoelectric module d_{33} increases abruptly to approach values reported for a single-domain nominally pure crystal [21]. Figure 72 also displays normalized frequency curves of the piezoelectric resonance signal for test samples oscillating in the z direction in two temperature ranges corresponding to different states of the domain structure. In these experiments, as in static measurements, intense resonance peaks also appeared in the vicinity of T_0 ~ 340 K and were conserved irreversibly during thermocycling. Long-term exposures (up to several weeks) of samples in an open state at room temperature recover the initial low values of the macroscopic static piezoelectric module and practically eliminate piezoelectric resonance (on the level of the instrumental signal-to-noise ratio) [232]. Similar dependences were observed in LN crystals doped with Zn^{2+} and B^{2+}.

In these crystals, however, effects are not so well defined and not reproducible [232]. An abrupt increase in d_{ss} in $LiNbO_3$:Gd crystals is accompanied by a substantial alteration of the near-surface etching relief, associated with the regular domain structure. For example, microrelief with a distinct orientation showing fine features of the regular domain structure is clearly seen in a sample that was examined with a KPD SMM-2000 atomic-force microscope before temperature measurements and that had only a weak natural unipolarity (Figure 73a).

For a sample that was etched immediately after dielectric measurements (at $T > 340$ K) and that had $d \approx (11\text{-}12) \times 10^{-12}$ C/N, in fact no such microrelief is observed (Figure 73b). This experiment gives direct evidence of the rearrangement of the domain structure in $LiNbO_3$:Gd single crystals in the range of temperatures around T_0. Therefore, the experimental results in [232] substantiate the supposition that initially polydomain $LiNbO_3$:Gd crystals have a rather labile domain structure, associated with point defects whose dynamics in low-frequency fields significantly contributes to the static dielectric constant

$\varepsilon_{s33}(T)$ of polydomain samples (Figure 69). When temperature elevates to $T_0 \sim 340$ K, the crystals behave anomalously; this anomalous thermal behavior indicates that the domain structure converts to a strongly unipolar state with its properties resembling the single-domain state.

This state is stable when $T > T_0$ and metastable at lower temperatures. The relaxation kinetics of the samples to recover the initial macroscopically unpolar state is governed greatly by the temperature and field prehistory; at room temperature, the relaxation times are up to several weeks and even months, which is responsible for the strong temperature hysteresis of anomalies observed in shorter thermocycles [232].

Lithium niobate single crystals having a labile domain structure at low temperatures can find important applications in periodic domain-structure technology in integral optics. Such a crystal is, in addition, a suitable test object for the study of relaxation of spontaneous polarization and the kinetics of domain reorientation in lithium niobate.

The inference that a strongly unipolar state is formed in the domain structure of a crystal doped as specified is consistent with photovoltaic effect data [64, 137] for nominally pure and doped crystals. In a polydo-main sample of a nominally pure crystal, the inertialess photoresponse changed its sign upon surface laser scanning; in a $LiNbO_3$:Gd crystal, the photoresponse sign remained unchanged, proving the high unipolarity of the crystal.

Since the $LiNbO_3$:Gd samples used in [64, 137] were prepared from an initially polydomain single crystal, we must suppose that the domain-structure uni-polarity found in the photovoltaic experiments is, likely, the result of photoinduced processes that occur in the electronic subsystem. Photoconductance measurements [234] showed that the external electric field in a $LiNbO_3$ crystal is shielded as a result of thermoactivated electron transitions to the conduction band from shallow traps with an activation energy of 0.2 eV. According to [235], this value was observed when the photoinduced optical inhomogeneity was studied in the temperature range where Δn_{st} = const (300-360 K). When $T > 370$ K, the effect is due to deep trapping levels with an activation energy of about 1 eV [235].

The bulk static conductance versus temperature in $LiNbO_3$:Gd samples at 300-340 K has an Arrhenius plot with an enthalpy of activation of about 0.22 eV (Figure 70) [232]; the conductance is anomalously high (more than two orders of magnitude higher than the conductance of nominally pure crystals). Presumably, conductance in $LiNbO_3$:Gd in the range 300-340 K is also controlled by shallow traps located near the bottom of the conduction band.

In was found in [98] that nonphotorefractive impurities (Mg) in $LiNbO_3$ can form shallow electron traps, e.g., a Mg^+ complex, which is a Mg^+ ion in a Li^+ site with an electron delocalized on several ions nearby [236]. This notably reduces the photorefractive effect through enhancing the recombination efficiency of photoexcited carriers without trapping them to deep levels. Thermostimulated luminescence studies [235] showed that such electron traps are thermolyzed at comparatively low temperatures ($T < 370$ K); that is, these are relatively shallow traps. The energy levels of local centers calculated from thermostimulated luminescence curves are 0.18–0.23 eV. The highest temperature luminescence peak falls in the temperature range 340–380 K; this range coincides with the temperature range in which dielectric anomalies were observed [232]. In [18, 30, 38, 64, 72, 77, 86, 87, 94, 131-136], it was shown that the structural quality of LN single crystals is improved when doping cations have charges intermediate between the matrix cations (Li^+, Nb^{5+}) and ionic radii that do not significantly distort the oxygen sublatice of the crystal (Mg, Zn, B, Gd) and when the dopant level falls in a certain range.

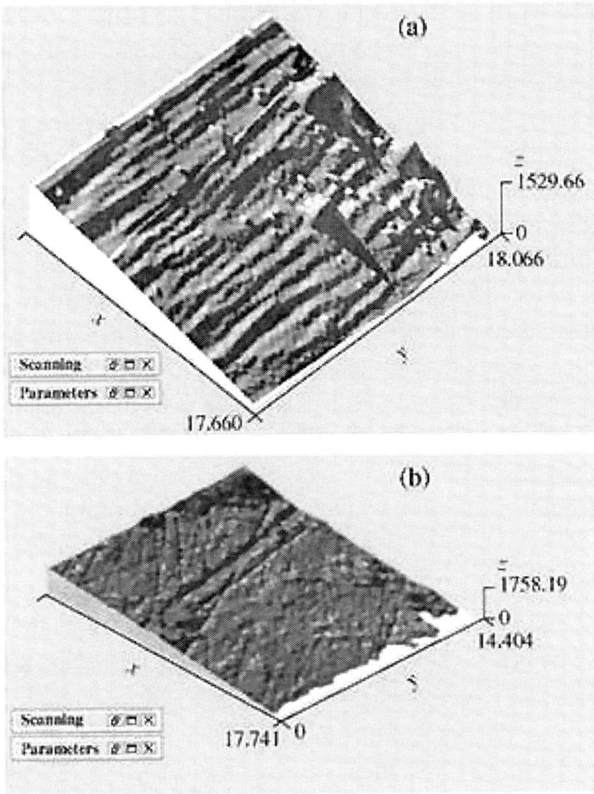

Figure 73. Near-surface etching relief for a $LiNbO_3$:Gd crystal (0.44 wt %, z direction) (a) before heating (d_{33}~ 0.2 ×10 C/N) and (b) after carrying out electrophysical measurements (d_{33} = (10-12)×10 C/N) (etching at room temperature, KPD SMM-2000 atomic force microscope).

The associated notable suppression of the photorefractive effect [131, 136] might indicate (by analogy with [132]) that the density of states near the bottom of the conduction band increases; that is, shallow electron traps on nonphotorefractive Gd^{3+} dopants appear. This must increase both photo-and dark conductance, which is responsible for the anomalous high conductance of $LiNbO_3$:Gd crystals in the region 300-340 K with the activation energy of about 0.2 eV.

It seems that shallow traps at $T > 340$ K become unstable, and their occupancy decreases dramatically. Electrons will drift from the regions rich in nonphoto-refractive Gd^{3+} (domain boundaries) toward the positive pole of spontaneous polarization and will be trapped by deep traps in the bandgap. These traps are, likely, antisite defects Nb_{Li}. Defects Nb_{Li} are deep electron traps; they generate polarons and bipolarons when trapping electrons [237]. This supposition is consistent with the finding that the activation energy in this temperature range changes from about 0.2 to 1 eV [235]. Moreover, there is evidence [238] that polaron conduction is the dominant conduction mechanism in nominally pure LN single crystals at above-room temperatures. Thus, in $LiNbO_3$:Gd single crystals at $T > T_0$, the static conductance decreases to values characteristic of a nominally pure crystal at the same temperature.

When shallow electron traps based on Gd^{3+} complexes (most of which are located at domain boundaries) lose electrons, the crystalline surroundings are additionally polarized by the field of a charged center.

It was found in [239] that the polarization field in LN decreases by two orders of magnitude when the temperature elevates from ambient values to about 440 K. Under additional polarization, the domain size in ferro-electrics can increase as a result of an easier polarizability. Given that the dopant concentration is at a sufficient level, a quasi-cooperative effect can appear with the formation of macroscopic domains whose sizes are comparable to the size of the sample [238]. This might be responsible for the dramatic enhancement of the unipolarity of a $LiNbO_3$:Gd crystal doped to the specified level near 340 K [232].

A change in the electron conduction mechanism must, to some extent, be responsible for the physical anomalies observed at 300-400 K in both doped and nominally pure crystals, regardless of the initial state of the domain structure. Particular values of the anomalies and the kinetics of processes involved are, probably, controlled by the ratio of the densities of states of low - and deep-lying traps and by the actual structure of samples.

17. Dielectric and Spectral Characteristics of Lithium Tantalate Polidomain Crystals

Lithium tantalate, like LN, is a well-studied ferroelectric [4, 8, 21]. However, many properties of crystals important for their applications need more careful consideration. Comparative studies of mono- and polydomain single crystals are of importance, because they offer a tool for highlighting the preparation conditions for a stable single-domain state, determining reproducible schedules for converting a crystal to a single-domain state, and evaluating the extent of conversion to the single-domain state.

The study of dielectric properties and relaxation of spontaneous polarization over a wide range of temperatures near the paraelectric transition carried out in [141, 240, 241] is essential for choosing process parameters in order to convert the crystal to a single domain.

The comparative Raman study of poly- and single-domain samples can appear a reliable tool of control over the polydomain state of crystals [241].

The temperature curves of the real part of the complex dielectric constant e_{33} for poly- and single-domain LT crystals measured at 1 kHz at various field amplitudes showed that for polydomain samples the Curie-Weiss law holds in the ferroelectric phase in the range T_c - T <*150* K and for field amplitudes no larger than 1 V/cm. Larger field amplitudes cause the £$_{3\ 3}$ versus temperature relationship to deviate from the Curie-Weiss law. Measurements carried out on single-domain samples in the heating mode starting at room temperature showed that the Curie-Weiss law in the range T_c - T < 150 K holds when the field amplitude is far larger than 1 V/cm (E_{meas} > 50 V/cm) [141, 240, 241].

For polydomain samples, the anomaly in the £$_{3\ 3}$ versus temperature curve (at E_{meas} > 1 V/cm) is associated with the dielectric contribution from the inelastic relaxation of domain walls in the ferroelectric phase at T_c - T < 150 K (where the coercive field is comparable to the measuring field). The field dependence of electric polarization is in this case substantially nonlinear.

Figure 74 illustrates the results of third-harmonic dielectric nonlinearity measurements in a polydomain sample (fundamental frequency, 200 Hz) at various field amplitudes [240, 241]. (To eliminate linear effects, U_{III} is normalized to the measuring field and to the electrode surface area).

These curves are characterized by two maxima, one at a low temperature and the other at a high temperature, at measuring field amplitudes larger than 1 V/cm. With an increase in the field amplitude, the low-temperature maximum shifts to lower temperatures; the high-temperature maximum, in fact, keeps its position; and $U_{III}/(ES)$ increases appreciably. The temperature range of the high-temperature maximum covers both the ferroelectric and the paraelectric phases of LT [240, 241].

The high-temperature maximum is due to the general nonlinear field dependence of induced polarization; a linear dependence would not have generated higher harmonics. When £$_{33}$ > 1 (i.e., when in the temperature range of a ferroelectric transition), it is likely that factors at the higher order terms of function *P(E)* (this function expresses the field dependence of electric polarization) have values sufficient for their contribution to induced polarization to be manifested in an experiment even when the measuring field amplitude is comparatively small. This is a normal lattice contribution to dielectric nonlinearity [141, 240, 241].

The abnormal contribution, which exists only in the ferroelectric phase and is related to the low-temperature third-harmonic dielectric nonlinearity peak (Figure 74), is due to spontaneous polarization P_s and the domain-boundary dynamics, i.e., to switching effects in part of the sample volume. This contribution is the greatest in the temperature range in which P_s is still high but the coercive field is comparable in its value to the measuring field amplitude [141, 240, 241]. In this case, when the measuring field amplitude increases, the anomalous contribution peak shifts to lower temperatures and increases substantially because of an increase in the switched volume fraction and a rise in P_s (Figure 74). It is expected that external displacing fields will suppress the anomalous dielectric nonlinearity in polydomain samples, which corresponds to a transition to the single-domain state. This expectation was supported by experiments in [141, 240, 241] (Figure 75).

Measurements on single-domain samples also showed the absence of a low-temperature third-harmonic amplitude peak. Thus, the experiments unambiguously indicate that the anomalous dielectric nonlinearity is related to switching processes in the measuring field.

The existence of a low-frequency dielectric dispersion branch, induced by displacements of domain boundaries in the measuring field, must be supposed in this case (as opposed to the high-frequency branch, associated with softening of the modes responsible for the phase transition). The domain dispersion branch must appear in the temperature range of the anomalous nonlinearity peak (given the proper measuring field amplitude).

Dielectric dispersion studies in the frequency range from 10 Hz to 10 MHz in [141, 240, 241] revealed this low-frequency branch. The relaxation time is $\tau \sim 10^{-4}$ s. Two relaxation processes are observed in Cole-Cole diagrams (Figure 76a).

A low-frequency relaxation process, like anomalous third-harmonic dielectric nonlinearity (low-temperature peak), is in fact not observed in a stationary ($\geq$500 V/cm) displacing field (Figures 75, 76b). These phenomena are also nonexistent in single-domain samples; this observation gives another piece of evidence for their domain character [141, 240, 241].

It may be inferred that dielectric spectra (for poly-domain LT samples) are spit into two vibrational branches. The high-frequency branch has a lattice character and is related to softening of the modes responsible for the ferroelectric transition. Raman spectra near the ferroelectric transition support this inference: instead of a separate soft mode, a whole continuum is soften with a poorly defined low-frequency maximum. The center of mass of this continuum shifts to the exciting line, and its intensity increases [242, 243].

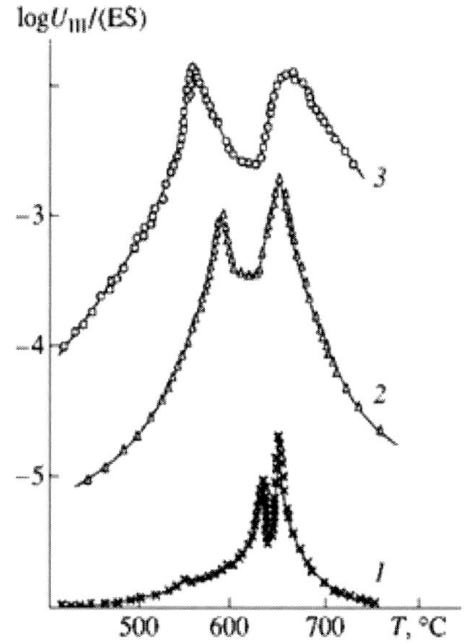

Figure 74. Plots of the third-harmonic amplitude vs. temperature for a polydomain LT single crystal at E_{meas} = (*1*) 5, (*2*) 50, and (*3*) 100 V/cm.

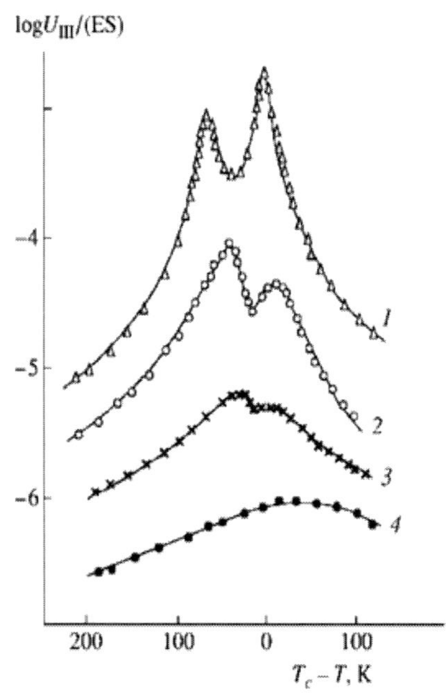

Figure 75. Plots of the third-harmonic amplitude vs. temperature in various displacing fields for a polydomain LT single crystal at E_{dis} = (*1*) 0, (*2*) 250, (*3*) 500, and (*4*) 1000 V/cm (E_{meas} = 50 V/cm).

This explains the abrupt rise in the dielectric constant, which is con tributed by several vibrations, rather than one; this, in particular, increases the probability of clear-cut normal nonlinearity in comparatively weak fields [141, 240, 241].

Their clear-cut manifestation in this case is likely related to the features of the domain structure of LT. In LN, macrodomains are observed along with micro-domains; macrodomains are comparable in their dimensions with a crystal (a result, the crystal has a certain spontaneous unipolarity [1, 8]). Conversely, an LT single crystal is completely penetrated by antiparallel domains (~0.2 μm in cross-sectional area [244]) oriented along the polar axis. Therefore, a polydomain LT crystal is characterized by the absence of electrooptical and pyroelectric effects [8] and by the absence of an electromechanical response (as shown by measurements in [141, 240, 241]). A similar type of domain structure suggests the existence of many domain boundaries, capable of relaxing in an electric field, which leads to the well-defined anomalous nonlinearity [141, 240, 241].

Given this type of domain structure, Raman spectra give the extent of the polydomain state of a sample with a high accuracy. The intensity of Raman lines in a single crystal is dictated directly by the components of the Raman tensor, which are proportional to the first derivative of the unit cell polarizability along the corresponding normal coordinate.

In polydomain samples, the components of the Raman tensor cannot be observed separately because of the misalignment of the polarizability ellipsoids of domains and scattering at domain boundaries. Measuring the intensity of the Raman lines forbidden for the chosen scattering geometry is a means for judging the single-domain state and for studying the kinetics of conversion to a single domain [240, 241].

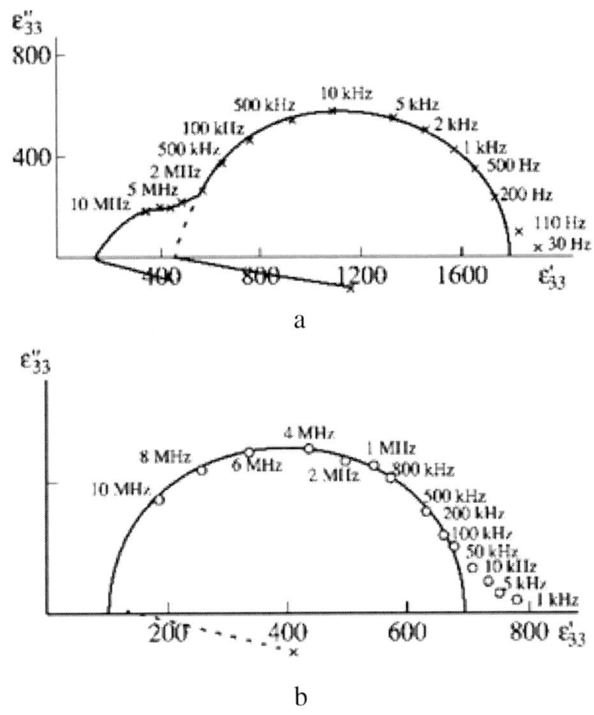

Figure 76. Cole-Cole diagrams for an LT single crystal: (a) $T = 582°C$, $E_{dis} = 0$; (b) $T = 620°C$, $E_{dis} = 1000$ V/cm. $E_{meas} = 50$ V/cm.

The unit cell of LT (like that of LN) contains two formula units and has space group C_{3v}^{6} [44]. Raman-active are four vibrations of symmetry A_1 and nine vibrations *E*. All of these vibrations experience LO-TO splitting as a result of their polar character.

The most significant differences between the spectra of mono-and polydomain samples for identical scattering geometries were found for the geometries in which TO vibrations of A_1 symmetry appear [219, 240, 241]. To determine the parameters of the corresponding Raman lines in optically uniaxial crystals, it is sufficient to study samples using scattering geometry *Y(ZZ)X*.

Figure 77 shows spectra for polydomain LT samples cut from various single crystals [219, 240, 241]. For both samples in orientation *Y(ZZ)X*, narrow lines appear in addition to the four quite intense lines associated with A_1 fundamentals and allowed by the selection rules. These narrow lines (marked by arrows) are due to *E* fundamentals and are forbidden for a fully single-domain crystal in the chosen scattering geometry.

The At the same time, experimental results allow one to appearance of forbidden lines is characteristic of all claim that the low-frequency relaxation process, as well polydomain LT single crystals. Therefore, Raman specas anomalous nonlinearity, is characteristic of polydotra measured using scattering geometry *Y(ZZ)X* offer an main samples [141, 240, 241]. These phenomena must efficient means for quantifying the degree to which an be intrinsic to any polydomain ferroelectric crystal. LT crystal is a single domain.

The strong forbidden line at 189 cm^{-1} associated with vibrational symmetry *E* is most suitable for determining the single-domain extent.

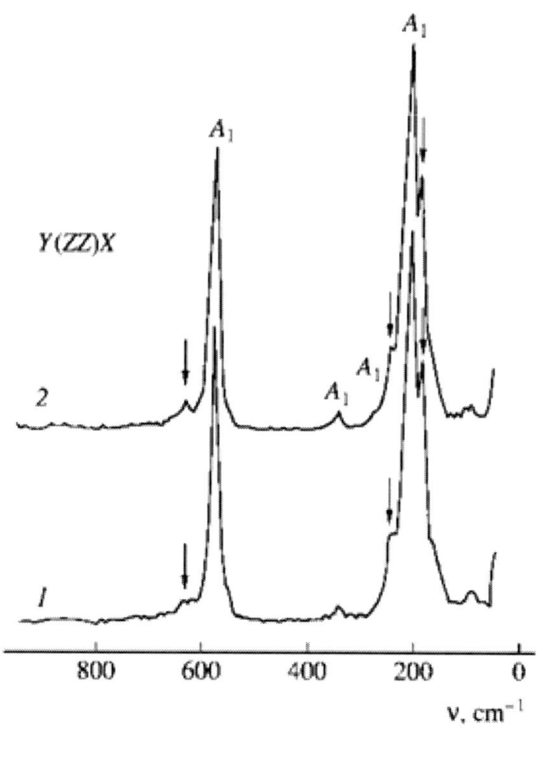

Figure 77. Raman spectra for LT samples cut from single crystals differing in the degree of conversion to a singledomain state: (*1*) sample *1* and (*2*) sample *2*.

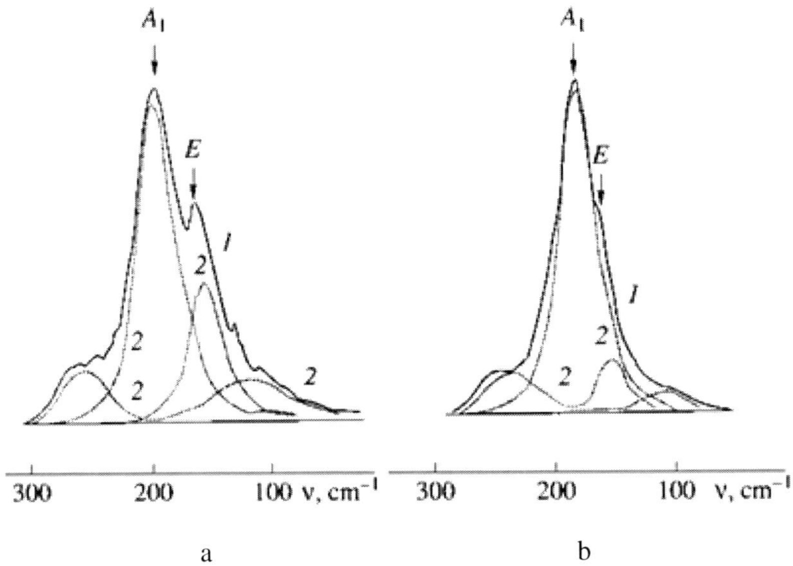

Figure 78. Raman line profiles for $LiTaO_3$ single crystals: (*1*) experimentally observed and (*2*) computer-processed. Panel (a): a poly-domain crystal, $I(A_1)/I(E) = 2.98$. Panel (b): a crystal converted to a single domain by an electric field, $I(A_1)/I(E) = 11.58$.

The intensity of this line changes appreciably after exposing a crystal near the ferroelectric transition to a stationary displacing field: the intensity tends to zero as a crystal converts to a single domain [240, 241]. This line is overlapped with a broader and strong line at 202 cm^{-1} associated with fundamentals A_1. Therefore, the intensity ratio of these lines $I(A_1)/ I(E)$ is a suitable criterion to judge the poly-domain state.

To quantify the intensity ratio exactly, the spectral line profiles at 189 and 202 cm^{-1} must be separated. Figure 80 [240, 241] shows fragments of observed Raman line profiles in the region of 100-300 cm^{-1} and computer-separated line profiles for a polydomain crystal after converting the crystal to a single domain.

The relative intensities of the lines at 202 and 189 cm^{-1}, characterizing the extent of conversion of the crystal to a single domain, are shown in the same figure.

The procedure allows both the single-domain extent to be evaluated and the kinetics of conversion to the single-domain state to be studied through recording narrow portions of Raman spectra (within the range where a crystal is converted to a single domain) with the electric field switched off for a short period [219, 240, 241]; this disorder is hardly modeled and severely affects the properties of crystals.

An approach to the design of ferroelectrics that does not involve the synthesis of new structures has been recognized; extant materials are modified with the goal of improving their optical parameters. The materials design based on an alteration of a structure order is especially topical, because extant technologies can be applied to design materials having basically novel properties.

It has been shown that vibrational spectroscopy is one of the most effective investigation tools for investigation a structure order in crystals. Vibrational spectra are very responsive to changes in the interactions between structure units in a crystal and, correspondingly, to various subtle structure rearrangements, in particular, the ones arising from compositional changes.

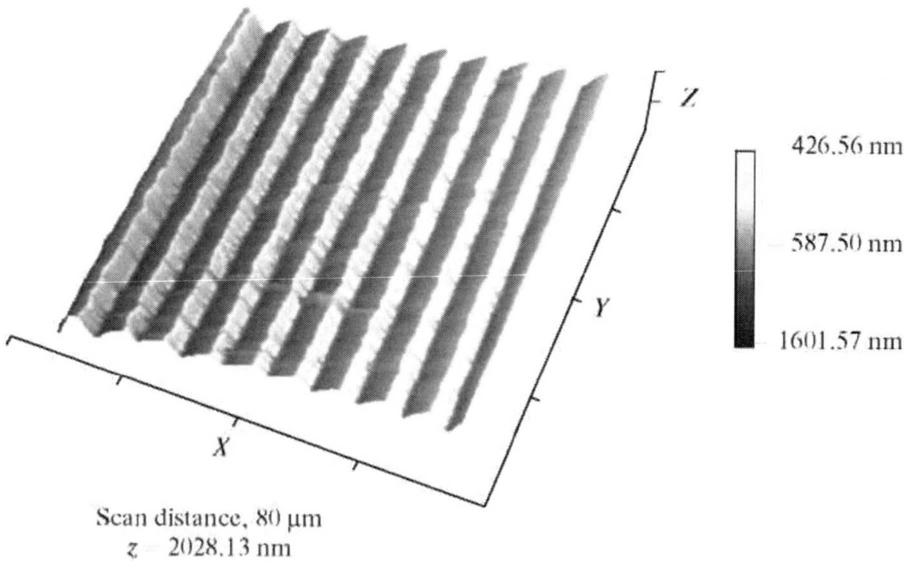

The image was obtained with a Nano-R2 atomic force microscope.

Figure 79. RDS in a $LiNbO_3$:Gd (0.44 wt % Gd) crystal grown under transient conditions (RDS period, 7.86 μm).

In addition, strong phonon interactions are observed in disordered and anharmonic structures such as LN and LT crystals; along with line broadening and weakening of fundamental lattice vibrations, phonon interactions can generate new vibrational lines due to many-phonon optical and acoustic vibrational states and to the mixing of one-phonon and two-phonon states. The parameters of relevant peaks can serve as a measure of the crystal-structure perfection and a suitable indicator for the detection and investigation of subtle structure-order features.

Raman studies included the detailed analysis of the structural features of the cation sublattice for real LN crystals differing in their chemical compositions (nominally pure and doped crystals). The following supposition has been substantiated: in the cation sublattice of off-stoichiometric crystals, clusterlike intrinsic and impurity defects form an ordered sublattice; this sublatice gives its own vibrational spectrum in the form of low-intensity extra lines that differ from the fundamental vibrational spectrum. No such defect structure exists in high-ordered stoichiometric crystals.

The peak observed in the region of 100-120 cm^{-1} in the Raman spectrum for an LN crystal and associated with the two-particle state of acoustic phonons with a null overall wave vector is responsive to the subtle features of the cation order. A stoichiometric, high-perfection crystal gives no Raman lines in the region of 100-150 cm^{-1}.

The absence of this peak can be taken as an experimental criterion to judge whether an LN crystal structure corresponds to the structure of a perfect stoichiometric crystal. It has been shown that the structure modification of complex oxide compounds by doping them with nonphotorefractive dopants could yield materials having improved optical properties, in particular, reduced photorefraction.

Techniques for determining the stoichiometry of LN ad LT crystals have been described. A model has been advanced to locate dopant cations in the structure proceeding from property versus concentration diagrams.

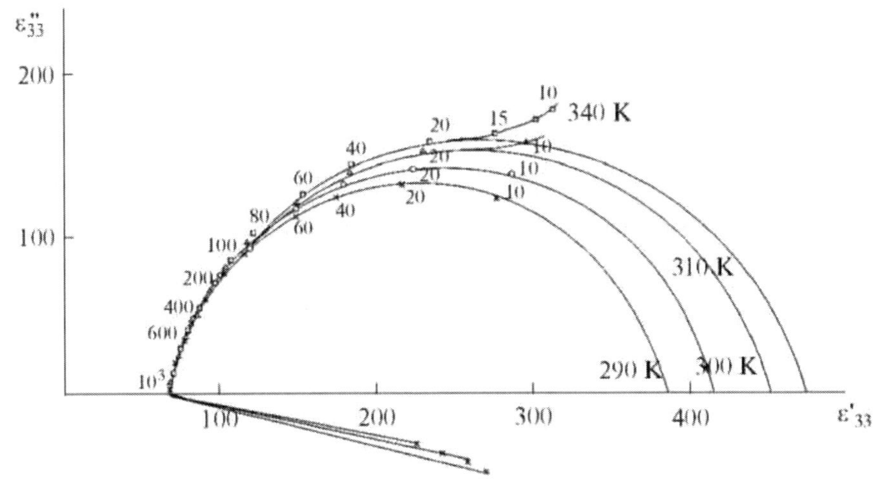

Figure 80. Cole-Cole diagrams for a $LiNbO_3$:Gd⟩ (0.44 wt % Gd, *Z*-cut) crystal at temperatures under 400 K. The frequencies are indicated at the curves in hertz.

Results have been presented concerning the stability of electrophysical and optical parameters in nominally pure and doped LN crystals in a significant temperature range of 300-350 K. It has been supposed that anomalies in the physical properties of LN observed in both doped and nominally pure crystals, regardless of the initial state of the domain structure, are to some extent related to a change in the electron conduction mechanism. Particular values of the anomalies and the kinetics of processes involved are dictated by the ratio of the densities of states for low- and deep-lying trapping centers and the real structure of samples.

We have presented the results of comparative studies of single-domain and polydomain LT single crystals. These results provide a basis for determining the formation conditions for a stable single-domain state, for developing reproducible modes of conversion to a single-domain state, and for evaluating the degree to which a crystal is a single domain from Raman spectra. The experimental results can be useful in the design of high-quality of optical crystals.

18. Electrophysical and Structural Properties of $LiNbO_3$⟨RE⟩ Single Crystals Grown under Steady-State and Transient Conditions

Lithium niobate crystals were reported to display a number of anomalies in their electrical conductivity and optical, dielectric, and pyroelectric properties in the temperature range of practical interest (300-400 K) [1-5, 8, 44, 95, 226, 227, 245, 247]. Most reports concerned with the anomalous temperature behavior of the physical characteristics of lithium niobate crystals pointed out a lack of quantitative reproducibility of results, which depended significantly on the thermal and field histories of samples. To gain insight into the origin of the anomalies, considerable research effort has focused on domain micro- and nanostructures and fine features of the structural order in $LiNbO_3$⟨RE⟩ (RE = rare earth) crystals grown under both steady-state and transient conditions. It has been shown that $LiNbO_3$⟨РЗЭ⟩ crystals grown under unstable conditions contained micron-scale regular domain structures (RDS's) with a variable or stable pitch and periodic nanostructures with a period from 10 to 100 nm.

The cation sublattice of rare-earth-doped lithium niobate crystals has a superstructural sublattice of clustered defects with a pitch equal to several translation periods [151, 152, 155].

Static and dynamic piezoelectric properties, dielectric dispersion, and electrical conductivity of rare-earth and magnesium (Gd, Tm, Gd, Mg) doped lithium niobate crystals grown under steady-state and transient conditions and having their micro-and nanodomain structures in various states are studied in the temperature range ~ 290-490 K and in a wide frequency range (0.5 to 10^6 Hz).

The growth of $LiNbO_3$⟨RE⟩ crystals under transient conditions leads to the formation of RDS's [151, 247-251]. Figure 79 shows a typical image of an RDS in a $LiNbO_3$⟨Gd⟩ crystal. Examination by atomic force microscopy revealed periodic nanostructures with a period from ~ 10 to 100 nm on the negative domain walls of RDS's in $LiNbO_3$⟨RE⟩ crystals [9]. Periodic structuring is not limited to the scale of 10-100 nm, which can be investigated with our atomic force microscopy facilities and procedures. These assumptions were validated by Raman scattering studies, which showed that the cation sublattice of the crystal had a superstructural sublattice of clustered defects. In the lithium niobate structure, clusters form near Nb_{Li} native defects and are arranged in ordered sublattices several translation periods in size, that is, have a pitch of 1-2 nm [252], which means that $LiNbO_3$⟨RE⟩ crystals contain periodic structures on length scales from ~ 1 nm to 100 |am [155].

At temperatures from ~ 330 to 380 K, the relative dielectric permittivity e_{33}(J) and conductivity of $LiNbO_3$⟨RE⟩ crystals have anomalies of a relaxation character [155]. At frequencies in the range 10 Hz to 19 kHz in weak electric fields, there is low-frequency dielectric dispersion which satisfies the Debye equation.

Figure 80 shows Cole-Cole diagrams for a $LiNbO_3$⟨Gd⟩ (0.44 wt % Gd) crystal. Raising the temperature leads to qualitative changes in the behavior of the dielectric dispersion. The changes show up in Cole-Cole diagrams as linear low-frequency portions. With increasing temperature, the linear portions become more pronounced and extended, and the relaxation time corresponding to the Debye process increases sharply. Above 400 K, the Debye dispersion decreases and disappears (Figure 81). In the temperature range ~ 290-410 K, two dispersion processes are manifest in the diagrams: the Debye process, represented by an arc of a circle, and a lower frequency process, represented by a linear portion. In Figure 81, these processes are denoted as *I* and *II*. As the temperature is raised from 290 to 340 K, the dispersion depth of the dispersion process *I* increases only slightly (Figure 82), whereas just above 340 K the dispersion depth drops precipitously (Figure 82).

It follows from the diagrams that, at temperatures from ~ 290 to 340 K and frequencies from ~0.5 Hz to 10 kHz, the dielectric dispersion in $LiNbO_3$:Gd is due to a single, Debye-type relaxation process (Figures 80, 82). In the temperature range T~ T_0 ~ 340-350 K, the dielectric characteristics vary over time, but the high-frequency dynamic permittivity (e') remains unchanged (Figure 82). Holding the sample at T_0 ~ 340 - 350 K for 4 h completely eliminates the type *I* (Debye) dispersion.

Studies of the complex permittivity dispersion in bias fields (0-25 kV/cm) (Figure 85) demonstrate that, even at room temperature, an applied bias field considerably reduces the dispersion depth of the dispersion process *I but* does not change its Debye character.

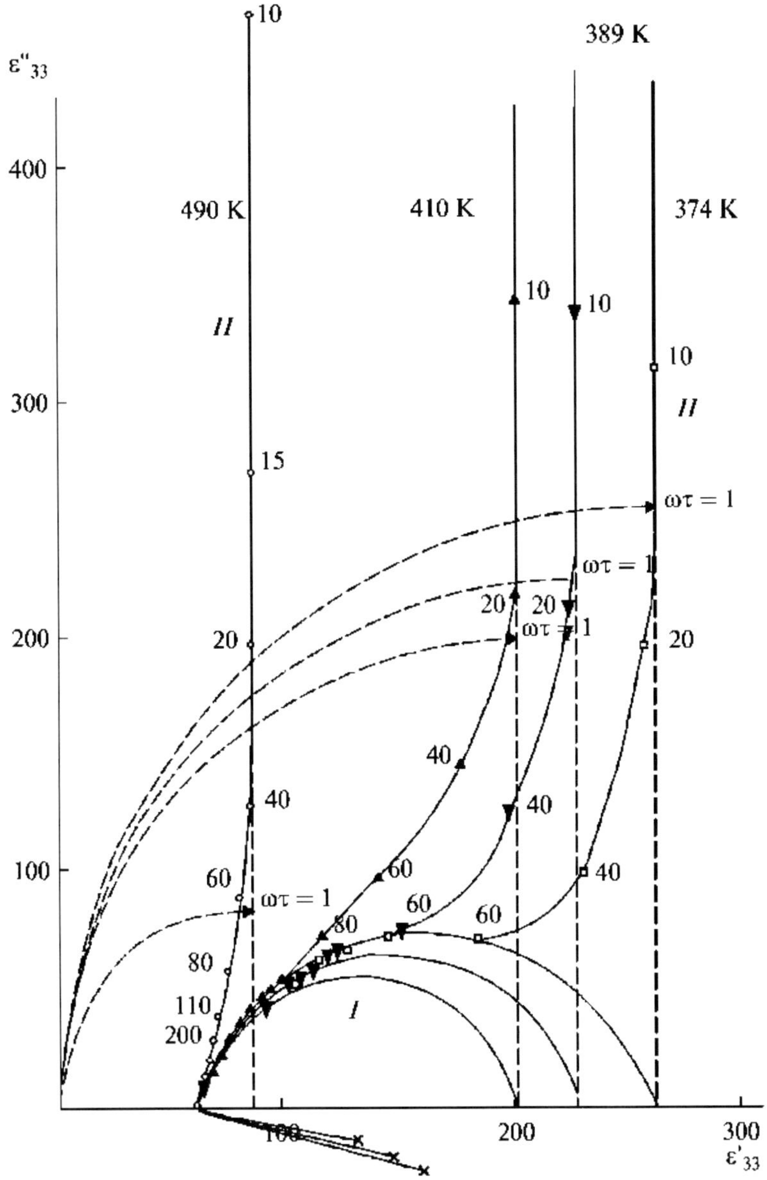

Figure 81. Cole-Cole diagrams for a $LiNbO_3$:Gd⟩ (0.44 wt % Gd, *Z*-cut) crystal at temperatures from 374 to 490 K. The frequencies are indicated at the curves in Hertz.

The low-frequency Debye-type dielectric dispersion seems to arise from both the spontaneous polarization relaxation and the relaxation of point defects that interact with periodic domain walls and periodic nanostructure boundaries.

Similar experiments were carried out with a $LiNbO_3$:Gd,Mg crystal, which also had well-developed micro- and nanodomain structures.

In the $LiNbO_3$⟨Gd,Mg⟩ sample, we identified a low-frequency relaxation process qualitatively similar to process *I* in the $LiNbO_3$:Gd: crystal (Figures 80, 82-84). Raising the temperature to ~ 400 K suppresses the type *I* dispersion.

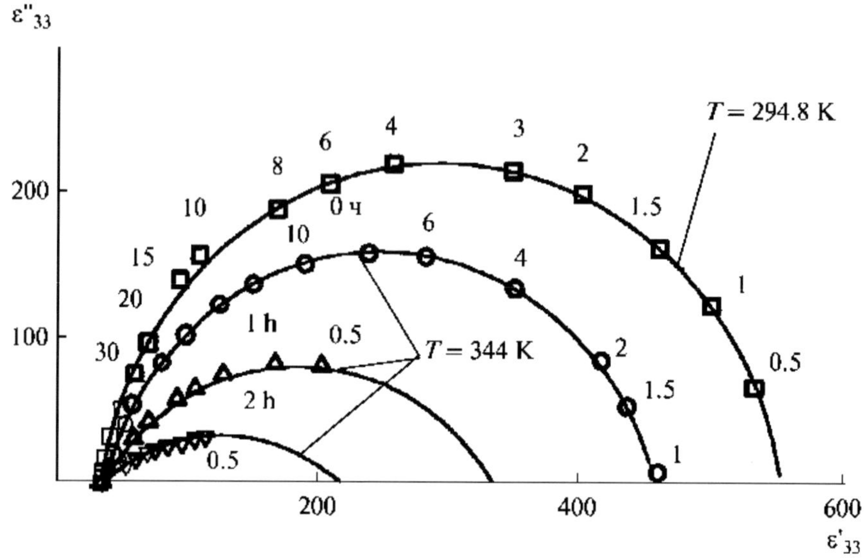

Figure 82. Cole-Cole diagrams for a $LiNbO_3$:Gd (0.44 wt % Gd, *Z*-cut) crystal at two temperatures. The holding time at $T = 344$ K was 0, 1, and 2 h. The frequencies are indicated at the curves in Hertz.

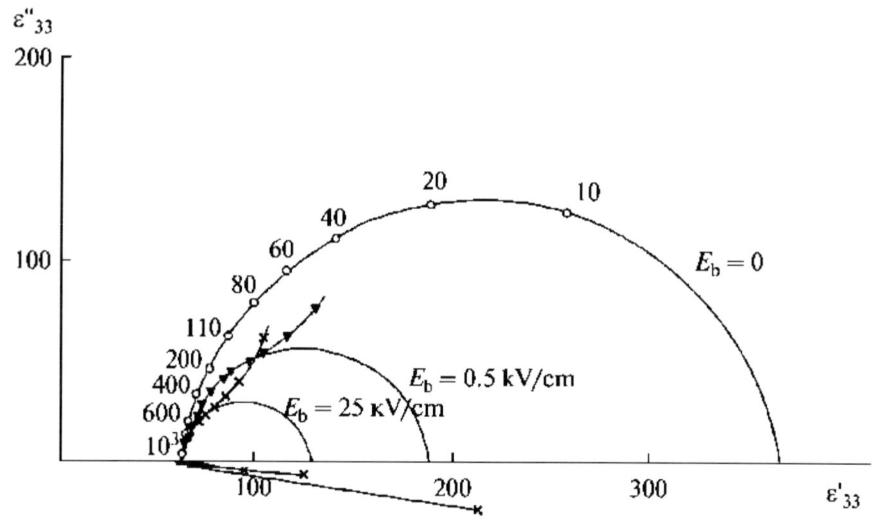

Figure 83. Complex permittivity dispersion in a $LiNbO_3$:Gd (0.44 wt % Gd, *Z*-cut) crystal in bias fields at $T = 296$ K. The frequencies are indicated at the curves in Hertz.

The arc of a circle in the Cole–Cole diagrams (Figures 80, 82-84) represents the dynamic relaxation of the spontaneous polarization related to periodically poled domain micro- and nanostructures and point defects in a weak measuring field with rather short relaxation times, ~ 0.1-0.01 s. An applied dc electric field causes a single-domain state to develop in a part of the sample, whose volume depends on field strength, and that part does not contribute to the dispersion, thereby reducing its depth (Figure 83). We studied temperature-dependent static and dynamic piezoelectric effects [152]. At temperatures below 340 K, our samples had low values of the d_{33} piezoelectric modulus, determined by natural unipolarity.

In the temperature range corresponding to the anomalies in dielectric properties and electrical conductivity (340-350 K), we observed a sharp increase in d_{33} to the level characteristic of single-domain crystals.

Long-term (up to several months) storage of the $LiNbO_3\langle Gd\rangle$ (0.44 wt % Gd) crystal in an open-circuit state at room temperature restored the low values of d_{33}. In the $LiNbO_3\langle Gd,Mg\rangle$ (0.37 wt % Gd, 0.72 wt % Mg) crystal, a state with a high degree of unipolarity persists for an extremely long time (up to 10 years) [152].

We measured the dielectric dispersion and conductivity of a$LiNbO_3$:Gd (0.52 wt % Gd) crystal grown under steady-state conditions, which had poorly developed domain micro- and nanostructures. At frequencies from 1 Hz to 1 kHz, we observed two relaxation processes: one of the Debye type (*I*) and the other represented by an inclined line (*II*) related to the conductivity of the sample (Figure 85a).

Even a slight increase in temperature eliminates the Debye process (Figure 86). Quantitatively, the type *I* dispersion depth in the crystals grown under steady-state conditions is substantially smaller than that in the crystals grown under transient conditions (cf. Figures 80 and 4-8). The relatively small dispersion depth attests to a rather strong unipolarity of the $LiNbO_3$:Gd (0.52 wt % Gd) crystal. Qualitatively, the conductivity data are similar to those for the crystals grown under transient conditions (cf. Figures 87 and 89b) [152]. Quantitatively, the conductivity of the crystals with a well-developed domain structure at temperatures from 290 to 340 K is higher by orders of magnitude (Figure 87). Heating reduces the static conductivity of the $LiNbO_3\langle Gd\rangle$ (0.52 wt % Gd) crystal by more than one order of magnitude, and there is no anomaly in conductivity during subsequent cooling (Figure 87a).

The dispersion behavior of $LiNbO_3\langle Tm\rangle$ (0.13 wt % Tm) crystals grown under steady-state conditions, which also have poorly developed micro- and nan-odomain structures, is qualitatively similar to that of $LiNbO_3\langle Gd\rangle$ (0.52 wt % Gd).

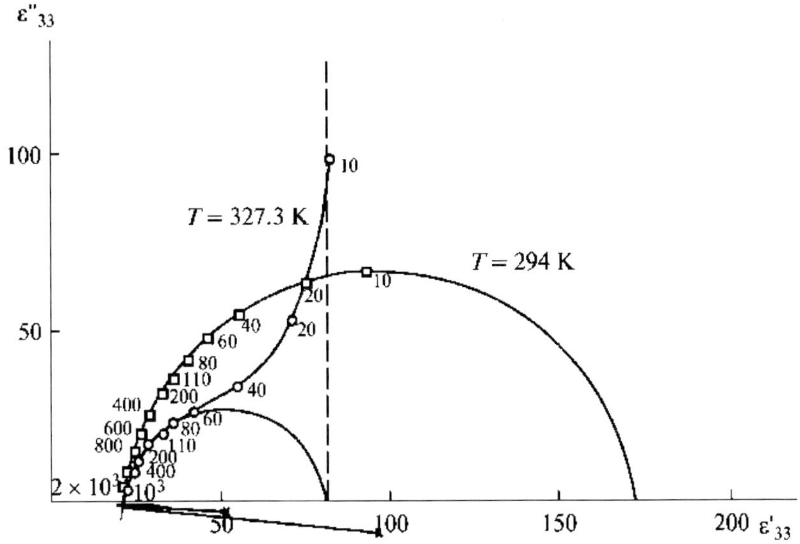

Figure 84. Complex permittivity dispersion in a $LiNbO_3$:Gd,Mg crystal (*Z*-cut). The frequencies are indicated at the curves in Hertz.

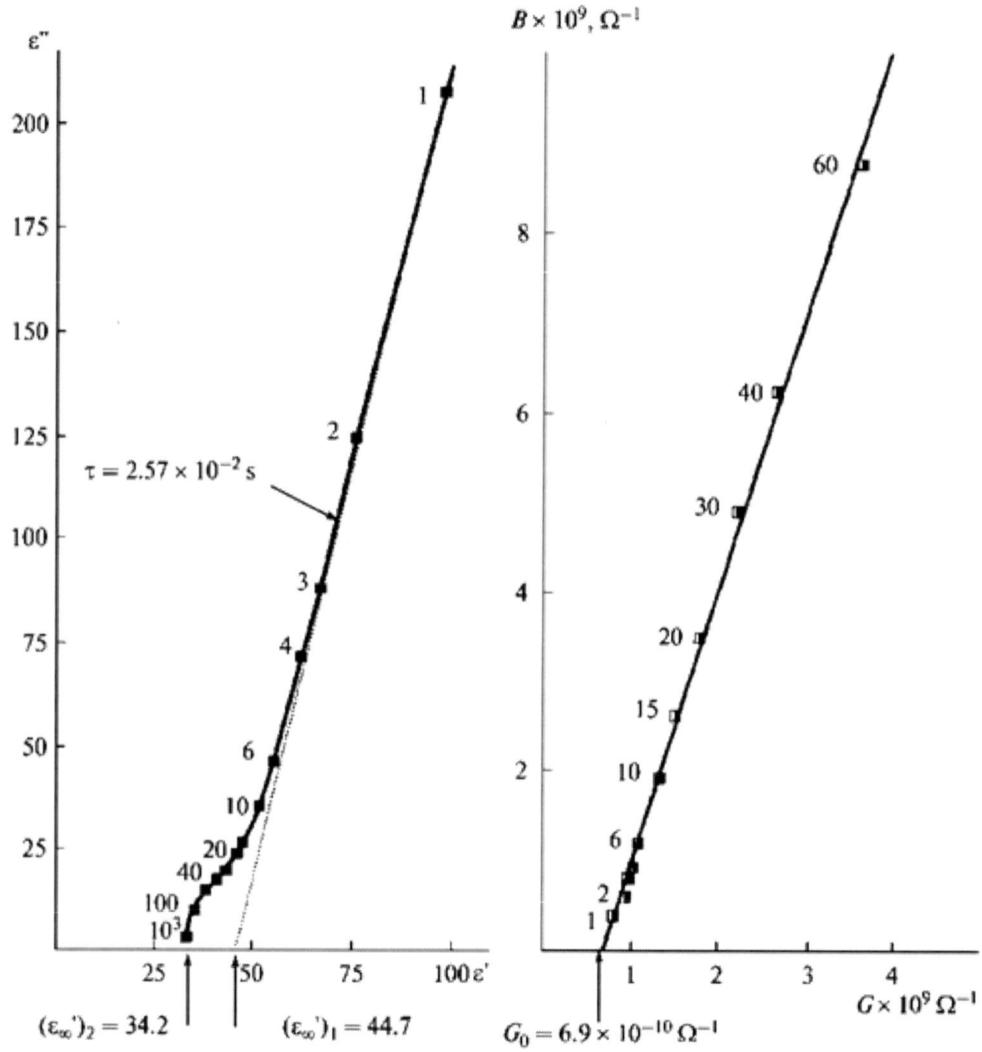

Figure 85. (a) Dielectric dispersion and (b) admittance diagram of a $LiNbO_3$:Gd (0.52 wt % Gd) crystal grown under steady-state conditions (T = 290 K). The frequencies are indicated at the curves in Hertz.

Their admittance diagrams are also qualitatively similar to those of $LiNbO_3$⟨Gd⟩. The diagrams were used to evaluate static conductivity as a function of temperature (Figure 88). In the first heating cycle (curve *I*), the $a_s(T)$ curve demonstrates a considerable decrease in static conductivity, which shows irreversible behavior during subsequent cooling (curve *II*).

The $a_s(T)$ data for $LiNbO_3$⟨Tm⟩ (0.13 wt % Tm) are qualitatively similar to those for $LiNbO_3$⟨Gd⟩ (0.52 wt % Gd) (Figures 9a, 10). It seems likely that the conductivity of the Tm-containing samples also has a maximum during the first heating, but at lower temperatures.

As in the case of the $LiNbO_3$⟨Gd⟩ (0.44 wt % Gd) crystals grown under transient conditions [152], the piezoelectric effect was studied by a static method. Figure 89 presents our results on the static piezoelectric effect in polydomain and single-domain $LiNbO_3$⟨Tm⟩ (0.13 wt % Tm) crystals.

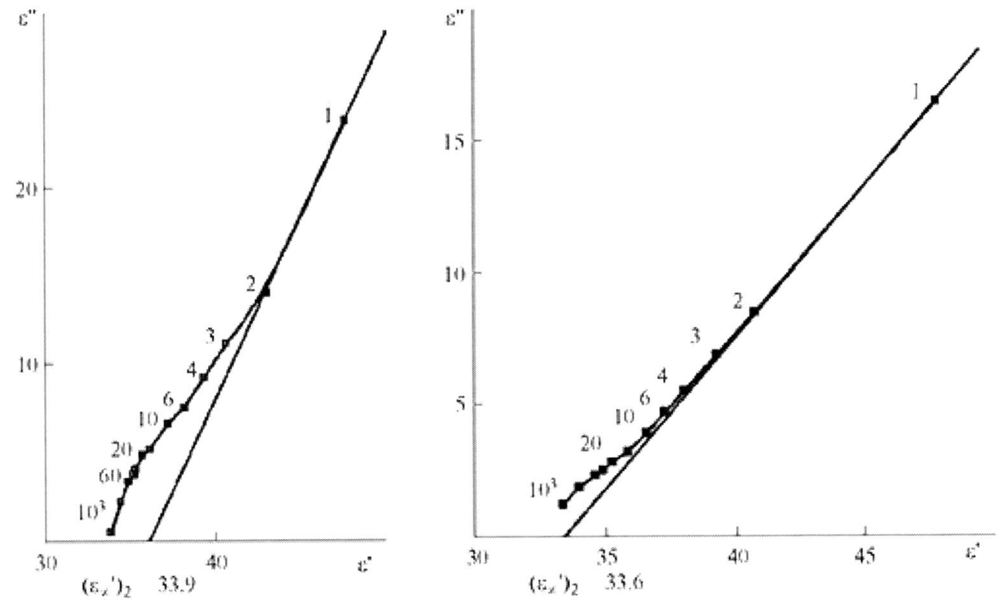

a b

Figure 86. Temperature effect on the dielectric dispersion in a $LiNbO_3$:Gd (0.52 wt % Gd) crystal grown under steady-state conditions: T = (a) 293, (b) 337 K. The frequencies are indicated at the curves in Hz.

The behaviors of the $LiNbO_3$⟨RE⟩ crystals grown under different conditions are qualitatively similar, but there are significant quantitative distinctions.

In particular, the initial room-temperature d_{33} of the $LiNbO_3$⟨Gd⟩ (0.44 wt % Gd) crystal grown under transient conditions, which has an RDS and a well-developed system of fractal nanostructures, is ~ (0.3-0.4) x 10^{-12} C/N, whereas that after a measurement cycle to temperatures $T > 340$ K is d_{33} ~ (10-12) x 10^{-12} C/N [10].

Whereas the polydomain $LiNbO_3$⟨Gd⟩ (0.52 wt % Gd) and $LiNbO_3$⟨Tm⟩ (0.13 wt % Tm) crystals grown under steady-state conditions exhibit a sizeable piezoelectric effect at room temperature, with d_{33} ~ 4.0 x $10–^{12}$ C/N, the d_{33} of the single-domain $LiNbO_3$⟨Tm⟩ crystal is ~12.8 x 10^{-12} C/N, like that of nominally undoped $LiNbO_3$ crystals [252].

The degree of unipolarity ξ of a $LiNbO_3$ crystal can be defined as

$$\xi = \frac{V^+ - V^-}{V^+ + V^-} = \frac{V^+ - V^-}{V_0},$$

where V^+ and V^- are the net volumes of all arbitrarily positive and negative domains, and their sum is the total volume of the sample, V_0. A single-domain crystalline sample has the maximum possible piezoelectric modulus d_{33}. If a crystal is ideally polydomain, max $d_{33\min}$ = 0. The same relations apply to ξ, so d_{33} is proportional to ξ and we have $\xi = d_{33} / d_{33\max}$, where d_{33} is the measured piezoelectric modulus of a sample in an intermediate (partially unipolar) state.

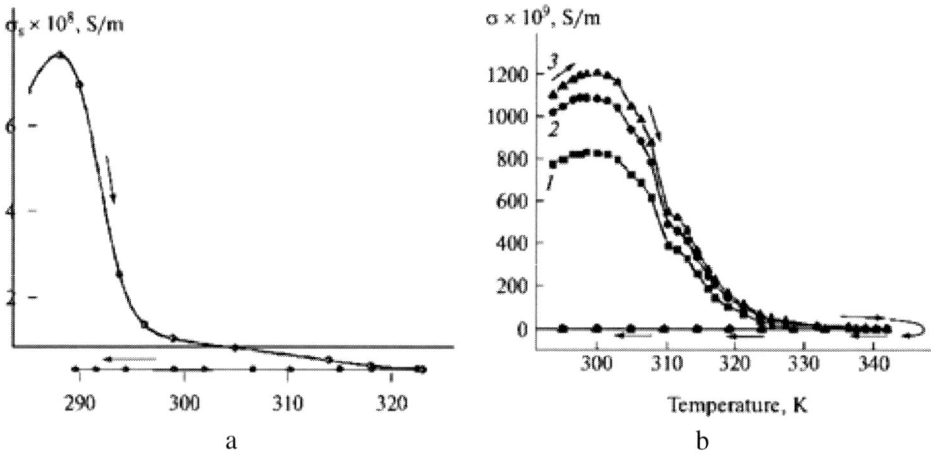

Figure 87. (a) Temperature dependence of static conductivity for a $LiNbO_3$⟨Gd⟩ (0.52 wt % Gd) crystal grown under steady-state conditions and (b) temperature dependences of conductivity at (*1*) 100 Hz, (*2*) 1 kHz, and (*3*) 10 kHz for a $LiNbO_3$⟨Gd⟩ (0.44 wt % Gd) crystal grown under transient conditions (heating–cooling cycles).

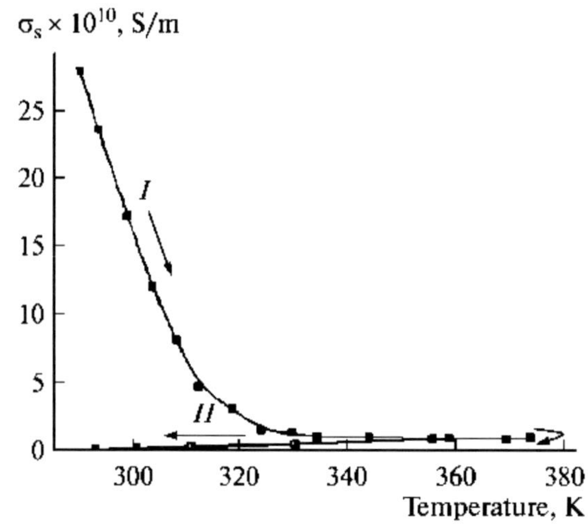

Data extracted from admittance diagrams.

Figure 88. Temperature dependence of static conductivity for a $LiNbO_3$⟨Tm⟩ (0.13 wt % Tm) crystal: (*I*) first heating, (*II*) subsequent cooling.

This means that the original degree of unipolarity in polydomain $LiNbO_3$⟨Tm⟩ is $\xi \approx 0.38$ (Figure 89).

In other words, the original degree of unipolarity of polydomain $LiNbO_3$⟨RE⟩ crystals with a poorly developed domain structure, grown under steady-state conditions, is higher than that of $LiNbO_3$⟨RE⟩ crystals grown under transient conditions, which have a well-developed domain structure. Heating to $T > 340$ K irreversibly increased the d_{33}*of* $LiNbO_3$⟨Tm⟩ to (8.5–8.6) x 10^{-12} C/N, which corresponds to a degree of unipolarity $\xi \approx 0.67$.

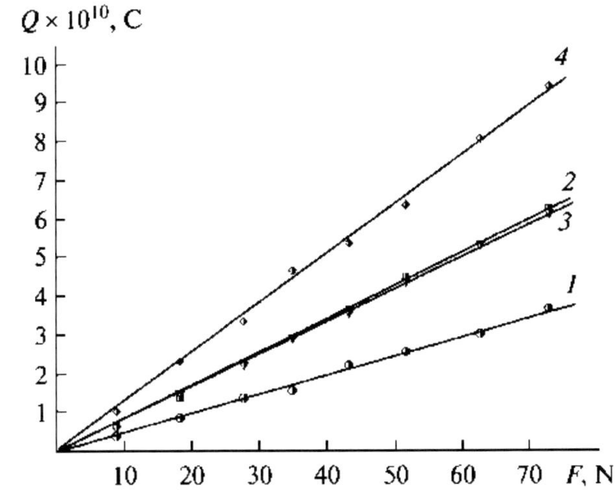

Figure 89. Static piezoelectric effect in $LiNbO_3$(Tm $^-$ (0.13 wt % Tm) samples: *(1)* T = 290 K, d_{33} =4.8 x 10 C/N; *(2)* T = 364 K, *d*33 = 8.6 x 10 C/N, heating in the first thermal cycle (Figure 10); *(3)* T = 290 K, *d*33 = 8.5 x 10 C/N, cooling in the first thermal cycle (Figure 10); *(4)* T = 290 K, *d*33 = 12.8 x 10 C/N, single-domain crystal.

This state persists for a long time (several weeks) after cooling to room temperature but is not entirely single-domain, in contrast to that of $LiNbO_3$:RE crystals with a well-developed domain structure, for example, $LiNbO_3$⟨Gd⟩ (0.44 wt % Gd), where after heating to T > 340 K d_{33} corresponds to that in an entirely single-domain crystal, and $\xi \approx 1$ [152].

Thus, the anomalies in various physical characteristics of $LiNbO_3$:RE crystals in the temperature range ~ 290 to 400 K depend significantly on the initial state of the micro- and nanodomain structures, and the magnitude of the anomalies and the kinetics of the underlying processes are probably determined by the temperature-dependent defect structure of the samples [152].

The static piezoelectric and dielectric properties and electrical conductivity of $LiNbO_3$ ⟨RE⟩ crystals grown under steady-state and transient conditions have been studied in the temperature range ~ 290-490 K.

The magnitude of the observed anomalies in their electrical characteristics and the kinetics of the underlying processes are determined by the development of the micro- and nanodomain structures of the samples.

The low-frequency Debye-type dielectric dispersion in the rare-earth-doped lithium niobate crystals arises from the spontaneous polarization relaxation and the relaxation of point defects bound to periodic domain walls and periodic nanostructure boundaries. The low-frequency Debye-type depth is determined by the development of the micro- and nanodomain structures and the degree of unipolarity of the crystalline sample.

The present results demonstrate that, when $LiNbO_3$⟨RE⟩ crystals are heated to T_0 ~ 340 K, their piezoelectric modulus d_{33} is relatively low and is determined by their natural unipolarity, which depends on the development of their domain micro- and nano- structures.

At temperatures above T_0, corresponding to the observed anomalies in the dielectric properties and conductivity of the crystals, their piezoelectric modulus d_{33} increases sharply.

The resultant degree of unipolarity in the crystalline $LiNbO_3\langle RE\rangle$ samples is determined by the initial development of the micro- and nanodomain structures and may reach the level characteristic of single-domain, nominally undoped lithium niobate crystals.

REFERENCES

[1] Kuzminov, Yu. S. *Electrooptical and Nonlinear Optical Lithium Niobate Crystal*; Nauka: Moscow, Russia, 1987; p. 263.
[2] Rauber, A. Chemistry and Physics of Lithium Niobate. *Current Topics in Materials Science;* Elsevier: Amsterdam, Netherlands, 1978; Vol. 1, p. 481.
[3] Abrahams, S. C. Properties of Lithium Niobate; *EMIS Datareviews Series No. 5 INSPEC;* The Institution of Electrical Engineers: London, Great Britain, 1989; p.
[4] Lines, M.; Glass, A. *Principles and Applications of Ferroelectrics and Related Materials*; Oxford Univ.: Oxford, Great Britain, 1977; p. 680.
[5] Kuzminov, Yu. S.; Osiko, V. V.; Prokhorov, A. M. *Quantum Electron*. 1980, 7, 1621-1627.
[6] Isupov, V. A. *Bull. Rus. Acad. Sciences Phys*. 1983, 47, 559-563.
[7] Svaasand, L. O.; Erikrund, M.; Nakken, G. *J. Cryst. Growth*. 1974, 22, 230-236.
[8] Kuzminov, Yu. S. *Lithium Niobate and Lithium Tantalate: Materials for Nonlinear Optics*; Nauka: Moscow, Russia, 1975; p. 337.
[9] Abrahams, S. K.; Reddy, J. M.; Bernstein, J. L. *J. Phys. Chem. Solids*. 1966, 27, 997-1001.
[10] Lerner, P.; Legras, C.; Dumas, J. P. *J. Cryst. Growth*. 1968, 231, 3-7.
[11] Donnerberg, H. J.; Tomlinson, S. M.; Catlow, E. R. A.; Schirmer, A. F. *Phys. Rev*. 1989, 40, 11909-11911.
[12] Bordui, P. F.; Norwood, R. G.; Jundt, D. H.; Fejer, M. M. *J. Appl. Phys*. 1992, 71, 875-880.
[13] Abrahams, S. C.; March, F. *Acta Crystallogr. Sect. B Struct. Crystallogr. Cryst. Chem*. 1986, 42, 61-64.
[14] Kuzminov, Yu. S.; Osiko, V. V. *Crystallogr. Rep*. 1994, 39, 530-535.
[15] Morozov, A. N.; Voronova, M. I.; Vyrelkin, V. P.; Makarevskaya, E. V.; Kugaenko, O. M.; Blistanov, A. A. *Crystallogr. Rep*. 1993, 38, 219-223.
[16] Malovichko, G. A., *Doctoral Dissertation*, Kiev, 1987.
[17] Kuzminov, Yu. S. *Crystallogr. Rep*. 1995, 40, 1034-1041.
[18] Palatnikov, M. N.; Sidorov, N. V.; Stefanovich, S. Yu.; Kalinnikov, V. T. *Inorg. Mater*. 1998, 34, p. 903-910.
[19] Balanevskaya, A. E.; Pyatigorskaya, L. I.; Shapiro, Z. I.; Margolin, L. N.; Bovina, E. A. *J. Appl. Spectrosc*. 1983, 38, 662-667.
[20] Balasanyan, R. N.; Gabrielyan, V. T.; Kokanyan, E. P. *Cryst. Rep*. 1990, 35, 1545-1550.
[21] *Acoustic Crystals: A Handbook*; Shaskolskaya, M. P.; Ed.; Nauka: Moscow, Russia, 1982; p. 632.
[22] Sangeeta, D.; Raipurkar, M. K.; Kothiyal, G. P.; Ghosh, B. *Indian J. Phys*. 1987, 61, 373-379.

[23] Srivastava, K. N.; Gangarh, J. R.; Rishi, M. V.; Singh, R. *Indian J. Pure Appl. Phys.* 1984, 22, 154-160.

[24] Balasanyan, R. N.; Polgar, K.; Erdei, Sh. *Crystallogr. Rep.* 1987, 32, 482-488.

[25] Scott, B. A.; Burns, G. *J. Amer. Ceram. Soc.* 1972, 55, 225-229.

[26] Chow, K.; McKnight, H. G.; Rothrock, L. R. *Mat. Res. Bull.* 1974, 9, 106-109.

[27] Grabmaier, B. C.; Wersning, W.; Koestler, W. *J. Cryst. Growth*, 1991, 110, 339-345.

[28] Born, E.; Willibald, E.; Hofmann, K.; Grabmaier, B. C.; Talsky, G.; *Abstracts of Papers, IEEE Ultrasonics Symposium*, 1998, p. 119.

[29] Arizmendi, L. *J. Appl. Phys.* 1988, 64, 4654-4659.

[30] Palatnikov, M. N.; Sidorov, N. V.; Stefanovich, S. Yu.; Kalinnikov, V. T. *Abstracts of Papers, 3 Mezhdunarodnaya konferentsiya: Kristally, rost, svoistva, real'naya struktura, primenenie* (3 Int. Conf. on Crystal Growth, Properties, Real Structure, and Application), Aleksandrov, 1997, vol. 1, p. 349.

[31] Krol, D. M.; Blasse, G. *J. Chem. Phys.* 1980, 73, 163-171.

[32] Foldvari, I.; Polgar, K.; Voszka, K.; Balasanyan, R. N. *Cryst. Res. Technol.* 1984, 19, 1659-1665.

[33] Gallagher, P. K.; O'Bryan, H. M. *J. Am. Ceram. Soc.* 1985, 68, 147-152.

[34] O'Bryan, H. M.; Gallagher, P. K.; Brandle, C. D. *J. Am. Ceram. Soc.* 1985, 68, 493-498.

[35] Fenske, M.; Kuzminov, Yu. S. *Preprint Gen. Phys. Inst.* 1998, 207, 45-43.

[36] Volk, T. R.; Rubinina, N. M. *Phys. Status Solidi*, 1988, 18, 437-441.

[37] Grabmaier, B. C.; Willibald, E.; Born, E. *Siemens Forsth. Entwicklungber.* 1988, 17, 159-170.

[38] Sidorov, N. V.; Palatnikov, M. N. and Kalinnikov, V. T. *Abstracts of Papers, 3rd Mezhdunarodnaya konferentsiya: Kristally, rost, svoistva, real'naya struktura, primenenie* (3 Int. Conf. on Crystal Growth, Properties, Real Structure, and Application), Aleksandrov, 1997, vol. 1, p. 375.

[39] Kostritskii, S. M.; Kanaev, I. F.; Malinovskii, V. K.; Novomlintsev, A. V.; Pugachev, A. M.; *Bull. Rus. Acad. Sciences Phys.* 1995, 59, 41-48.

[40] Balasanyan, R. N.; Gabrielyan, V. T.; Kokanyan, E. P.; Feldvari, I. *Crystallogr. Rep.* 1990, 35, 1540-1548.

[41] Dyakov, V. A.; Luchinskii, G. V.; Rubinina, N. M.; Kholodnykh, A. I. *Tech. Phys.* 1981, 51, 1557-1661.

[42] Zakharov, A. M. *Phase Diagrams of Binary and Ternary Systems;* Metallurgiya: Moscow, Russia, 1990; p. 240.

[43] *Compounds of Variable Composition*; Ormont, B. F.; Ed.; Khimiya: Leningrad, Russia, 1969; p. 520.

[44] Sidorov, N. V.; Volk, T. R.; Mavrin, B. N.; Kalinnikov, V. T. *Lithium Niobate: Defects, Photorefraction, Raman Spectra, Polaritons;* Nauka: Moskow, Russia, 2003; p. 255.

[45] Rosenman, G.; Skliar, A.; Arie, A. *Ferroelectr. Rev.* 1999, 1, 263–326.

[46] Zhang, Q.-R.; Feng, X.-Q. *Phys. Rev. B* 1991, 43, 12019–12024.

[47] Furukawa, Y.; Sato, M.; Nitanda, F.; Ito, K. *J. Cryst. Growth.* 1990, 99, 832–836.

[48] Choubey, R. K.; Sen, P.; Sen, P. K.; Bhatt, R.; Kar, S.; Shukla, V.; Bartwal, K. S. *Opt. Mater.* 2006, 28, 467–472.

[49] Hu, L. J.; Chang, Y. H.; Yen, F. S.; Lin, S. P.; Lin, I.-N.; Lin, W. Y. *J. Appl. Phys.* 1991, 69, 7635–7639.

[50] Chen, Y. I.; Guo, J.; Lou, C. B.; Yuan, J. W.; Zhang, W. L.; Chen, S. L.; Huang, Z. H.; Zhang, G. Y. *J. Cryst. Growth*. 2004, 263, 427–430.

[51] Palatnikov, M. N.; Sidorov, N. V.; Kalinnikov, V. T. *Stahl Eisen*. 2000, 10, 54–60.

[52] Palatnikov, M. N.; Sidorov, N. V.; Kalinnikov, V. T. *Ferroelectric Solid Solution Based on Oxide Compounds of Niobium and Tantalum: Synthesis, the Search of Structure Ordering and of Physical Characteristics;* Nauka; Saint-Petersburg, Russia, 2001; p. 255.

[53] Sidorov, N. V.; Mavrin, B. N.; Chufyrev, P. G.; Kalinnikov, V. T. *The Phonon Spectra of Single Crystals of Lithium Niobate*; Kola Science Center of Russain Academy of Sciences: Apatity, Russia, 2012; p. 214.

[54] Voronko, Yu. K.; Kudrjavcev, A. B.; Osiko, V. V.; Sobol, A. A.; Sorokin, E. V. *Phys. Solid State*. 1987, 29, 1348–1355.

[55] Voronko, Yu. K; Kudrjavcev, A. B.; Sobol, A. A.; Sorokin, E. V. *Proceedings of IOFAN* 29 (1991) 50–100.

[56] Voronko, Yu. K.; Kudrjavcev, A. B.; Osiko, V. V.; Sobol, A. A.; Sorokin, E. V. *Proceedings of FIAI* 2 1987, 34–36.

[57] Bordui, P. F.; Norwood, R. G.; Bird, C. D.; Carella, J. T. *J. Appl. Phys*., 1995, 78, 4647–4650.

[58] Katz, M.; Route, R. K.; Hum, D. S. *Opt. Lett*., 2004, 29, 1775–1778.

[59] Hum, D. S.; Route, R. K.; Miller, G. D.; Kondilenko, V.; Alexandrovski, A.; Huang, J.; Urbanek, K.; Byer, R. L.; Fejer, M. M. *J. Appl. Phys*. 2007, 101, 093108.

[60] Tian, L.; Gopalan, V.; Galambos, L. *Appl. Phys. Lett*. 2004, 85, 4445–4447.

[61] Schaufeler, R. F.; Weber, L. L. *Phys. Rev*. 1966, 152, 705-710.

[62] Kaminov, I. P.; Johnston, W. D. *Phys. Rev*. 1967, 160, 519-528.

[63] Johnston, W. D.; Kaminov, I. P. *Phys. Rev*. 1968, 165, 1045-1054.

[64] Sidorov, N. V.; Palatnikov, M. N.; Serebryakov, Yu. A.; Lebedeva, E. L.; Kalinnikov, V. T. *Inorg. Mater*. 1997, 33, 496-525.

[65] Nippus, M. J. *Nature Res. A Phys. Sci*. 1976, 31, 231-239.

[66] Kojima, S. *Jpn. J. Appl. Phys*. 1993, 32, 4343-4348.

[67] Gorelik, V. S. *Tr. Fiz. Inst. im. P. N. Lebedeva, Akad. Nauk. SSSR*. 1982, 132, 15-19.

[68] Okamoto, Y.; Wang, P.-C.; Scott J. F. *Phys. Rev. B Condens. Matter*. 1985, 32, 6787-6797.

[69] Claus, R.; Borstel, G.; Wiesendanger, E.; Steffan, L. J. *Nature Res. A Phys. Sci*. 1972, 27, 1187-1198.

[70] Shuller, E.; Claus, R.; Falge, H. J.; Borstel, G. J. *Nature Res. A Phys. Sci*. 1977, 32, 47-59.

[71] Klimenko, V. A.; Korotkov, P. A.; Felinskii, G. S. *Opt. Spektrosk*. 1983, 54, 476-484.

[72] Sidorov, N. V.; Palatnikov, M. N.; Kalinnikov, V. T. *Abstracts of Papers, 3 Mezhdunarodnaya konferentsiya: Kristally, rost, svoistva, real'naya struktura, primenenie* (3rd Int. Conf. on Crystal Growth, Properties, Real Structure, and Application), Aleksandrov, 1997, vol. 1, p. 333.

[73] Obuhowskij, V. V.; Ponath, H.; Strizhewsij, V. L. *Phys. Status Solidi B*. 1970, 41, 837-842.

[74] Mavrin, B. N.; Abramovich, T. E.; Sterin, Kh. E. *Phys. Solid State*. 1972, 14, 1810-1820.

[75] Kondilenko, N. I.; Korotkov, P. A. *Opt. Spektrosk*. 1982, 52, 554-559.

[76] Baran, T. J.; Botto, I. L.; Nuto, F.; Kumada, N.; Kinomura, N. *J. Mat. Sci. Lett.* 1986, 5, 671-679.
[77] Sidorov, N. V.; Palatnikov, M. N.; Kalinnikov, V. T. *Opt. Spektrosk.* 1982, 1, 38-45.
[78] Donnerberg, H. J.; Tomlinson, S. M.; Catlow, C. R. A. *J. Phys. Chem. Solids.* 1991, 52, 201-210.
[79] Lyi, N.; Kitamura, K.; Izumi, F.; Yamamoto, I. K.; Hayashi, T.; Asano, H.; Kimura, S. *J. Solid State Chem.* 1992, 101, 340-350.
[80] Menta, A.; Navrotski, A.; Kumara, N.; Kinomura, N. *J. Solid State Chem.* 1993, 102, 213-219.
[81] Kumara, N.; Ozawa, N.; Muto, F.; Linomura, N. *J. Solid State Chem.* 1985, 57, 267-275.
[82] Tsivilev, R. P.; Fedulov, S. A.; Nezamaeva, M. F. *Bull. Rus. Acad. Sciences Inorg. Mater.* 1970, 6, 1539-1548.
[83] Agulyanskii, A. A.; Serebryakov, Yu. A.; Korobeinikov, L. S.; Balabanov, Yu. I.; Agulyanskaya, L. A.; Kalinnikov, V. T. *Rus. J. Gen. Chem.* 1986, 4, 734-741.
[84] Shimada, S.; Kodaira, K.; Matsushita, T. *Thermochim. Acta.* 1978, 23, 135-139.
[85] Bocharova, N. G., *Cand. Sci. Dissertation*, Moscow, 1986.
[86] Sidorov, N. V.; Serebryakov, Yu. A.; Lebold, V. V. *J. Appl. Spectrosc.* 1992, 56, 319-327.
[87] Sidorov, N. V.; Serebryakov, Yu. A. *Vibr. Spectrosc.* 1994, 6, 215-221.
[88] Semenov, A. E.; Cherkasov, E. V. *Rus. J. Phys. Chem. A.* 1980, 54, 2600-2611.
[89] Gorelik, V. S.; Reznik, L. G.; Umarov, B. S.; Gabrielyan, V. T. *Phys. Solid State.* 1983, 25, 1836-1842.
[90] Kustova, G. N.; Yurichenko, E. N. *Modern Vibrational Spectroscopy of Inorganic Compounds;* Nauka; Novosibirsk, Russia, 1990; p. 84.
[91] Zotov, N.; Boysen, H.; Frey, F.; Metzger, E. *J. Phys. Chem. Solids.* 1995, 55, 145-148.
[92] Neur Gamkar, R. R.; Lim, T. E.; Staples, E. J. *Ferroelectrics.* 1980, 27, 63-78.
[93] Andropov, P.; Kimura, S.; Sawada, T.; *Abstracts of Papers, 16 Congr. Int. Union Crystallogr.,* Beijing, 1992.
[94] Sidorov, N. V.; Serebryakov, Yu. A. *Abstracts of Papers, 2 Mezhdunarodnaya konferentsiya: Real'naya struktura i svoistva atsentrichnykh kristallov* (2nd Int. Conf. on Real Structure and Properties of Acentric Crystals), Aleksandrov, 1995, p. 338–356.
[95] Volk, T. R.; Wohlecke, M. *Lithium Niobate: Defects, Photorefraction, and Ferroelectric Switching*; Springer: Berlin, Germany, 2008; p. 78.
[96] Sidorov, N. V.; Yanichev, A. A.; Chufyrev, P. G.; Mavrin, B. N.; Palatnikov, M. N.; Kalinnikov, V. T. *Dokl. Chem.* 2009, 428, 492-496.
[97] Sturman, B. I.; Fridkin, V. M. *Photovoltaic Effect in Media without Symmetry Center and Related Phenomena;* Nauka: Moscow, Russia, 1992; p. 208.
[98] A. A. Blistanov, V. M. Lyubchenko and A. N. Goryunova. *Crystallogr. Rep.* 43 (1), 86 (1998).
[99] Maksimenko, V. A.; Syuy, A. V.; Karpets, Yu. M. *Photoinduced Processes in Lithium Niobate Crystals*; Fizmatlit: Moscow, Russia, 2008; p. 95.
[100] Palatnikov, M. N.; Shcherbina, O. B.; Efremov, V. V.; Sidorov, N. V.; Kalinnikov, V. T. *Inorg. Mater.* 2010, 46, 479-289.
[101] Goncharov, A. F.; Denisov, V. N.; Mavrin, B. N.; Podobedov, V. B. *J. Exp. Theor. Phys.* 1990, 94, 321-329.

[102] Sidorov, N. V.; Antonycheva, E. A.; Syuy, A. V.; Palatnikov, M. N. *Crystallogr. Rep.* 2010, 55, 765-773.

[103] Antonycheva, E. A.; Syuy, A. V.; Sidorov, N. V.; Chufyrev, P. G.; Yanichev, A. A. *J. Appl. Spectrosc.* 2010, 77, 89-95.

[104] Blistanov, A. A. *Crystals of Quantum and Nonlinear Optics;* MISIS: Moscow, Russia, 2000; p. 431.

[105] Huidnard, J.-P. *Photorefractive Materials and Their Applications*; Springer: Berlin, Germany, 2007; p. 421.

[106] Chernaya, T. S.; Maksimov, B. A.; Volk, T. R.; Rubinina, N. M.; Simonov, V. I. *J. Exp. Theor. Phys. Lett.* 2003, 73, 110-120.

[107] Chernaya, T. S.; Volk, T. R.; Verin, I. A.; Simonov, V. I. *Crystallogr. Rep.* 2008, 53, 573-578.

[108] Xin, F.; Zhang, G.; Bo, F.; Sun, H.; Kong, Y.; Xu, J.; Volk, T. R.; Rubinina, N. M. *J. Appl. Phys.* 2010, 107, 033113-033119.

[109] Rodriguez, M.; Cantelar, E.; Cusso, F.; Domingo, C. *Opt. Express.* 2010, 18, 5449-5458.

[110] Jermann, F.; Simon, M.; Kratzig, E. *J. Opt. Soc. Am.* 1995, 12, 2066-2071.

[111] Nevado, R.; Lifante, G. *Appl. Phys. A* 2001, 72, 725-738.

[112] Nevado, R.; Sada, C.; Segato, F.; Caccavale, F.; Kling, A.; Soares, J.; Cantelar, E.; Cusso, F.; Lifante, G. *Appl. Phys. B.* 2001, 73, 555-564.

[113] Zhang, Y.; Xu, Y. H.; Li, M. N.; Zhao, Y. Q. *J. Cryst. Growth.* 2001, 233, 537-543.

[114] Zhao, L.; Wang, X.; Wang, B.; Wen, W.; Zhang, T. *Appl. Phys. B.* 2004, 78, 769-776.

[115] Sidorov, N. V.; Yanichev, A. A.; Chufyrev, P. G.; Palatnikov, M. N.; Mavrin, B. N.; Bormanis, K. *Integr. Ferroelectr.* 2009, 109, 48-56.

[116] Semenov, A. E.; Cherkasov, E. V. *Rus. J. Phys. Chem. A.* 1980, 54, 2600-2615.

[117] Ridah, A.; Bourson, P.; Fontana, M. D.; Malovichko, G. *J. Phys. Condens. Mater.* 1997, 9, 9687-9696.

[118] Sun, M.-L.; Chia, C.-T.; Hu, M.-L.; Chang, J. Y.; Tse, W.-S.; Yang, Z. P.; Cheng, H. H. *Appl. Phys.* 2004, 78, 355-362.

[119] Sandler, V. A.; Sidorov, N. V.; Palatnikov, M. N. In: *Dielectric Crystals: Symmetry and Properties;* Kalinnikov, V. T.; Ed.; Izdatelstvo KSC RAN: Apatity, Russia, 2010; Vol. 2, p. 186.

[120] Syui, A. V.; Litvinova, M. N.; Goncharova, P. S.; Sidorov, N. V.; Palatnikov, M. N.; Krishtop, V. V.; Likhtin, V. V. *Tech. Phys.* 2013, 83, 109-118.

[121] Palatnikov, M. N.; Sidorov, N. V.; Biryukova, I. V.; Shcherbina, O. B.; Kalinnikov, V. T. *Inorg. Mater. Appl. Res.* 2011, 2, 93-108.

[122] M. N. Palatnikov, *Doctoral Dissertation* (Apatity, 2010).

[123] Abdi, F.; Aillerie, M.; Fontana, M.; Bourson, P.; Volk, T.; Maximov, B.; Sulyanov, S.; Rubinina, N.; Wohlecke, M. *Appl. Phys. B.* 1999, 199, 795-803.

[124] Umarov, M.; Gruzinenko, V.; Vtyurin, A. N.; Khodzhibaev, A. *Komponenty Tekhnol.* 2010, 6, 138-145.

[125] Anikev, A. A.; Gorelik, V. S.; Umarov, B. S. *Preprint Phys. Inst. RAS.* 1984, 154, 3-7.

[126] Atabaev, Sh.; Polivanov, Yu. N. *Phys. Solid State.* 1987, 29, 1165-1178.

[127] Barker, A. S.; London, R. *Phys. Rev.* 1967, 158, 433-440.

[128] Claus, R.; Borwel, G.; Wiesendanger, E.; Steffan, L. J. *Nature Res.* 1972, 27, 1186-1192.

[129] Anikev, A. A.; Gorelik, V. S.; Umarov, B. S. *Preprint Phys. Inst. RAS*. 1984, 38, 8-12.
[130] Anikev, A. A.; Sidorov, N. V.; Serebryakov, Yu. A. *J. Appl. Spectrosc*. 1992, 56, 670-679.
[131] Sidorov, N. V.; Serebryakov, Yu. A. *Ferroelectrics*, 1994, 160, 101-110.
[132] Palatnikov, M. N.; Sidorov, N. V.; Kalinnikov, V. T. *Rus. J. Non-ferrous Met*. 2000, 10, 54-61.
[133] Serebryakov, Yu. A.; Sidorov, N. V.; Palatnikov, M. N. *Ferroelectrics*, 1995, 167, 181-189.
[134] Serebryakov, Yu. A.; Sidorov, N. V.; Palatnikov, M. N. *Inorg. Mater*. 1992, 28, 1988-1991.
[135] Sidorov, N. V.; Palatnikov, M. N.; Serebryakov, Yu. A. *Ferroelectrics*, 1996, 188, 31-42.
[136] Sidorov, N. V.; Serebryakov, Yu. A. *Abstracts of Papers, 2 Mezhdunarodnaya konferentsiya: Real'naya struktura i svoistva atsentrichnykh kristallov* (2nd Int. Conf. on Real Structure and Properties of Acentric Crystals), Aleksandrov, 1995, p. 327.
[137] Zanadvorov, P. M.; Lebedeva, E. L.; Lebold, V. V. *Abstracts of Papers, 13 Konferentsiya po fizike segneto-elektrikov* (13th Conf. on Physics of Ferroelectrics), Tver, 1992, vol. 2, p. 29.
[138] Zanadvorov, P. M.; Lebedeva, E. L.; Moldavskaya, V. M.; Stepanov, Yu. A. *Phys. Solid State*. 1986, 26, 2823-2832.
[139] Kokanyan, E. P.; Lebedeva, E. L.; Moldavskaya, V. M. *Phys. Solid State*. 1986, 26, 2572-2578.
[140] Zanadvorov, P. M.; Lebedeva, E. L.; Kokanyan, E. P.; *Phys. Solid State*. 1986, 30, 2015-2028.
[141] Sandler, V.; Igoshin, I.; Serebryakov, Yu.; Palatnikov, M. *Collected, Abstracts, 8th Eur. Meet. on Ferroelectricity*, Nijgemen, 1995.
[142] Dmitriev, V. G.; Tarasov, L. V. *Applied Nonlinear Optics*; Fizmatlit: Moscow, Russia, 2004; p. 512.
[143] Akhmanov, S. A.; Nikitin, S. Yu. *Physical Optics*; Izdatelstvo MSU: Moscow, Russia, 2004, p. 656.
[144] Antipov, V. V.; Blistanov, A. A.; Sorokin, N. G.; Chizhikov, S. I. *Crystallogr. Rep*. 1985, 30, 734-745.
[145] Ito, H.; Takyu, C.; Inaba, H. *Electron. Lett*. 1991, 27, 1221-1234.
[146] Magel, G. A.; Fejer, M. M.; Byer, R. L. *Appl. Phys. Lett*. 1990, 56, 108-120.
[147] Shur, V. Ya.; Rumyantsev, E. L.; Bachko, R. G. *Phys. Solid State*. 1999, 41, 1831-1840.
[148] Naumova, I. I. *Crystallogr. Rep*. 1994, 39, 1119-1130.
[149] I. I. Naumova, N. F. Evlanova, O. A. Gilko, and Lavrishchev, J. *Cryst. Growth* 181, 160-164 (1997).
[150] Naumova, I. I.; Gilko, O. A. *Crystallogr. Rep*. 1996, 41, 712-713.
[151] Bermudez, V.; Serrano, M. D.; Dieguez, E. *J. Cryst. Growth*. 1999, 200, 185-193.
[152] Palatnikov, M. N.; Shcherbina, O. B.; Biryukova, I. V.; Sidorov, N. V. *Ferroelectrics*. 2008, 374, 41-47.
[153] Palatnikov, M. N.; Sandler, V. A.; Sidorov, N. V.; Guryanov, A. V.; Kalinnikov, V. T. *Phys. Solid State*. 2000, 42, 1456-1468.

[154] Palatnikov, M. N.; Shcherbina, O. B.; Biryukova, I. V. *Vestn. Kolsk. Nauchn. Tsentr.* 2010, 3, 40–46.

[155] Palatnikov, M. N.; Shcherbina, O. B.; Sidorov, N. V. *Crystallogr. Rep.* 2010, 55, 811 – 814.

[156] Palatnikov, M. N.; Shcherbina, O. B.; Efremov, V. V. *Inorg. Mater.* 2010, 46, 418–423.

[157] Sidorov, N. V.; Yanichev, A. A.; Chufyrev, P. G.; Mavrin, B. N.; Palatnikov, M. N.; Kalinnikov, V. T. *Dokl. Chem.* 2009, 428, 492-496.

[158] Sidorov, N. V.; Syui, A. V.; Palatnikov, M. N. *Opt. Spectrosc.* 2011, 110, 864–870.

[159] Sidorov, N. V.; Syui, A. V.; Palatnikov, M. N.; Kalinnikov, V. T. *Dokl. Phys. Chem.* 2011, 437, 47–49.

[160] Huidnard, J.-P. *Photorefractive Materials and Their Applications*; Springer: Berlin, Germany, 2007; p. 365.

[161] V. V. Obukhovskii, *Doctoral Dissertation* (Kievsk, Gos. Univ., 1989).

[162] Palatnikov, M.; Pikoul, O.; Sidorov, N.; Makarova, O.; Bormanis, K. *Ferroelectrics.* 2012, 436, 19-28.

[163] Palatnikov, M. N.; Biryukova, I. V.; Masloboeva, S. M.; Makarova, O. V.; Kravchenko, O. E.; Yanichev, A. A.; Sidorov, N. V. *Inorg. Mater.* 2013, 49, 765-773.

[164] Pikoul, O. Yu.; Alekseeva, L. V.; Povkh, I. V.; Stroganov, V. I.; Rudoi, K. A.; Tolstov, E. V.; Krishtop, V. V. *Izv. Vyssh. Uchebn. Zaved.* 2004, 12, 53-68.

[165] Pikoul, O. Y. *J. Appl. Crystallogr.* 2010, 43, 949-956.

[166] Razdobarin, A. G.; Basun, S. A.; Bursian, V. E.; Sochava, L. S.; Evans, D. R. *Phys. Solid State.* 2010, 52, 656-664.

[167] Sidorov, N. V.; Palatnikov, M. N.; Kalinnikov, V. T. *Dokl. Phys. Chem.* 2011, 441, 215-218.

[168] Sidorov, N. V.; Syui, A. V.; Palatnikov, M. N.; Kalinnikov, V. T. *Dokl. Phys. Chem.* 2011, 437, 47-49. (similar link 159).

[169] Zhong, G. G.; Jin, J.; Wu, Z. K. *Opt. Lett.* 1994, 19, 933-944.

[170] Gabrielyan, V. T.; Lebedeva, E. L.; Pirozerski, A. L.; Normatov, S. A. *Ferroelectrics.* 2002, 281, 151-159.

[171] Sweeney, K. L.; Halliburton, L. E.; Bryan, D. A.; Rice, R. R.; Gerson, R.; Tomachke, H. E. *J. Appl. Phys.* 1985, 57, 1036-1048.

[172] Fedorova, E. P.; Aleshina, L. A.; Sidorov, N. V.; Chufyrev, P. G.; Yanichev, A. A.; Palatnikov, M. N.; Voskresenskii, V. M.; Kalinnikov, V. T. *Inorg. Mater.* 2010, 46, 247-252.

[173] Stoiber, R.; Morse, S. *Microscopic Identification of Crystals*; The Ronald Press Company: New York, US, 1972; p. 278.

[174] Veiras, F. E.; Garea, M. T.; Perez, L. I. *Appl. Opt.* 2012, 51, 3081-3090.

[175] Mamedov, N.; Shim, Y.; Yamamoto, N. *Jpn. J. Appl. Phys.* 2005, 44, 754-760.

[176] Dumitrascu, L.; Dumitrascu, I.; Dorohoi, D. O.; Subbarao, E. C.; Shirane, G.; Jona, F. *J. Appl. Cryst.*, 2009, 42, 878-884.

[177] Pikoul, O. Y. *J. Appl. Cryst.* 2010, 43, 949-954.

[178] Rudoi, K. A.; Nabatov, B. V.; Stroganov, V. I.; Konstantinova, A. F.; Alekseeva, L. V.; Evdishchenko, E. A.; Kidyarov, B. I. *Crystallogr. Rep.* 2003, 48, 300-304.

[179] Geday, M. A.; Glazer, A. M. *J. Appl. Cryst.*, 2002, 35, 185-190.

[180] Gao, Ch. Y.; Xia, H. R.; Xu, J. Q.; Si, Sh. Ch.; Zhang, H. J.; Wang, J. Y.; Song, H. L. *Cryst. Res. Tech.*, 2007, 42, 1126-1131.

[181] Wang, P. Q. *Opt. Lett.* 2012, 37, 4392-4394.
[182] Kar, S.; Rajeev, B.; Gurvinderjit, S.; Gaurav, G.; Bartwal, K. S. *Proceeding of DAE-BRNS National Laser Symposium IIT,* Kharagpur, December 2003.
[183] Shtukenberg, A. G.; Punin, Y. O. *Optical Anomaliesin Crystals*; Science: Saint-Petersburg, Russia, 2004; p. 262.
[184] Kokhanchik, L. S.; Palatnikov, M. N.; Shcherbina, O. B. *J. Surf. Invest.* X-ray Synchrotron Neutron Tech., 2010, 9, 42-48.
[185] Simon, M.; Jermann, F.; Kraetzig, E.; Volk, T. *Phys. Stat. Soll. A.* 1995, 149, 723-728.
[186] Iyi, N.; Kitamara, K.; Izumi, F.; Yomanato, J. K.; Hayashi, T.; Asano, H.; Kimura, S. *J. Solid State Chem.* 1992, 101, 340-352.
[187] Gunter, P.; Huignard, J. P. *Photorefractive Materials and Their Applications.* Basic Effects; Springer: Paris, France, 2007; p. 640.
[188] Guilbert, L. *Opt. Express.* 2009, 17, 10782-10785.
[189] Simon, M.; Jermann, F.; Krätzig, E. *Opt. Mater.* 1994, 3, 243-250.
[190] Antonycheva, E. A.; Sidorov, N. V.; Syuy, A. V.; Chufyrev, P. G.; Yanichev, A. A. *Inorg. Mater. Appl. Res.* 2010, 5, 36-40.
[191] Goulkov, M.; Imlau, M.; Woike, Th. *Phys. Rev. B.* 2008, 77, 235110-235118.
[192] Volk, T.; Rubinina, N.; Wöhlecke, M. *J. Opt. Soc. Am. B.* 1994, 11, 1681-1687.
[193] Furukawa, Y.; Kitamura, K.; Takekawa, S.; Miyamoto, A.; Terao, M.; Suda, N. *Appl. Phys. Lett.* 2000, 77, 2494-2496.
[194] Malovichko, G. I.; Grachev, V. G.; Kokanyan, E. P.; Schirmer, O. F.; Betzler, K.; Gather, B.; Jermann, F.; Klauer, S.; Schlarb, U.; Wöhlecke, M. *Appl. Phys. A.* 1993, 56, 103-108.
[195] Furukawa, Y.; Sato, M.; Kitamura, K.; Yajima, Y.; Minakata, M. *J. Appl. Phys.* 1992, 72, 3250-3255.
[196] Palatnikov, M. N.; Sidorov, N. V.; Biryukova, I. V.; Chyfyrev, P. G.; Kalinnikov, V. T. *Inorg. Mater. Appl. Res.* 2003, 10, 258-365.
[197] Parfianovich, I. A.; Penzina, E. E., *Electronic Color Centers in Ionic Crystals; Vostochno-Sibirskoe Knizhnoe Izdatelstvo*: Irkutsk, Russia, 1977; p. 208.
[198] Bollman, W.; Gernand, M. *Phys. Status Solidi A.* 1972, 9, 301-308.
[199] Soroka, V. B.; Khromova, N. N.; Klyuev, V. P. *J. Appl. Spectrosc.* 1974, 20, 541–543.
[200] Vartanyan, E. S.; Ovsepyan, R. K.; Pogosyan, A. R.; Timofeev, A. L. *Phys. Solid State.* 1984, 26, 2418–2423.
[201] Mironov, S. P.; Akhmadulin, I. Sh.; Golenishchev-Kutuzov, V. A.; Imitative, S. A. *Phys. Solid State.* 1995, 37, 3179–3181.
[202] Palatnikov, M. N. *Inorg. Mater.* 2008, 44, 538–541.
[203] Palatnikov, M. N.; Biryukova, I. V.; Sidorov, N. V. *J. Cryst. Growth.* 2006, 291, 390–397.
[204] Hong, Xi. Z.; Guan, X. Ch.; Xuebin, Ch.; Yuheng, X. *J. Chim. Ceram. Soc.* 1991, 19, 523-530.
[205] Feng, X.-Q.; Tang, T. B. *J. Phys.: Condens. Mater.* 1993, 5, 2423-2430.
[206] Zhiling, K. K. *Bull. Rus. Acad. Sciences Phys.* 1997, 61, 327-334.
[207] Prieto, C.; Zaldo, C.; Fessler, R.; Dexpert, H.; Sanz-Garcia, J. A.; Dieguez, E. *Phys. Rev. B.* 1991, 43, 2594-2608.
[208] Volk, T. R.; Rubinina, N. M. *Phys. Solid Stat.* 1991, 33, 1192-1201.
[209] Wang, H.; Wen, J.; Li, B.; Wang, H. *Phys. Status Solidi A.* 1990, 118, 47-53.

[210] Minoz Santinste, J. E.; Mocalik, B.; Garsia, S. *J. Phys. Rev. B.* 1993, 47, 88-95.
[211] McDonald, P. B.; Tam, C. P.; Mok, Y. W. *J. Chem. Phys.* 1972, 56, 1007-1019.
[212] Dishler, B.; Herington, J. R.; Rauber, A.; Schneider, J. *Solid State Commun.* 1973, 12, 737-741.
[213] Lorenzo, A.; Jaffezic, H.; Roux, B.; Boulon, G.; Garsia-Sole, J. *Appl. Phys. Lett.* 1995, 67, 3735-3746.
[214] Grachev, V. G.; Malovichko, G. I.; Troitskii, V. V. *Phys. Solid State.* 1987, 29, 607-615.
[215] Rebonta, L.; Soares, J. C.; Da Silva, M. F.; Sanz-Garcia, J. A.; Diequez, E.; Agullo-Lopez, F. *Appl. Phys. Lett.* 1989, 55, 120-129.
[216] Palatnikov, M. N.; Sidorov, N. V.; Skiba, V. I. *Inorg. Mater.* 2000, 36, 315-324.
[217] Nassau, K.; Lines, M. E. *J. Appl. Phys.* 1970, 41, 533-548.
[218] Grabmaier, B. C.; Otto, F. *J. Cryst. Growth.* 1986, 79, 682-696.
[219] Kalinnikov, V. T.; Palatnikov, M. N.; Sidorov, N. V. *Vestn. Kolsk. Nauchn. Tsentr.* 2003, 48, 1-31.
[220] Shimura, F.; Fujino, Y. *J. Cryst. Growth.* 1977, 38, 239-249.
[221] Fukuda, T.; Hirano, H. *J. Cryst. Growth.* 1976, 35, 127-134.
[222] Kobayashi, T.; Muto, K.; Kai, J.; Kawamori, A. *J. Magn. Reson.* 1979, 34, 459-468.
[223] Parsonage, N.; Staveley, L. *Disorder in Crystals*; Oxford University: Oxford, Great Britain, 1978; p. 926.
[224] Kamentsev, V. P.; Nekrasov, A. V.; Pedko, B. B. *Bull. Rus. Acad. Sciences Phys.* 1983, 47, 791-806.
[225] Roitberg, M. B.; Novik, V. K.; Gavrilova, N. D. *Crystallogr. Rep.* 1969, 14, 938-946.
[226] Pedko, B. B.; Lebedev, E. V.; Kislova, I. L.; Volk, T. R. *Phys. Solid State.* 1998, 40, 337-341.
[227] Blistanov, A. A.; Makarevskaya, E. V.; Geraskin, V. V. *Phys. Solid State.* 1978, 20, 2575-2583.
[228] Bagdasarov, Kh. S.; Bogdanov, M. Ya. *Phys. Solid State.* 1987, 29, 2380-2391.
[229] Pedko, B. B.; Lebedev, E. V. *Bull. Rus. Acad. Sciences Phys.* 1997, 61,321-332.
[230] Belabaev, K. G.; Gabrielyan, V. T.; Sarkisov, V. Kh. *Crystallogr. Rep.* 1973, 18, 198-210.
[231] Pedko, B. B.; Rudyak, V. M.; Shabalin, A. L. *Bull. Rus. Acad. Sciences Phys.* 1990, 54, 1154-1163.
[232] Palatnikov, M. N.; Sandler, V. A.; Sidorov, N. V.; Guryanov, A. V.; Kalinnikov, V. T. *Phys. Solid State.* 2000, 42, 1453-1464.
[233] Ming, N.; Hong, J.; Feng, D. *J. Mater. Sci.* 1982, 17, 1633-1642.
[234] Pashkov, V. A.; Solovev, N. M.; Uyukin, E. M. *Phys. Solid State.* 1979, 21, 1879-1884.
[235] Pashkov, I. A.; Solovev, N. M.; Angert, N. B. *Phys. Solid State*, 1979, 21, 92-103.
[236] Sweeney, K. L.; Halliburton, L. E.; Bryan, D. A. *J. Appl. Phys.* 1985, 57, 1036-1042.
[237] Schirmer, O. F.; Thieman, O.; Wohlecke, M. *J. Phys. Chem. Solids.* 1991, 52, 185-193.
[238] Akhmadullin, I. Sh.; Golenishchev-Kutuzov, V. A.; Migachev, S. A.; Mironov, S. P. *Phys. Solid State.* 1998, 40, 1307–1312.
[239] Kovalevich, V.; Shuvalov, L.; Volk, T. *Phys. Status Solidi,* 1978, 45, 249-251.
[240] Palatnikov, M.; Sandler, V.; Serebryakov, Yu. *Ferroelectrics.* 1996, 175, 183-194.
[241] Palatnikov, M. N.; Sandler, V. A.; Serebryakov, Yu. A. *Inorg. Mater.* 1995, 31, 96-102.

[242] Smolenskii, G. A.; Bokov, V. A.; Isupov, V. A.; Krainik, N. N.; Pasynkov, R. E.; Shur, M. S. *Ferroelectrics and Antiferroelectrics*; Nauka: Leningrad, Russia, 1971; p. 476.

[243] Sushchinskii, M. M. *Tr. Fiz. Inst. im. P.N. Lebedeva,* Akad. Nauk. SSSR. 1982, 132, 3-8.

[244] Borisov, V. N.; Pereverzeva, L. P. *Phys. Solid State*. 1985, 27, 3112-3120.

[245] Kamentsev V. P.; Nekrasov A. V.; Pedko, B. B. *Bull. Rus. Acad. Sciences Phys.* 1983, 47, 791–793.

[246] Roitberg, M. B.; Novik, V. K.; Gavrilova, N. D. *Crystallogr. Rep*. 1969, 14, 938–943.

[247] Bagdasarov, Kh. S.; Bogdanov, M. Ya. *Phys. Solid State*. 1987, 29, 2380–2387.

[248] Palatnikov, M. N.; Shcherbina, O. B.; Kazakov, A. A. *Inorg. Mater*. 2008, 44, 305–310.

[249] Naumova, I. I. *Crystallogr. Rep*. 1994, 39, 1119–1122.

[250] Ming, N.; Hong, J.; Feng, D. *J. Mater. Sci*. 1982, 17, 1663–1670.

[251] Voskresenskii, V. M.; Starodub, O. R.; Sidorov, N. V. *Crystallogr. Rep*. 2011, 56, 221–226.

[252] *Akoustic Crystals: A Handbook*; Shaskolskaya, M. P.; Ed.; Nauka: Moscow, Russia, 1982; p. 632).

In: Oxide Electronics and Functional Properties ... ISBN: 978-1-63321-499-6
Editor: Alexander Pergament

Chapter 3

SPUTTER DEPOSITED NANOLAMINATES CONTAINING GROUP IVB (Ti, Zr, Hf)-OXIDES: PHASE STRUCTURE AND NEAR BAND GAP OPTICAL ABSORPTION BEHAVIOR

Carolyn Rubin Aita*

Department of Chemistry and Biochemistry, University of Wisconsin-Milwaukee, Milwaukee, Wisconsin, US

ABSTRACT

A nanolaminate architecture was used to produce Group IVB transition metal oxide films with unique nanocrystal phase structures and tailored optical behavior at the fundamental optical absorption edge (FOAE). Five nanolaminate systems, ZrO_2-Al_2O_3, HfO_2-Al_2O_3, TiO_2-Al_2O_3, ZrO_2-TiO_2, and HfO_2-TiO_2, were grown by reactive sputter deposition on unheated substrates with a dissimilar oxide surface, SiO_2. Despite the far-from-equilibrium growth environment, bulk phase diagrams of the corresponding pseudobinary systems served as guides for predicting cation mixing at bilayer interfaces. However, unexpected and technologically-important structures resulting from finite crystal size effects were prevalent in the transition metal oxide nanocrystals, captured in a nanolaminate structure which limited nanocrystal growth. Elevated temperature and pressure phases, metastable at ambient conditions, were the rule rather than the exception. The optical absorption coefficient at the FOAE in all nanolaminates was successfully described by a persistence rather than an amalgamation model, although in some cases (TiO_2-Al_2O_3, ZrO_2-TiO_2, and HfO_2-TiO_2), accommodation was made for the electronic influence of one oxide on the other at a bilayer interface. The onset of the FOAE, i.e., the optical band gap, did not move across the energy spectrum between the extrema set by the individual constituents. Rather it was determined by the constituent with the lowest energy band gap. However, the energy at which significant absorption occurs can be tailored by adjusting the amount of each constituent in a bilayer.

* aita@uwm.edu.

I. INTRODUCTION

A nanolaminate film is a multilayer stack of different materials sequentially deposited on a substrate. These materials are "laminated" together at the atomic level. The thickness of an individual layer is in the nanometer scale range. The films that are the subject of this Chapter contain a bilayer repeat unit, Λ, schematically shown in Figure 1. One or more of the constituent layers is the transition metal oxide TiO_2, ZrO_2, or HfO_2, either partnered with another Group IVB oxide or with amorphous Al_2O_3. These oxides are wide band gap solids, and have found many optical and electronic applications as such. This Chapter addresses the relationship between the phase structure of the constituent oxides at the atomic level and the consequent ultraviolet optical absorption behavior of the nanolaminate vicinal to its electronic band gap.

The nanolaminates in this Chapter were grown by reactive sputter deposition. Figure 2 shows a schematic diagram of a reactor. Several reviews [1, 2] address this growth technique specifically in regard to nanolaminate deposition, and so it is only briefly summarized here. Reactive sputter deposition involves film growth on a substrate in contact with a low pressure glow discharge containing a reactive gas. Atoms (M) and simple molecules (MO_x) are ejected from a target surface under ion (R^+) bombardment and travel through the discharge to the substrate. The flux that arrives at the substrate consists of neutral species in ground and electronic excited states and a small population of ionic species, as well. This energetic flux arrives faster than the adsorbed flux can equilibrate with the substrate. Film growth is, in essence, quenching of these species into structures that may not predicted by equilibrium thermodynamics, that is, structures not found on a bulk binary oxide phase diagram at the growth temperature and pressure. Herein lies the beauty of reactive sputter deposition. It is ideally suited for growth of artificial but technologically important non-equilibrium structures including high melting point phases near room temperature, metastable phases, and nanometer-scale layered structures with controlled interfaces. This technique, in one its several modifications (diode, magnetron, triode, pulsed excitation), is widely used for oxide film deposition.

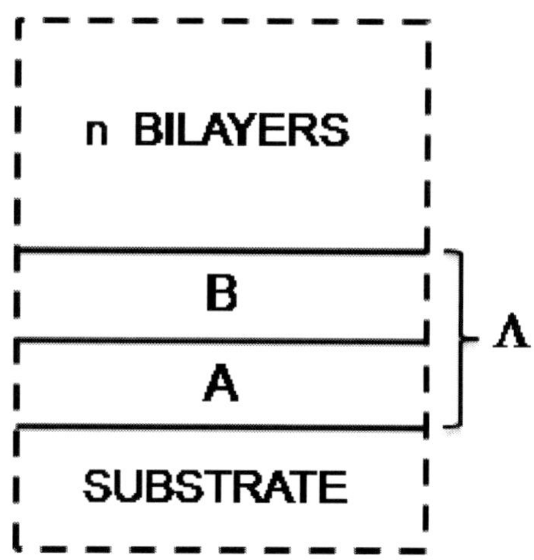

Figure 1. A nanolaminate with a bilayer repeat unit, Λ.

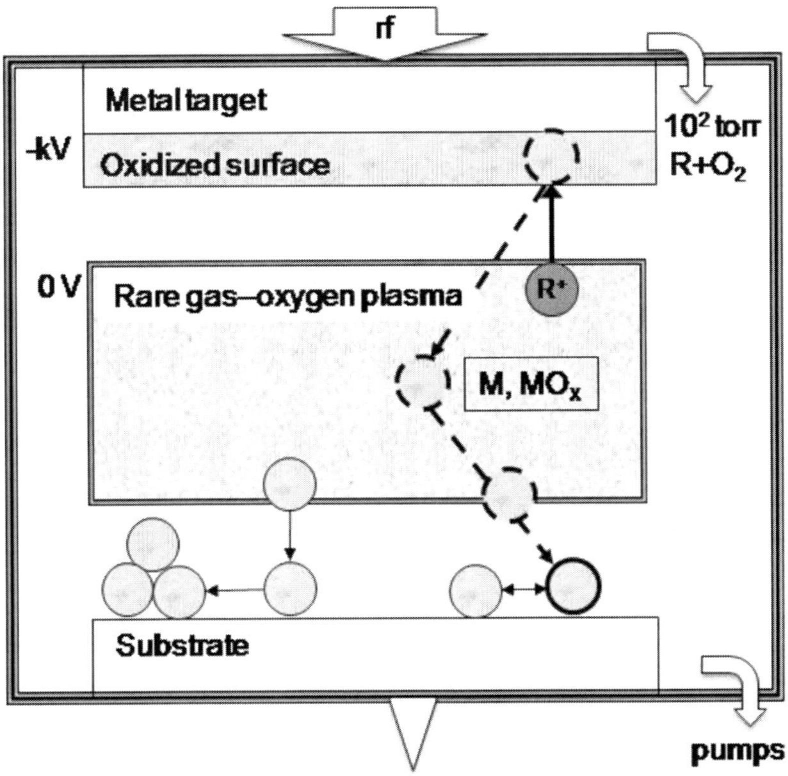

Figure 2. A sputter deposition reactor showing basic processes of film growth.

The nanolaminates discussed in this chapter were grown in an automated reactor by sequentially passing unheated substrates at the anode under sputtered metal targets (Ti, Zr, Hf, or Al) at the cathode. Kinetic processes at each electrode and in the gas phase determine phase formation in the film. Momentum transfer from incident ions leads to synthesis and reduction reactions at the target surface that determine the chemistry of the species sputtered into the plasma volume. Collisional and radiative processes in the plasma volume lead to gas phase ionization, excitation, association, and dissociation of the sputtered species. Adsorption, surface diffusion, bulk diffusion, and desorption of the sputtered flux at the substrate surface ultimately lead to cluster formation, coalescence, and ultimately, film formation. Although the physics and chemistry of the process is at first glance daunting, understanding and careful manipulation of the kinetics at each electrode enables the production of a film with desirable and reproducible properties and behavior tailored for a specific purpose.

The substrates on which the nanolaminates were grown are fused SiO_2 and crystalline Si from which the nascent Si-oxide had not been removed. These substrates present a noncrystalline oxide growth surface onto which the first nanolayer of the film is deposited. The fundamental physiochemical issue is, therefore, that of the growth of an overlayer on a dissimilar oxide under kinetically constrained conditions.

To address oxide growth under these conditions, we look to Felner [3] who introduced the concept of "structural complexity" in relation to vitreous oxides formed at room temperature. This concept is used here to understand nanocrystalline versus amorphous growth of a nanolaminate's constituents. Oxides with structural complexity have either bond flexibility giving rise to polymorphs with the same chemistry and very similar free energy of formation (type I), or form mixed valence compounds with vernier, block, or infinitely adaptive physical structures that accommodate changes in stoichiometry (type II). Oxides with structural complexity in general do not crystallize into single phase structures at their earliest growth stages. Al_2O_3 exhibits type I structural complexity; it is an amorphous constituent in the nanolaminates. Ti-oxides exhibit type II structural complexity. Consequently TiO_2 forms an intergrowth structure in the nanolaminates.

On the other hand, oxides that do not have structural complexity such as HfO_2 and ZrO_2 readily grow from the earliest stages as single phase nanocrystals. However, there are two important features that experimental results have found about the crystal structures of these oxides. First, their nanometer-scale crystallites are not the equilibrium phases predicted from the corresponding bulk phase diagrams, they are metastable phases. The term "metastable" is used here to designate a phase that is not the bulk equilibrium structure of a large crystal at standard temperature and pressure (STP). In general, metastable phases form in the nanolaminates due to various finite crystal size effects.

There are several ways in which the nanoscale size of a crystallite can affect its structure. The finite crystal size effect considers a balance between the surface and volume contributions to the Gibbs free energy of formation [4]. As the crystallite size becomes smaller, the ratio of the surface-to-volume free energy becomes larger, thus effecting a phase change. The Gibbs–Thomson effect considers the increased hydrostatic pressure above a curved surface compared to a flat surface [5]. Hence, high pressure phases can form in tiny crystallites. A surface dipole repulsion mechanism results in an expanded lattice and ultimately a phase change to a structure of higher symmetry with decreasing crystal size [6-9]. All of these effects are observed, and sometimes in combination, in the nanocrystalline transition metal oxides that are discussed in this Chapter.

A second feature of the crystallites in the nanolaminates is that they have a preferred orientation along the direction of arriving flux. This 'texturing" can be clearly observed by comparing the nanolaminates' diffraction patterns to those from bulk standards. Yet, often no in-plane structural relationship exists between nanocrystals. Bauer [10] has shown that this morphology is characteristic of Volmer-Weber growth (growth from initial three dimensional nuclei) when there is weak adsorbate-substrate interaction compared to the interaction among the adsorbed species. In that case, the low energy orientation of a crystallite is one in which the most densely packed planes are parallel to the substrate (or the underlayer), that is, these planes are stacked along the direction of the arriving flux. Hence, there is a thermodynamic driving force for a preferred orientation at the very inception of stable nucleus formation.

In Section II of this Chapter, the nanolaminates are classified based on the propensity for interfacial mixing and the growth of two-cation phases. This classification is important for understanding the as-grown interfacial structure, as well as interfacial cation-mixed structures that develop after intentional (and unintentional) post-deposition annealing. Although the size effects enumerated above make it difficult to predict intralayer binary oxide phases on the basis of bulk binary oxide phase diagrams, these diagrams are very useful for predicting interfacial mixing and the formation of new interfacial structures in the nanolaminates.

The nanolaminates that are the subject of this Chapter are wide electron band gap solids with transparency throughout the visible region of the optical spectrum. One of their uses is as variable band gap materials across the ultraviolet spectral region. To this end, optical behavior near the fundamental optical absorption edge (FOAE) is determined and transitions across the optical band gap are modeled. The formalisms used to determine the optical band gap and model the FOAE structure are given in Section III.

In addition, the transition metal dioxides discussed here are "d^0 oxides" [11]. The interband transitions that define the onset of the FOAE in these oxides are from O *2p* electron states at the top of the valence band to metal *nd* electron states at the bottom of the conduction band, where n = 3, 4, and 5 for Ti, Zr, and Hf, respectively. In their fully oxidized state, all valence electrons in these oxides fill metal-oxygen bonding orbits. Hence, there are "zero" electrons that are not in metal-oxygen bonding configurations. These *d* electron states form two "split-off" bands (that is, split-off from *(n+1) s* states at slightly higher energy in the conduction band). The localized nature of *d* electron states permits the use of either an ionic model or a band model (or anything in between) to describe the initial band-to-band transition in these oxides. An important consideration with respect to the nanolaminates is how this rather pristine bonding picture is disrupted when two dissimilar oxides are brought into intimate contact. Does the nature of each oxide persist in the nanolaminate or is there amalgamation to some degree, especially when the bilayer thickness is ultrathin? How does intralayer and interlayer structure enter into the picture? These questions are answered with respect to individual nanolaminates in Sections IV and V. Lastly, overarching conclusions drawn from the research are presented in Section VI.

II. Materials Classification

The nanolaminate compositions are next divided into three classes based on the miscibility of their pseudobinary constituents in bulk crystals. The classes are characterized by different thermodynamic driving forces for interfacial cation mixing and physical atomic registry (nanoscale heteroepitaxy or pseudomorphism). Interfaces play an increasingly important role in determining overall film properties as individual layer thicknesses decrease.

This classification relies on equilibrium thermodynamics, that is, information obtained from a pseudobinary phase diagram. Thermodynamics tells where a system wants to end up when rate considerations are no longer of consequence. Although the kinetic constraints imposed by a high arrival flux and low surface diffusion leave very little time for atomic arrangement into thermodynamically-favored structures, Tromp and Hannon [12] found that collective phenomena such as self-assembly in individual critical nuclei become possible on this timescale.

The classification scheme is as follows.

Class I: The binary oxides have complete immiscibility. There is no driving force for interfacial physical registry, such as heteroepitaxy, or chemical mixing. ZrO_2-Al_2O_3 and HfO_2-Al_2O_3 are class I nanolaminates.

Class II: The binary oxides have limited miscibility without a common end-member lattice. There is a driving force for interfacial chemical mixing but not for physical registry. ZrO_2-TiO_2, HfO_2-TiO_2, and TiO_2-Al_2O_3 are class II nanolaminates.

Class III: The binary oxides have complete miscibility. There is a driving force for both interfacial physical registry and chemical mixing, e.g., ZrO_2-HfO_2 nanolaminates.

This Chapter is concerned with classes I and II nanolaminates. The reason for placing a specific nanolaminate in one of these classes is given in Sections IV and V.

III. FORMALISMS FOR OPTICAL BEHAVIOR ANALYSIS

Spectrophotometry was used to determine optical transmission (T) and reflection (R) for the nanolaminates in the 190 to 1100 nm wavelength (λ) range. Transmission data was taken in single beam mode. An aluminum mirror with a reflectivity greater than 90% over the wavelength range examined here was used as a standard for reflection data. The absorption coefficient, α, was calculated for a film of thickness, x, in the short wavelength region where the contribution to α from reflection at the film-substrate interface are negligible, using the expression [13],

$$T = (1\text{-}R)\exp(-\alpha x). \tag{1}$$

The functional dependence of α on E is indicative of the types of optical transitions at the fundamental optical absorption edge (FOAE). In general, α depends on oscillator strength, *f*, and a density of states factor, D(E),

$$\alpha = C f D(E), \tag{2}$$

where C is a constant, and the real part of the index of refraction is assumed to vary only slightly over the energy range of interest. In the case of parabolic valence and conduction band edges with an optical band gap E_G, $D(E) = (E-E_G)^{1/2}$ for direct interband transitions [14], and $D(E) = (E-E_G)^2$ for indirect interband transitions [14] in the case of an atomically ordered solid, or non-direct interband transitions in the case of a disordered solid [15].

Three common assumptions for the behavior of *f* [16] appear in the literature. (1) If *f* is independent of E over the energy range of interest, then α is proportional to D(E). (2) α can also be expressed in terms of the momentum matrix element M_p for the transition. Since *f* is proportional to $|M_p|^2/E$, Eq. (2) shows that if M_p is independent of E over the energy range of interest, then αE is proportional to D(E). (3) α can also be expressed in terms of the electric dipole matrix element M_d for the transition. Since *f* is proportional to $E|M_d|^2$, Eq. (2) shows that if M_d is independent of E over the energy range of interest, then α/E is proportional to D(E).

By examining the α, αE, and α/E versus E curves with assumed forms for D(E), it is theoretically possible to decide which, if any, of the quantities *f*, M_p, or M_d, is approximately constant over the energy range of interest. However, in practice the differences between assumptions may be very small. The first assumption, α ~ D(E), is used in the following sections when FOAE behavior for specific nanolaminates is presented. Since the other two assumptions occasionally appear in the literature, we analyzed [17] the case of single layer, 0.2 μm-thick HfO_2 films to determine how close data obtained from these formalisms are to each other. It was not possible to determine which formalism offered a better fit to the experimental data since all assumed forms for *f* worked equally well over the energy range of interest. Furthermore, linear regression analysis showed only a small difference in E_G between forms over the 6.21 eV $\leq E \leq$ 6.54 eV region in which interband (O *2p* → Hf *5d* electron transitions) absorption initiated, with E_G = 5.51eV for $\alpha^{1/2}$ versus E, bracketed by E_G = 5.44 eV for $(\alpha/E)^{1/2}$versus E, and E_G = 5.55 eV for $(\alpha E)^{1/2}$ versus E.

The above discussion applies to a homogeneous film with one type of absorber. However, individual contributions to α from each constituent must be considered in the case of a nanolaminate. Volmer-Weber growth produces morphological roughness that precludes interfacial specular reflection, which greatly simplifies the analysis. But even if interfacial smoothness were the case, the thickness of individual layers for most of the architectures discussed in this Chapter is much smaller than one-quarter wavelength of the incident radiation in that material. Each constituent in a bilayer is therefore exposed to the same electric field, and the film appears as a single dielectric slab to the incident radiation. In other words, the nanolaminates do not behave as classical optical multilayer stacks with transmission and specular reflection at each interface, but as composite materials.

The absorption coefficient for a composite film of constituents A and B with total thickness, x, is written,

$$\alpha = (v_A\alpha_A + v_B\alpha_B), \tag{3}$$

where v_A and v_B are the path lengths through which radiation travels in each constituent, and α_A and α_B are the absorption coefficients of the pure solids, A and B. Equation (1) becomes,

$$T = (1\text{-}R)\ \exp[-(v_A\alpha_A + v_B\alpha_B)\ x]. \tag{4}$$

Expanding Eq.(4),

$$T = (1\text{-}R)\exp[-(v_A\alpha_A/x + v_B\alpha_B/x)], \tag{5}$$

where the quantities v_A/x and v_B/x are the fractional path lengths for each constituent. (These quantities will be replaced by the volume fraction of each constituent in Sections IV and V.)

Equations (3-5) are especially useful when α of one pure constituent is known and the other is unknown but of interest. For example, Eqs. (3-5) were applied to a ZrO_2-Al_2O_3 nanolaminate in Section IV.A to determine the two initial interband transitions in pure tetragonal ZrO_2 and the band gap of this illusive yet technologically-important phase [18].

The Beer-Lambert-Bouguer law [19], expressed by Eqs (1), (4), and (5) can be expanded to account for any number, i, of constituents, in a nanolaminate of known total thickness where,

$$\alpha = \Sigma_i\ (v_i\alpha_i). \tag{6}$$

Equation (6) was used to separate individual intralayer and interfacial contributions to the FOAE in class II nanolaminates, as discussed in Sec. V.

IV. CLASS I NANOLAMINATES: ZrO_2-Al_2O_3 AND HfO_2-Al_2O_3

The pseudobinary phase diagrams of the class I nanolaminates show that the constituent oxides are immiscible. Furthermore, there is mutual limited solid solubility of the metal atoms of one partner in the oxide partner. Class I nanolaminates are distinguished by extremely limited interfacial mixing in the as-grown state. One value of class I nanolaminates is that they can be used to isolate scientifically-interesting and technologically-useful metastable transition metal oxide phases that are formed by a finite crystal size effect, as shown next.

A. ZrO_2-Al_2O_3

Bulk pure ZrO_2 at atmospheric pressure crystallizes in three polymorphs: monoclinic (m-ZrO_2) stable to ~ 1448 K, tetragonal (t-ZrO_2) stable to 2633 K, and cubic (c-ZrO_2) stable to the liquidus at 2953 K [20]. However, t-ZrO_2 occurs at room temperature in small crystallites, and c-ZrO_2 occurs at room temperature by doping with M^{3+} ions [20].

Films on fused SiO_2 and the nascent oxide surface of <111>-Si were studied by x-ray diffraction (XRD) and high resolution transmission electron microscopy (HTREM) [18, 21-24]. The as-grown structure consisted of nanocrystalline ZrO_2 and amorphous Al_2O_3. The ZrO_2 phase composition changed from t-ZrO_2 to t+m-ZrO_2 to m-ZrO_2 with increasing ZrO_2 layer thickness. ZrO_2 nanocrystal growth was oriented such that t-ZrO_2 {111} and m-ZrO_2 (11-1) planes grew perpendicular to the direction of the incoming flux, that is, parallel to the substrate in the absence of layer roughening. These are the closest-packed orientations, and are expected when there is weak adsorbate-substrate interaction compared to the interaction among adsorbed species [10].

ZrO_2 nanocrystal size in the growth direction was calculated from XRD line broadening using the Scherrer equation [25]. Figure 3a shows the average t-ZrO_2 (<r(t)>) and m-ZrO_2 (<r(m)>) nanocrystal size as a function of ZrO_2 layer thickness for a series of films in which the ZrO_2 layer thickness ranged from 5 to 30 nm and Al_2O_3 layer thickness was held constant at 4 nm. Several observations can be made. First, <r(t)> saturated at 6.0±0.2 nm, whereas <r(m)> continued to increase as a function of ZrO_2 layer thickness. Second, m-ZrO_2 was not present in the thinnest ZrO_2 layers, that is, in the smallest nanocrystals. Third, the appearance of m-ZrO_2 was concurrent with the saturation of t-ZrO_2 nanocrystal size.

HTREM showed that each oxide layer was a separate entity, having incoherent interfaces with adjacent layers [23, 24]. For ZrO_2 layers that were less than a critical thickness, r_c (discussed below), nanocrystals were approximately rectangular in shape with a height r in the growth direction and a base 2r in the substrate plane. Figure 3b [24] shows a t-ZrO_2 nanocrystal in the first layer deposited on the nascent oxide of <111> Si, topped by the first amorphous Al_2O_3 layer.

The behavior shown in Figure 3a is consistent with a finite crystal size effect responsible for the formation of t-ZrO_2. Figure 4 shows a hemispherical cap nanocrytal making a contact angle Θ with the underlying substrate. There are four energy terms to consider: g is the volume free energy of the nanocrystal, γ_{fv} is the surface free energy of the nanocrystal exposed to the vapour, γ_{sv} is the surface free energy of the substrate exposed to the vapour, and γ_{fs} is the surface free energy at the nanocrystal/substrate interface. The value of $\Theta = \pi/2$

occurs when $\gamma_{sv}=\gamma_{fs}$, that is, the nanocrystal neither wets nor balls-up on the substrate. That case applies to nanocrystals whose dimension in the substrate plane is twice that of their dimension perpendicular to the substrate plane. The change in the Gibbs free energy accompanying a phase change that preserves this geometry is given by $\Delta G=Ar^3\Delta g+Br^2\Delta\gamma_{fv}$, where A and B are geometric factors. At the point of transformation, $\Delta G = 0$, and the corresponding critical dimension, $r = r_c$ can be calculated from a balance of the volume and surface energy terms. For the nanocrystal geometry here, such an end-point thermodynamics calculation [22] yielded,

$$r_c=3.79[1-(T/1448\ K)]^{-1}(nm). \tag{7}$$

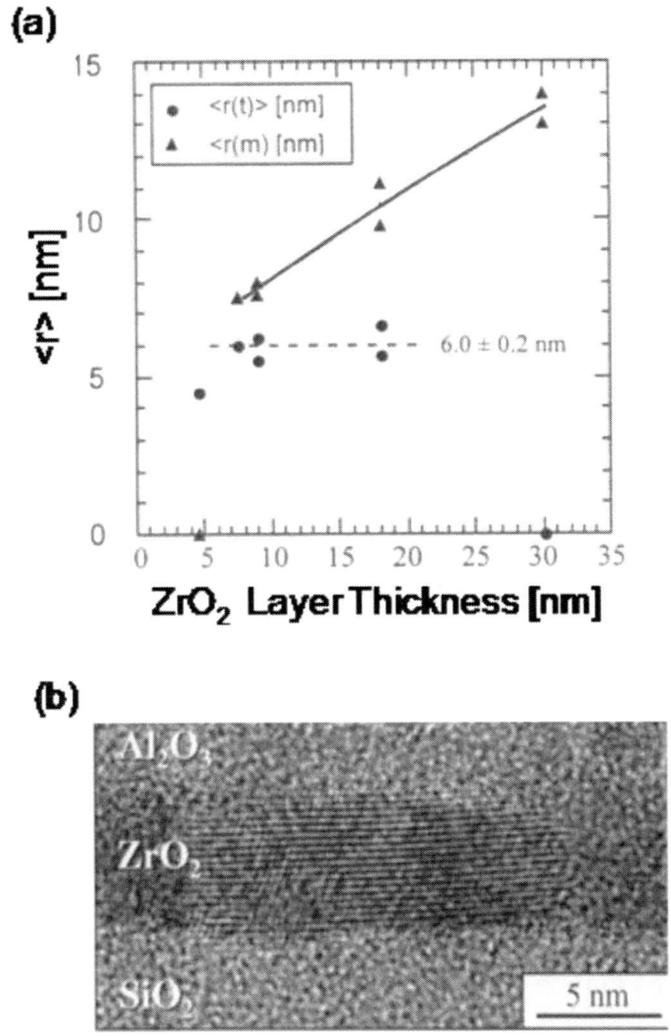

Figure 3. (a) The average t-ZrO_2 (<r(t)>) and m-ZrO_2 (<r(m)>) nanocrystal size as a function of ZrO_2 layer thickness in ZrO_2-Al_2O_3 nanolaminates on fused SiO_2. The Al_2O_3 layer thickness was 4 nm. (b) A t-ZrO_2 nanocrystal and amorphous Al_2O_3 in the first bilayer grown on the nascent oxide of <111> Si [24].

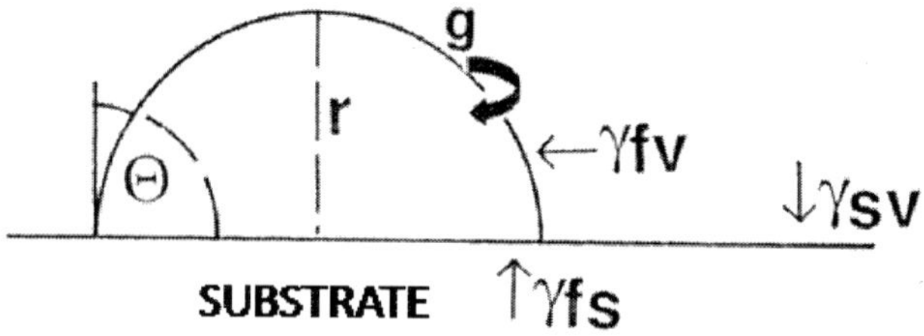

Figure 4. A hemispherical cap nanocrytal making a contact angle Θ with the underlying substrate. g, γ_{fs}, γ_{sv}, and γ_{fv} are the volume free energy, and the interfacial film-substrate, substrate-vapor, and film-vapor free energies, respectively.

A value of r_c = 6.2 nm calculated from Eq. (7) at the growth temperature is in excellent agreement with the XRD experimental data presented in Figure 3a. These results demonstrate that solely t-ZrO_2 can be produced if a ZrO_2 layer thickness $<r_c$ is used. The function of the Al_2O_3 layers was to terminate t-ZrO_2 nanocrystal growth before reaching the critical size for transformation to m-ZrO_2.

We emphasize that pure t-ZrO_2 cannot be produced in a single layer film or in a bulk solid at STP. By "capturing" t-ZrO_2 an Al_2O_3 matrix, the FOAE of this illusive solid can be experimentally determined [18]. t-ZrO_2 is a d^0 oxide [11] in which characteristic features on the FOAE are determined at the nearest-neighbor coordination level, that is in the framework of a cluster model, by charge transfer within a $(ZrO_8)^{-12}$ cluster from an O *2p* to a Zr *4d* electron state [26-28]. Tauc [29] generalized this presumption by stating that interband optical transitions that can be described by wave functions localized over distances on the order of the lattice constant are relatively unchanged by disorder.

Figure 5 shows the FOAE calculated from experimental transmission and reflection data using Eq. (1) for a nanolaminate consisting of fifteen 5 nm t-ZrO_2-5 nm Al_2O_3 bilayers. Experimental data obtained for a single layer Al_2O_3 film is also shown. The third curve in Figure 5, α versus E for t-ZrO_2, was obtained using Eq. (3) using α(nanolaminate) = 0.5α (t-ZrO_2) + 0.5α (Al_2O_3).

The functional dependence of α on E was next examined to determine the nature of the transitions across the band gap near the FOAE of t-ZrO_2. Two transitions were identified. Figure 6a shows an initial indirect interband transition, affirming *ab initio* calculations for this solid [27, 28, 30-32]. Linear regression analysis yields an E_i = 5.22 eV. This initial indirect transition is followed by a direct interband transition with an onset at E_d = 5.87, shown in Figure 6b.

B. HfO_2-Al_2O_3

Bulk pure HfO_2 at atmospheric pressure crystallizes in three polymorphs: monoclinic (m-HfO_2) stable to 2000 K, tetragonal (t-HfO_2) stable to 2870 K, and cubic (c-HfO_2) stable to the liquidus at 3073 K [33-35]. This phase evolution with increasing temperature is similar to that of ZrO_2, the difference being that the phase transitions occur at a higher temperature in HfO_2. In addition, bulk m-HfO_2 transforms to a sequence of orthorhombic phases with increasing pressure [34-37], one of which becomes important in the following discussion.

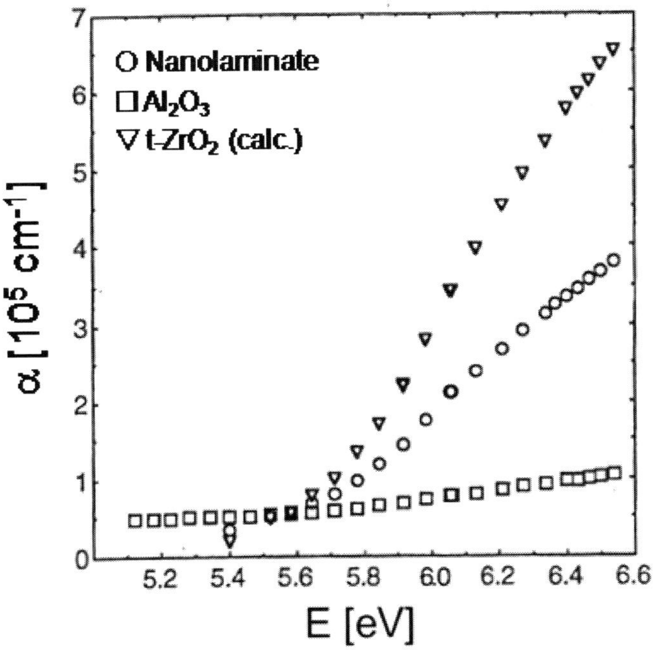

Figure 5. α versus E experimentally determined for a nanolaminate containing 15 bilayers of 5 nm ZrO_2-5 nm Al_2O_3, a single layer Al_2O_3 film, and calculated for t-ZrO_2 [18].

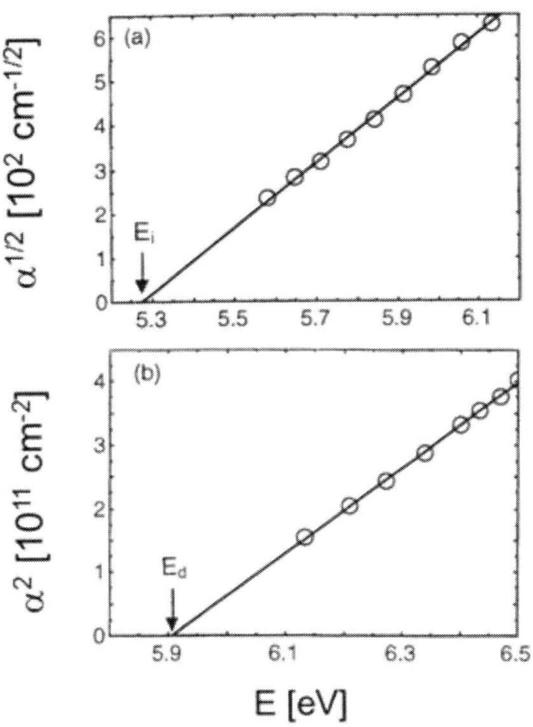

Figure 6. Two interband transitions at the fundamental optical absorption edge of t-ZrO2: (a) an initial indirect interband transition with an onset at $E_i = 5.22$ eV, and (b) a direct interband transition with an onset at $E_d = 5.87$ eV [18].

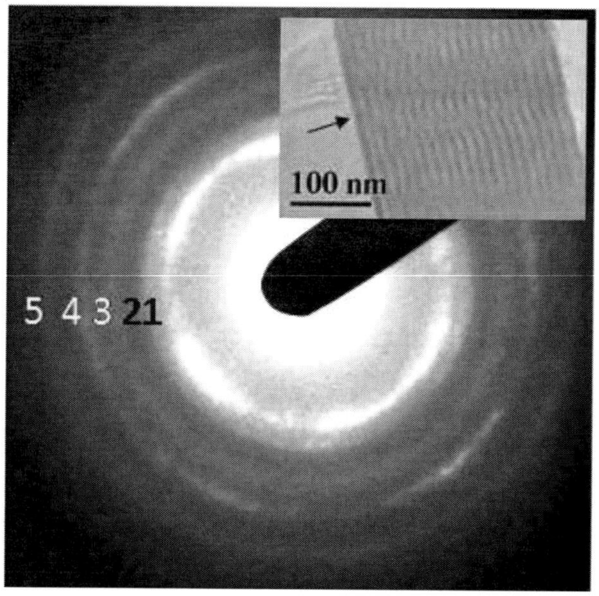

Figure 7. A selected area diffraction pattern across the cross-sectional thickness of a nanolaminate containing 19 bilayers of 7.3 nm HfO_2-5.2 nm Al_2O_3 bilayers grown on the nascent oxide of <111> Si. The film is shown in the insert (arrow indicates growth direction) [38].

The major question with respect to structure in HO_2-Al_2O_3 nanolaminates centers around the initial phase or phases that are formed. Can a direct comparison with ZrO_2-Al_2O_3 be made, where the nucleated phase was t-ZrO_2 not m-ZrO_2which is the STP phase, or does m-HfO_2 grow at nanocrystal inception? HfO_2-Al_2O_3 nanolaminates on fused SiO_2 and the nascent oxide surface of <111>-Si were studied by XRD and HTREM [38] to answer these questions, with surprising results.

Whereas the presence of t-ZrO_2 was readily observed even in ultrathin ZrO_2 layers by conventional Bragg-Brentano XRD of ZrO_2-Al_2O_3 nanolaminates [18, 21, 22, 24], t-HfO_2 could not be detected in analogous studies [38]. However, HTREM told a different story. Figure 7 is a selected area diffraction pattern (SAD) taken from the entire cross-sectional thickness of a nanolaminate with nineteen 7.3 nm HfO_2-5.2 nm Al_2O_3 bilayers, shown in the insert. The radial position of the rings was determined relative to diffraction spots from the <111> Si substrate. All but ring 3 could be indexed to t-HfO_2. Nor could ring 3 be indexed to m-ZrO_2. Lattice fringes obtained from high resolution images of many regions enabled identification of the mystery ring. As it turns out, not just one but two non-monoclinic phases were initially formed, t-HfO_2 and a high pressure phase, denoted "o I" in the literature [34-37].

Figure 8 shows a HRTEM image of a nanocrystalline HfO_2 layer between two amorphous Al_2O_3 layers. Two nanocrystals are outlined with boxes A and B and diffractograms of these areas were obtained. Analysis [38] of spacing and relative angles showed that area A enclosed a t-HfO_2 nanocrystal. Area B enclosed an o-HfO_2 nanocrystal that was in the process of transforming to m-HfO_2. The o-HfO_2→m-HfO_2 transition is a twinning operation, with geometry schematically shown in Figure 9. This transition can be readily analyzed by crystallographic group theory [36, 38].

A question is why only the tetragonal phase is initially formed in nanocrystalline ZrO_2, whereas both the tetragonal and orthorhombic phases are initially formed in nanocrytalline HfO_2. Although ZrO_2 and HfO_2 are in the broad view so similar and are referred to as "sister materials", there are electronic differences [35, 39] between them that might drive the structural differences observed here and warrants further investigation.

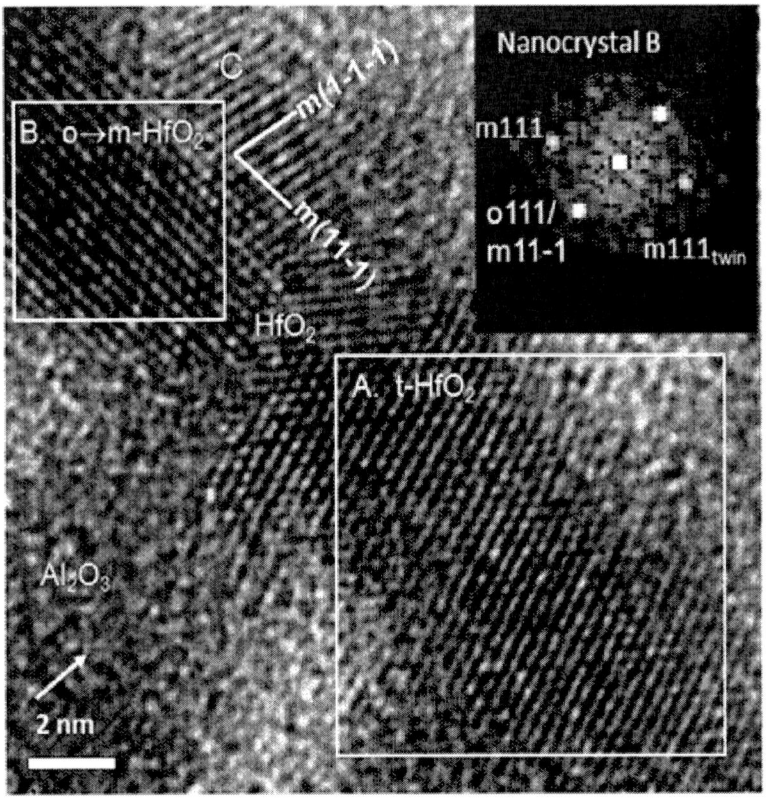

Figure 8. HRTEM image of a nanocrystalline HfO_2 layer between two amorphous Al_2O_3 layers in the nanolaminate shown in Figure 7. Two nanocrystals are outlined. Area A encloses a t-HfO_2 nanocrystal. Area B encloses an o-HfO_2 nanocrystal undergoing a transformation to m-HfO_2. Inserted diffractogram shows the transformation [38].

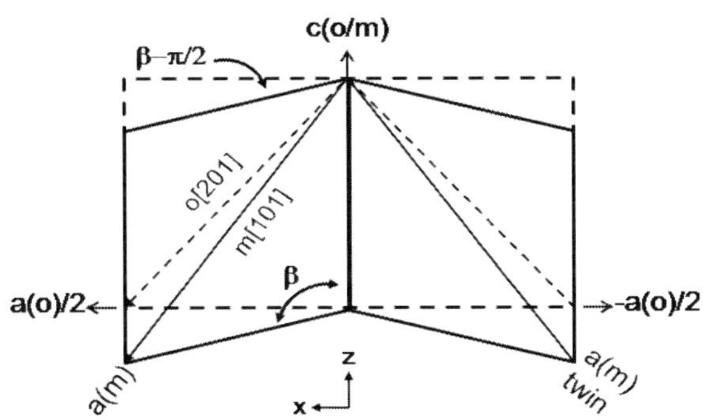

Figure 9. Geometry of the o-HfO_2 → m-HfO_2 twinning transition viewed along the [010] axes of both unit cells. β is the characteristic non-π/2 angle of the monoclinic unit cell.

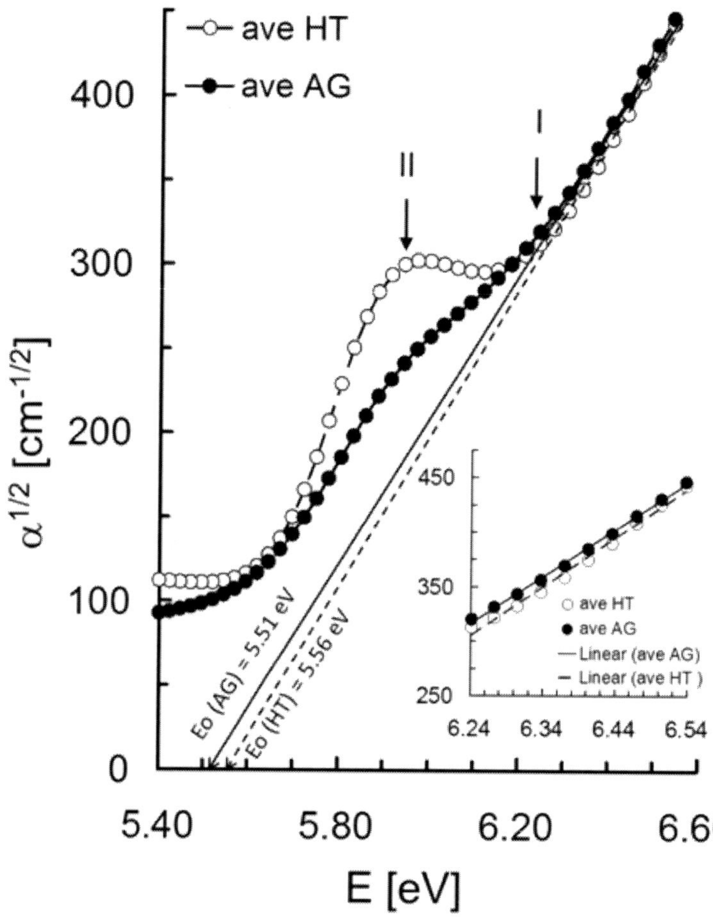

Figure 10. $\alpha^{1/2}$ versus E for as-grown and high temperature annealed single layer m-HfO_2 films grown on fused SiO_2 showing an indirect band gap (I) and a polaronic feature (II) [44].

The main current interest in HfO_2 is for its electronic and optical behavior. HfO_2 is a wide band gap solid with a high dielectric constant and high refractive index. Hence it has attracted much recent attention as a replacement gate dielectric for SiO_2 in complementary metal-oxide-semiconductor devices [40]. However, electronic states close in energy to the FOAE form intragap charge trap states. In particular, a feature at a slightly higher energy than O defect states is observed as a broad low energy shoulder on the FOAE [41-44] or a loss peak in the dielectric function [45, 46]. In order to understand how this undesirable feature can be prevented, we first consider its presence on the FOAE on single layer HfO_2 films.

Single layer as-grown HfO_2 films were annealed at 1273 K in furnace air for 24 h. XRD showed that that the as-grown films were m-HfO_2 whose crystal structure was refined after annealing. Figure 10 shows $\alpha^{1/2}$ versus E for both film states. Two features are indicated. Feature I, expanded in the insert, represents an indirect band gap associated with a O *2p*→ Hf *5d* electron transition between Γ→B points on the m-HfO_2 reciprocal lattice. (Aside: alternatively, the transition can be expressed using a cluster model; these equivalent descriptions being a manifestation of Tauc's rule [29]). The value of the indirect gap is indicated on Figure 10. Feature I does not change with annealing, again demonstrating the primacy of nearest neighbor configurations in determining interband transitions in d^0 oxides.

Now regard undesirable feature II. Note that (1) the position of Feature II does not change with annealing, and (2) its intensity increases as the film crystal structure is refined, indicating that this feature is not associated with a defect. Rather, it is intrinsic to the 7-fold Hf–O coordination in well-ordered m-HfO_2, that is, material in which the Hf-O coordination in which O exists in ordered alternate rows in 3-fold and 4-fold coordination with Hf [34]. The polaronic origin of feature II was proposed by several investigators [38, 41, 43, 47]. This proposal is supported by recent theoretical calculations showing that both electrons and holes can be self-trapped in a perfect monoclinic HfO_2 lattice, producing self-trapped small polarons [48-50].

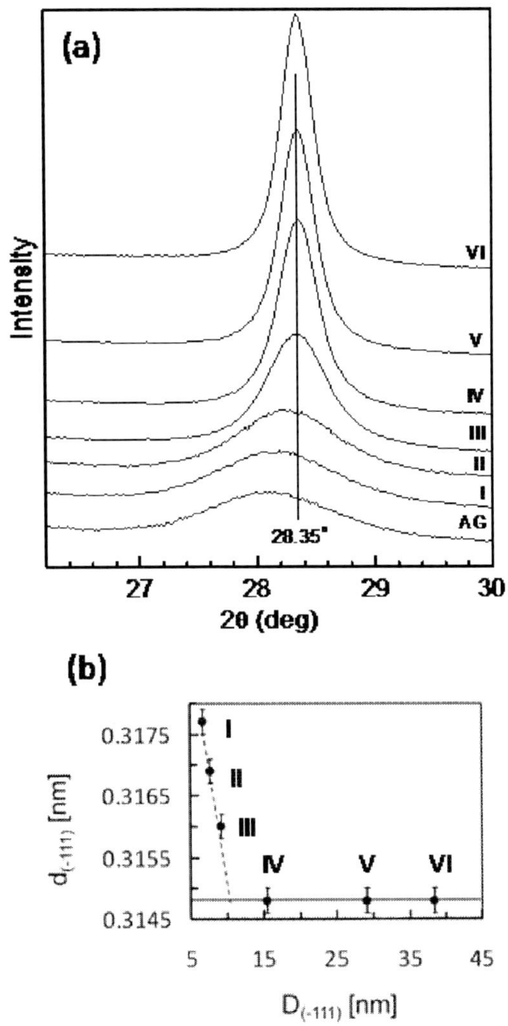

Figure 11. (a) XRD intensity of the m-HfO_2 (-111) peak for six different nanocrystal sizes. Line designates the peak position for this reflection in bulk m-HfO_2. (b) m-HfO_2 (-111) interplanar spacing, d(-111), versus nanocrystal size, D(-111) [51].

Further experiments using a series of single layer films with a range of nanocrystal sizes clearly showed the development of the pre-gap band with increasing volume/surface area ratio [51]. These experiments also led to the discovery of a finite crystal size effect in m-HfO_2.

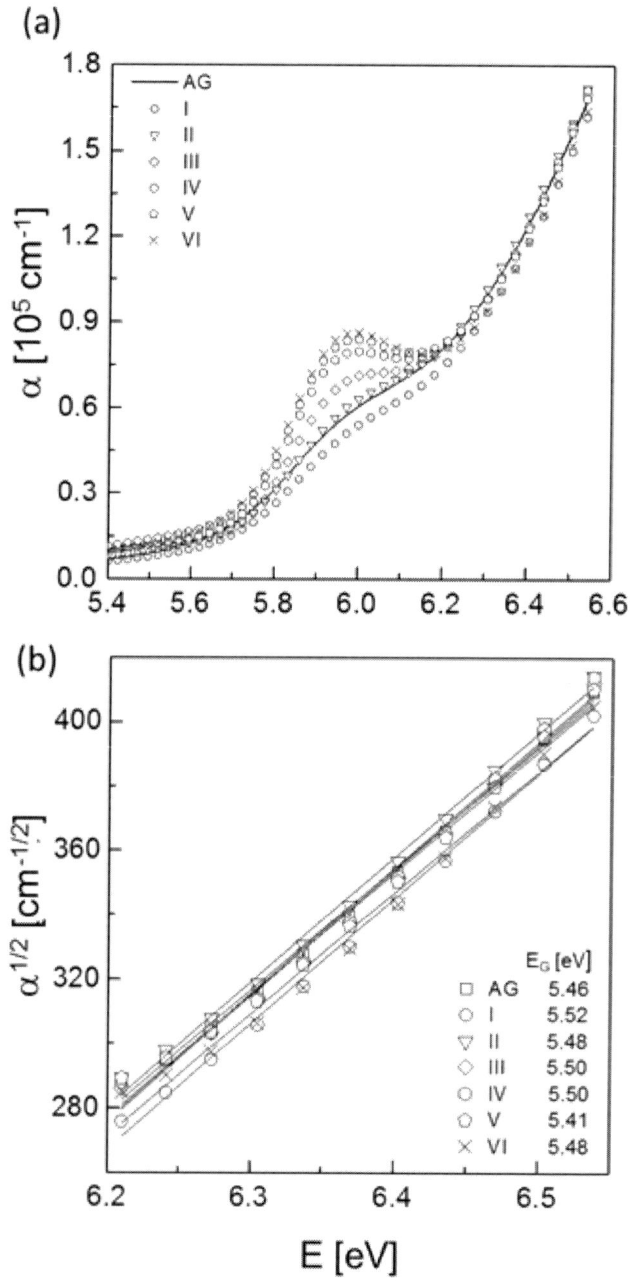

Figure 12. (a) α vs. E, and (b) indirect interband transition for the single layer m-HfO_2 films with nanocrystal sizes shown in Figure 11 [51].

Figure 11a shows the XRD intensity of the m-HfO_2 (-111) peak for six different average nanocrystal sizes. A line designates the peak position for this reflection in bulk m-HfO_2. Figure 11b shows the m-HfO_2 (-111) interplanar spacing, $d_{(-111)}$, calculated from the peak shift, graphed as a function of nanocrystal size, $D_{(-111)}$. For the smallest nanocrystals, $d_{(-111)}$ decreases linearly with increasing $D_{(-111)}$, but this effect ultimately saturates. The value of $D_{(-111)}$ at which the saturation occurs is determined by extrapolation of the regression analysis line through the data in which $d_{(-111)}$ decreases with increasing $D_{(-111)}$ (shown as a dashed line in Fig. 11b) to the standard value of $d_{(-111)}$ (shown as a solid line). A value of $D_{(-111)}$= 10.7 nm is obtained as the critical dimension above which $d_{(-111)}$ becomes insensitive to nanocrystal size. After careful elimination of other causes, it was concluded [51] that the lattice expansion observed in m-HfO_2 was due to dipole-dipole repulsion at a nanocrystal's surface.

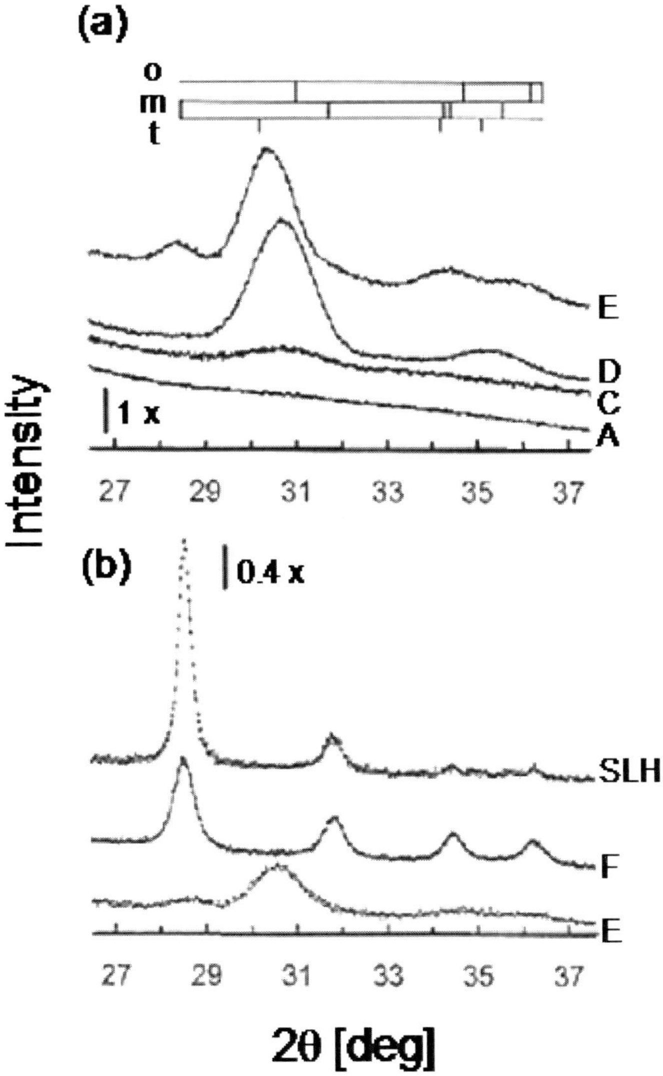

Figure 13. XRD pattern for a nanolaminate with 15, 7.3 nm HfO_2-5.2 nm Al_2O_3 bilayers after annealing for 1 h in air at 573 K (A), 773 K (B), 973 K (C), 1173 K (D), and 1273 K (E), 24 h at 1273 K (F) [52].

Figure 12a graphs α(E) versus E for this set of single layer films. A rapid rise in α(E) for E ≥ 6.24 eV is equivalent to feature I, discussed above. Figure 12b shows again that this interband transition is indirect, and as seen in Figure 10, is independent of nanocrystal size. The pre-gap band that initiates at ~ 5.65 eV and reaches maximum intensity at 5.94 eV is equivalent to the aforementioned undesirable feature II in Figure 10. Figure 12a shows that the spectral position of the band is unaffected by nanocrystal size but its strength increases as nanocrystal size, hence volume/surface area, increases. That is, the number of absorption states in the pre-gap band increases with increasing film perfection.

Finally, note that lattice expansion at small nanocrystal size (Figure 11b) is consistent with the inhibition of small polaron formation (Figure 12a) due to increased nearest-neighbor distance. Alternatively, one can envision HfO_2 as becoming more covalent as its lattice expands, a phenomenon observed for other ionic materials [9], and therefore less likely to support small polaron formation. This result is also consistent with the fact that an analogous pre-gap band has not been observed in m-ZrO_2 which is, in fact, less ionic than HfO_2 [39].

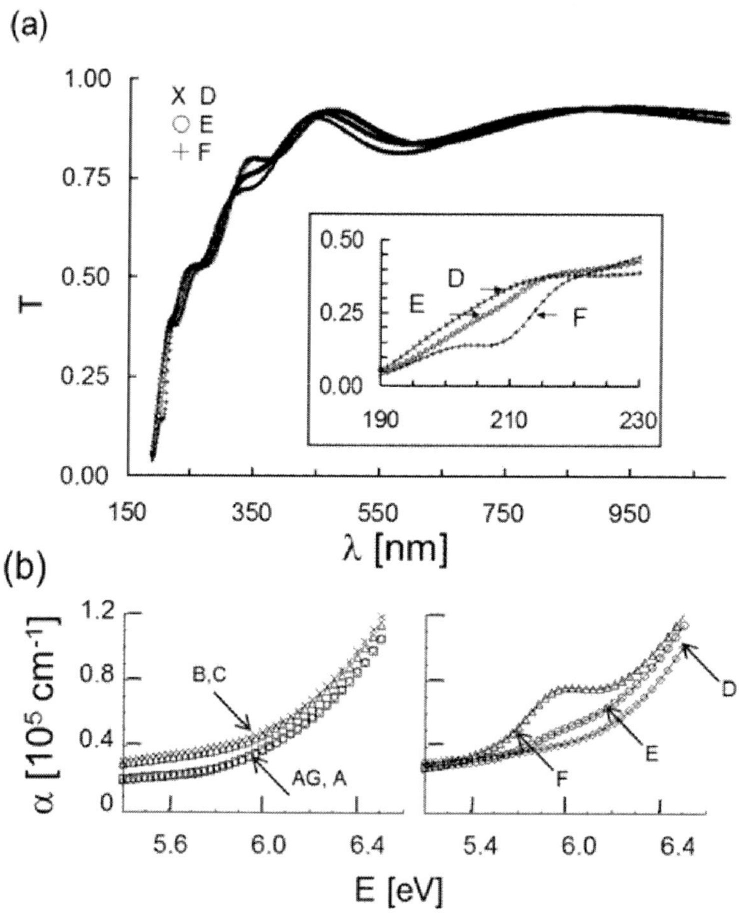

Figure 14. (a) Transmission vs. wavelength and (b) α vs. E for a HfO_2-Al2O_3 nanolaminate (Figure 13) after sequential annealing showing the development of a polaron band [52].

With the single layer HfO_2 film results in mind, a reasonable hypothesis is if the non-monoclinic phases of HfO_2 can be isolated, then the undesirable pre-band absorption band can be suppressed. The FOAE of the HfO_2-Al_2O_3 nanolaminate whose structure is presented in Figures 7 and 8 was examined to test this hypothesis [52]. In addition, to determine the nanolaminate's phase stability and relate it to changes in the FOAE, this film was annealed for 1 h in air at 573 K (A), 773 K (B), 973 K (C), 1173 K (D), and 1273 K (E), followed 24 h at 1273 K (F). The film was examined *ex situ* by XRD and spectrophotometry between annealing steps. Figure 13 shows the structural evolution of the nanolaminate. The bars at the head of Figure 13 indicate the position of XRD peaks of standards for the monoclinic, tetragonal, and orthorhombic HfO_2 phases. States A and B (not shown in Figure 13a) do not yield diffraction peaks. State C shows a broad, low intensity peak with contributions from both t-HfO_2 and o-HfO_2. This peak further develops and shifts towards t-HfO_2 in states D and E. In addition, two peaks unequivocally attributed to m-HfO_2 appear in the pattern of state E. For comparison, State F is shown in Figure 13b on a contracted intensity scale, along with state E and a well-crystallized single layer HfO_2 film. It can be seen that after a high temperature-long term anneal, the nanolaminate's structure is polycrystalline m-HfO_2.

Figure 14a shows transmission versus wavelength data for states D, E, and F, when phase changes from t + o-$HfO_2 \rightarrow$ m-HfO_2 are occurring in the nanolaminate. The insert in Figure 14a enlarges the data at the FOAE. The difference in transmission between these three states is clearly observed. Whereas only a slight decrease in transmission in state E occurs as m-HfO_2 observed along with the non-monoclinic phases, a large dip in transmission occurs in state F, as non-monoclinic phases are banished from the film.

Figure 14b shows α versus E for all films. These data tell a complementary story to Figure 14a, and proves our hypothesis to be correct. Undesirable absorption feature II (Figure 10) is not present in any of the states except state F, although a pre-gap increase in α state E indicates that the formation of feature II is incipient, and coincident with the appearance of m-HfO_2 nanocrystals.

V. CLASS II NANOLAMINATES: TiO_2-Al_2O_3, ZrO_2-TiO_2, AND HfO_2-TiO_2.

In theory, a nanolaminate's physiochemistry can be tailored for a specific application by simple architectural changes. However, a basic concern involves the electronic nature of the interfaces, in addition to miscibility issues that place the nanolaminates in the specific classes outlined in Section III. In the case of the class I nanolaminates, ZrO_2–Al_2O_3 and HfO_2–Al_2O_3, the electronic interaction between the oxides is extremely localized. The strength of the FOAE is determined by a contribution from each constitiuent oxide based on its volume fraction in a bilayer, that is, the absorption behavior of each end-member persists [53, 54] in the composite. At the other extreme, constituent oxides of class III nanolaminates become electronically amalgamated [53, 54] at their interfaces. A well-known example of amalgamation not discussed in this Chapter but widely researched is the case of ZrO_2–Y_2O_3 nanolaminates, in which bilayer interfaces disappear as the constituents form a cubic solid solution of graduated chemistry and electronic behavior [1]. Metallurgically speaking, most oxide pairs have neither complete miscibility nor immiscibility, and their optical absorption behavior is expected to fall somewhere between persistence and amalgamation. The three examples provided in this section give three different examples of optical absorption that is regulated by both the persistence of the individual oxide constituents and amalgamation at their interfaces. This model is denoted "modified persistence" in the following text.

A. TiO_2-Al_2O_3

Bulk TiO_2–Al_2O_3 ceramics have historically been used as structural refractory materials because of their low thermal expansion, thermal shock resistance, and high temperature stability [55-57]. Recent interest in TiO_2–Al_2O_3 ceramics extends to thin films and nanocomposites that depend upon the materials' response to light, including tailored refractive index films [58-60] transparent dielectrics [60], and photocatalytic applications [61, 62].

TiO_2–Al_2O_3 nanolaminates present a different challenge than the other class II oxide systems discussed here because not only do fully oxidized Ti and Al atoms vastly differ in radii (0.68 and 0.50 Å for Ti^{+4} and Al^{+3}, respectively [57]), these ions are not isovalent. Consequently, bulk TiO_2 and Al_2O_3 have no common lattice structure at any temperature or pressure [57], form no solid solutions, and only one high temperature, enthalpy-stabilized [63] compound, Al_2TiO_5. However, there is a strong affinity for Ti–O–Al bond formation across a TiO_2–Al_2O_3 interface [64, 65] with local structural adjustments to satiate charge imbalance [66, 67]. Even in the absence of actual physical mixing that extends beyond a few bonding units, the electronic changes due to the influence of one component on another can affect the FOAE.

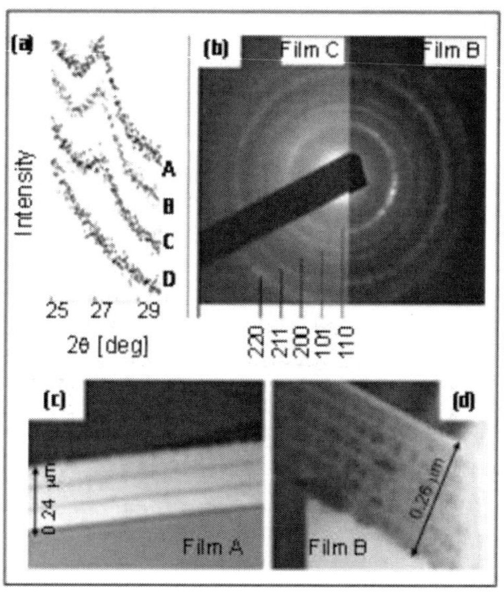

Figure 15. Structural aspects of TiO_2-Al_2O_3 nanolaminates (Table 1). (a) XRD peak from all nanolaminates. (b) SAD patterns showing rutile rings in films B and C. (c) TEM image of film A. (c) Under focused TEM image of film B showing columnar structure [68, 69].

Nanolaminates with various architectures grown on fused SiO_2 and the nascent oxide surfaces of <100> and <111>-Si were studied by XRD, HRTEM, and Raman microscopy [68, 69]. Results from a series of films with whose bilayer architectures are given in Table 1 are presented next.

Table 1. TiO_2-Al_2O_3 nanolaminate architecture

Film	Thickness (nm)	TiO_2/Al_2O_3 layer thickness (nm)	No. of bilayers	TiO_2 VF	Ti-O-Ti VF
A	237	72/7	3	0.91	0.88
B	258	36/7	6	0.85	0.70
C	250	18/7	10	0.68	0.56
D	240	9/7	15	0.51	0.35

Figure 15a shows the only XRD peak obtained from the nanolaminates. This peak, present in films A, B, and C, is indexed as rutile (r) TiO_2. Figure 15b shows SAD patterns taken from cross-sectional HRTEM images of films B and C. All of the rings are indexed to rutile. From Figures 15a and b, it can be seen that the TiO_2 layers are nanocrystalline and the Al_2O_3 layers are amorphous.

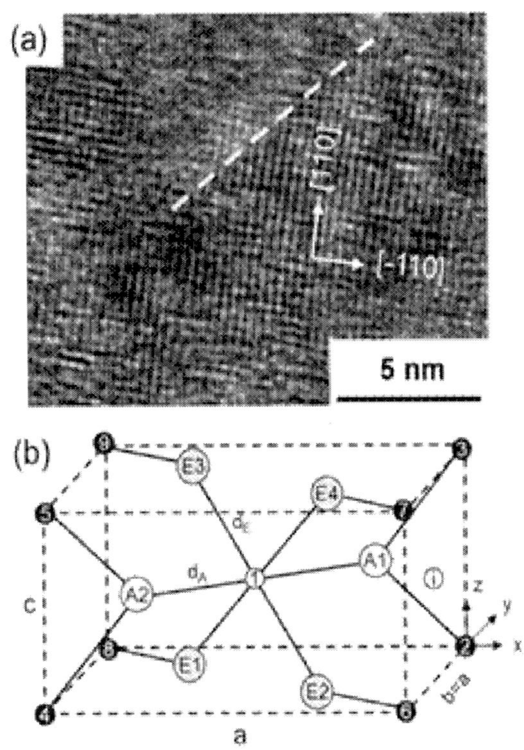

Figure 16.(a) HREM image of nanomosaic rutile TiO_2. Zone axis of the domain at the center of the image is [001]. Dashed line indicates a domain boundary. (b) Rutile unit cell. Ti atoms 1 – 9 are at allowed lattice sites. An interstitial Ti, labeled "i" is shown on the (100) plane. Equitorial and apical O atoms and the bonding configurations are indicated [72].

A low magnification image of film A is presented in Figure 15c, showing three bilayers of Λ=72 nm TiO_2/7 nm Al_2O_3 each. TiO_2 layers are light and Al_2O_3 layers are dark in this image. Figure 15d is a z-contrast image of film B showing six bilayers of Λ=36 nm TiO_2/7 nm Al_2O_3 each. This image was obtained at severe underfocus and shows in a greatly exaggerated manner differences in density and atomic contrast in film B. TiO_2 layers are dark and Al_2O_3 layers are light in this image. In addition, the TiO_2 intralayer columnar structure can be resolved.

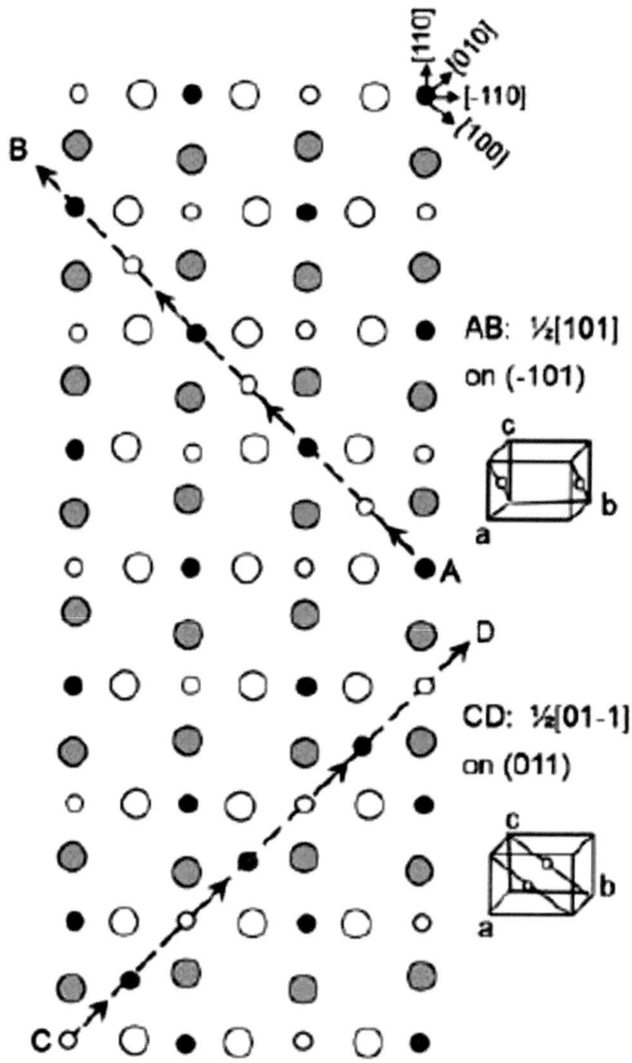

Figure 17. Ti and O sublattices projected onto rutile (001) plane. Large circles indicate O atoms. Small circles indicate Ti atoms. Light and dark shaded atoms of either species are removed from each other by c/2. Two ½<011>{011} stacking faults are shown along dashed lines AB and CD. Ti atoms are removed from the base of the arrows and placed in interstitial positions at the tip of the arrows. Side drawings show Ti interstitial positions in the TiO_2 unit cell [72].

HREM images of the TiO_2 layers in nanolaminates A, B, and C revealed a mosaic structure consisting of rectangular domains, shown in Figure 16a. Further examination showed that this nanomosaic structure was present in ultrathin single layer TiO_2 films, as well. This structure was generated by the "mistake" attachment of Ti at interstitial instead of allowed sites on growing rutile nuclei [70-72], schematically shown in the rutile unit cell in Figure 16b. As a result, ½<011>{011} stacking faults are created in an ideal rutile lattice. A ½<011>{011} stacking fault is equivalent to an antiphase boundary on the Ti sublattice, shown schematically in Figure 17. The O sublattice is unchanged across the fault and there is no change in the overall film stoichiometry. These faults are bounded by partial edge dislocations with ½<011> Burgers vectors. As an aside, annealing studies [72] show this nanostructure to be very robust. Without the high temperature required for dislocation climb, ½<011>{011}-type faults inherent to nanomosaic rutile provide thermal stability against massive crystallite growth.

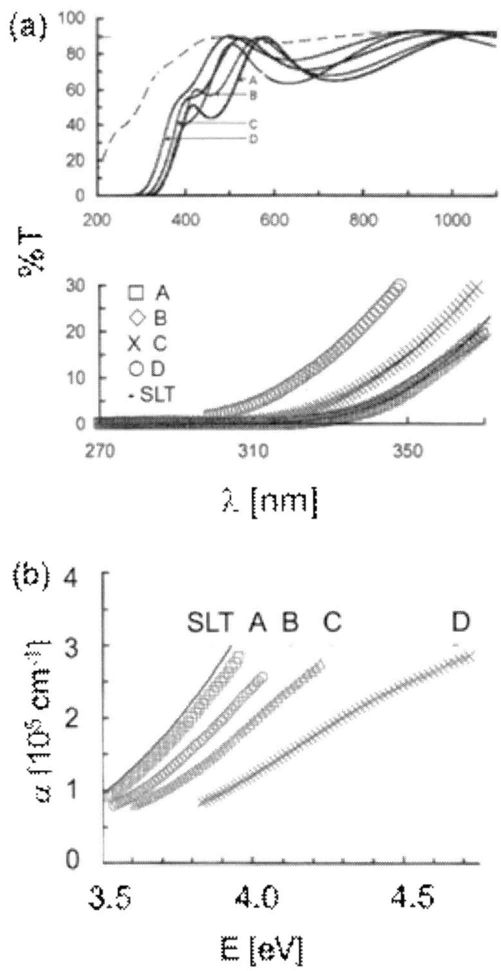

Figure 18. (a) Transmission vs. wavelength for TiO_2-Al_2O_3 nanolaminates grown on fused SiO_2 (Table 1). (b) α vs. E at the fundamental optical absorption edge [69].

Figure 18a shows transmission versus wavelength for the TiO_2-Al_2O_3 nanolaminates in Table 1 grown on fused SiO_2. Data for a 195 nm-thick single layer TiO_2 film and a 210 nm-thick Al_2O_3 film are also included for comparison. Figure 18b shows α versus E at the FOAE for the nanolaminates and single layer TiO_2. $\alpha^{1/2}$ versus E is shown in Figure 19. The values of E_G calculated from regression analysis are recorded in the insert.

It is clear that the apparent blueshift of α with decreasing TiO_2 layer thickness is not caused by a shift of E_G to higher energy but by diminished absorption at the FOAE with increasing Al_2O_3 volume fraction. Furthermore, the values of E_G are close or identical to that for SL TiO_2, 2.95eV. At first glance TiO_2-Al_2O_3 appears to exhibit similar behavior to ZrO_2-Al_2O_3 and HfO_2-Al_2O_3 in which each component weighted by its volume fraction contributes as a separate entity to α. That is, the two-component persistence model of Eq. 3 can be applied,

$$\alpha = VF(TiO_2)\alpha(TiO_2) + VF(Al_2O_3)\alpha(Al_2O_3). \tag{8}$$

Al_2O_3 is highly transmitting over the energy range of interest (Figure 18a) and Eq. (8) becomes,

$$\alpha = VF(TiO_2)\alpha(TiO_2). \tag{9}$$

If Eq. (9) correctly describes α in the case of the nanolaminates, then dividing each of the curves in Figure 19 by the TiO_2 volume fraction should bring them into coincidence with that of single layer TiO_2. $\alpha/VF(TiO_2)$ versus E is graphed in Figure 20a. It can be seen that $VF(TiO_2)$ does not work as a normalization factor. Equation (9) does not apply to the TiO_2-Al_2O_3 nanolaminates; α is weaker in strength than predicted by a simple persistence model.

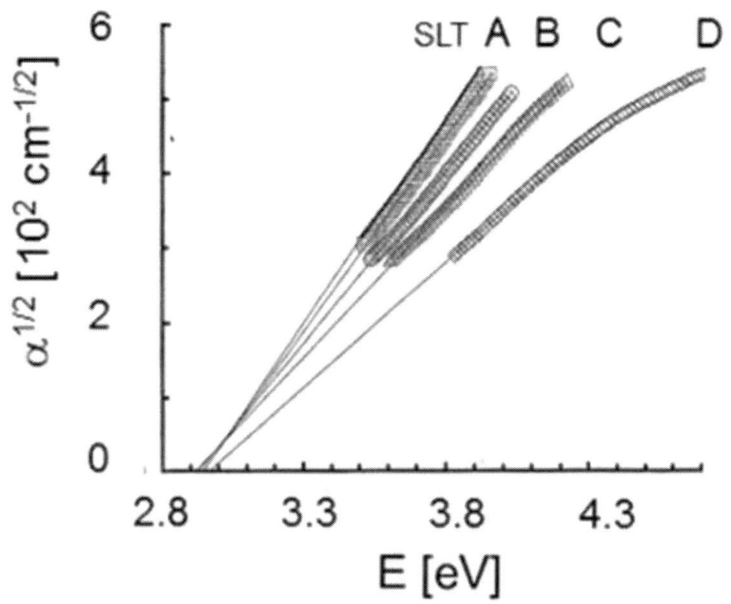

Figure 19. $\alpha^{1/2}$ vs. E for TiO_2-Al_2O_3 nanolaminates grown on fused SiO_2 (Table 1). E_G is recorded in the insert [69].

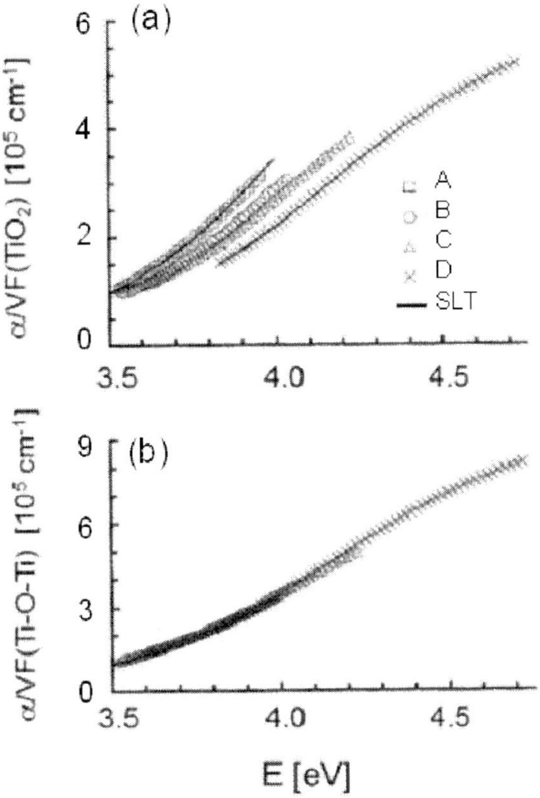

Figure 20. (a) α/VF(TiO_2) vs. E and (b) α/VF(Ti–O–Ti) vs. E, where VF(Ti–O–Ti) < VF(TiO_2) for TiO_2-Al_2O_3 nanolaminates grown on fused SiO_2 (Table 1) [69].

A reasonable hypothesis for the diminished strength of α in TiO_2-Al_2O_3 compared to a persistence model prediction is that Al_2O_3 has an electronic influence on the TiO_2 joint density of states at an interface even though no extended metallurgical bonds exist. Other investigators [64-67] have suggested that Al_2O_3 distorts the electronic configuration of TiO_2 even if the interface between these oxides is abrupt. The effect in the nanolaminates is that volume fraction of TiO_2 that electronically behaves like nanomosaic rutile with respect to interband transitions is reduced.

A modified persistence model is proposed in which the FOAE onset is determined by the volume fraction of Ti–O–Ti linkages with nanomosaic rutile (average) nearest neighbor bonding. But this volume fraction is less than the actual TiO_2 layer thickness, i.e., VF(Ti–O–Ti) < VF(TiO_2). Figure 20b shows α versus E curves brought into coincidence with single layer TiO_2 by the fit of a single parameter, VF(Ti–O–Ti), whose values are given in Table 1. Comparing VF(TiO_2) and VF(Ti-O-Ti) given in Table 1, it can be seen that the volume fraction of TiO_2 that behaves electronically like nanomosaic rutile is 97% of the actual amount of nanomosaic rutile present in film A, 83% in film B, 77% in film C, and only 63% in film D.

The volume fraction of TiO_2 that does not behave like nanomosaic rutile is denoted "Δ". It is reasonable to expect that Δ is close to the metallurgical interface of each bilayer because that is where the proximity to Al_2O_3 is most likely to exert an influence on TiO_2. The remaining volume fraction of a TiO_2 bilayer, (1- Δ), is responsible for the onset of optical absorption in TiO_2-Al_2O_3 nanolaominates.

B. ZrO_2-TiO_2

The large negative enthalpies of formation of zirconium titanate compounds and a solid solution with extensive stoichiometry [73-75] demonstrates the chemical driving force for Zr-O-Ti bond formation in preference to Zr-O-Zr or Ti-O-Ti bonds. This situation is exactly the reverse of the bonding preferences in TiO_2-Al_2O_3. At ambient pressure, however, all crystalline polymorphs of $(Zr,Ti)O_2$ have α-PbO_2-type orthorhombic lattices (*Pbcn* space group). This lattice type differs from any ambient pressure polymorph of the end-member oxides, ZrO_2 and TiO_2.

The bulk standard temperature and pressure (STP) polymorph of ZrO_2 has a baddeleyite-type monoclinic structure. TiO_2 undergoes the following polymorphic changes upon compression at room temperature: rutile → α-PbO_2-type orthorhombic, → baddeleyite monoclinic [76, 77]. The intriguing feature about the bulk pseudobinary ZrO_2-TiO_2 system is that the structural path followed by $(Zr,Ti)O_2$ with increasing Zr content is the same as that of TiO_2 under compression, i.e., STP m-ZrO_2 and high pressure m-TiO_2 are isostructural. Consideration of the reported unit cell volumes for the monoclinic phases of ZrO_2, $ZrTiO_4$, and TiO_2 [73, 76, 77] shows that Vegard's rule is obeyed.

Nanolaminates with various architectures grown on fused SiO_2 and the nascent oxide surfaces of <100> and <111>-Si were studied by XRD, Raman microscopy, and spectrophotometry [78-82]. Results from a series of films with whose bilayer architectures are given in Table 2 are presented next.

Table 2. ZrO_2-TiO_2 nanolaminate architecture

Film	Thickness (nm)	ZrO_2/TiO_2 layer thickness (nm)	No. of bilayers	ZrO_2 VF
A	204	1.5/1.5	68	0.50
B	204	4.0/1.5	37	0.73
C	308	10.8/1.5	25	0.88
D	354	16.2/1.5	20	0.91
E	200	4.0/4.0	25	0.50
F	204	4.0/8.0	17	0.33
G	200	4.0/36.0	5	0.10

Film A, with Λ =1.5 nm ZrO_2-1.5 nm TiO_2, did not yield XRD peaks. However, a major peak attributed to diffraction from (11-1) planes of an extensive m-$(ZrTi)O_2$ solid solution was observed as the ZrO_2 layer thickness increased to 4 nm in film B with Λ =4.0 nm ZrO_2-1.5 nm TiO_2. This solid solution is isomorphic with two high pressure, fixed composition phases, m-TiO_2 which naturally occurs in rocks [76, 77], and m-$ZrTiO_4$, which is produced by applying high hydrostatic pressure to orthorhombic $ZrTiO_4$ powder [73]. The formation of a high pressure phase nanocrystal in the film can be understood in terms of the Gibbs-Thomson effect [5], that is, vapor pressure enhancement, p/p_o, above a nanocrystal of radius, r, given by

$p/p_o = \exp(2\gamma\Omega/rkT)$, where γ is the surface energy, Ω is the volume of a $(Zr,Ti)O_2$ unit, k is Boltzmann's constant, and T is absolute temperature, as demonstrated in References 81 and 82.

Non-resonant Raman spectra [81, 82] for films A and B reveal an interfacial structure. Specific features between 700 and 900 cm^{-1} that are unambiguously attributed to Ti-O-Zr linkages [83-88]. A feature at 180 cm^{-1} in film B is attributed to m-$(Zr,Ti)O_2$ [81, 82]. The Raman spectrum and absence of XRD peaks for film A suggest that initial deposition of a Zr flux onto a TiO_2 underlayer produces amorphous Zr-doped TiO_2 with overall rutile short range order, denoted "r-TiO_2:Zr". A Zr^{+4} ion dopant in r-TiO_2:Zr retains its coordination number (CN) of 8 while Ti^{+4} ions in the surrounding TiO_2 host have a CN of 6 characteristic of rutile [89, 90]. As the Zr flux continues to arrive at the growth interface, the TiO_2 host matrix becomes saturated with Zr dopant and an adaptive mixed cation structure with a "flexible" CN develops [88], denoted "a-$(Zr,Ti)O_2$". A cation in a-$(Zr,Ti)O_2$ has an average CN = 6 but the CN of Zr is > 6 and the CN of Ti < 6.

Beginning the deposition of the next TiO_2 layer in films A and B switches the structure back to amorphous r-TiO_2:Zr as the arrival flux changes from Zr back to Ti. However, the continued arrival of Zr flux when depositing film B leads to the formation of m-$(Zr,Ti)O_2$ nanocrystals. The average cation CN in m-$(Zr,Ti)O_2$ is 7. A schematic drawing of the inhomogeneous chemical structure of a ZrO_2-on-TiO_2 bilayer interface in film B is shown in Figure 21 [81, 82].

So far we have not discussed cation mixing at a TiO_2-on-ZrO_2 interface. Diffusion of the arriving Ti flux into an underlying ZrO_2 layer is expected in view of the ballistic nature of sputter deposition coupled with the thermodynamic driving force for cation mixing in this pseudobinary. However, low temperature post-deposition annealing studies of both a single ZrO_2-TiO_2 bilayer [91] and ZrO_2-TiO_2 nanolaminates [79] show that the formation of a mixed cation interface initiates by diffusion of Zr into TiO_2 not by diffusion of Ti into ZrO_2. Because of this asymmetric diffusivity, we surmise that the mixed cation region at a ZrO_2-on-TiO_2 interface is wider than the mixed cation region at a TiO_2-on-ZrO_2 interface.

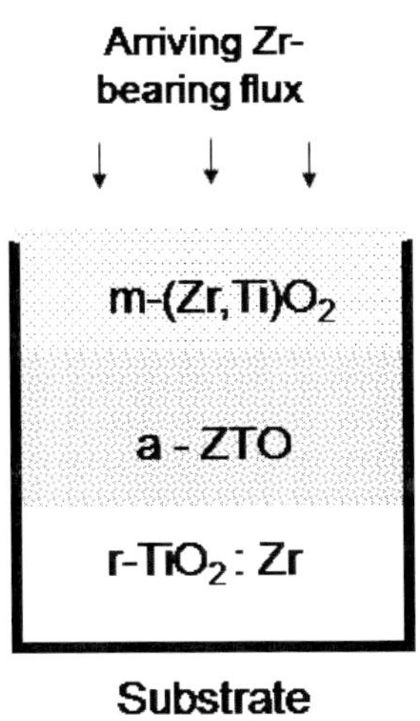

Figure 21. The inhomogeneous chemical structure of the ZrO2-on-TiO2 bilayer interface in ZrO_2-TiO_2 nanolaminates, exemplfied by film B (Table 2) [82].

Films with thicker ZrO_2 layers have interfaces that are the same as film B. The question is how does microstructure develop as the ZrO_2 layer thickness is further increased and the prevalent bonding changes from Zr-O-Ti to Zr-O-Zr. Film C (Λ=10.8 nm ZrO_2-1.5 nm TiO_2) showed a dominant reflection from c-ZrO_2 {220} planes. The dominant reflection changed to c-ZrO_2 (200) planes as the ZrO_2 layer thickness was increased to 16.2 nm (film D). The change in texture from c-ZrO_2 {220} to {200} can be explained in terms of a growth-controlled [92], rather than nucleation-controlled [93] model. The {220} planes are the only low-index planes that do not carry a net surface charge in the fluorite lattice, and it has been suggested [94, 95] that this factor results in favorable nucleation energetics. However, the {200} planes have the lowest surface energy in the fluorite c-ZrO_2 lattice [96]. Bauer [10] has shown that lowest surface energy planes form the habits of nanocrystals, even when the crystallites have nonequilibrium growth shapes such as columns. It is also well known that nanocrystals oriented with the fast-growth direction perpendicular to the substrate (here <100>) shadow and prevent the growth of neighbors with different alignment. We conclude that c-ZrO_2 {200} nanocrystals in film D grew at a faster rate, geometrically shadowed, and ultimately extinguished the growth of other cubic orientations.

ZrO_2–TiO_2 nanolaminates with different overall chemistry are good candidates for a family of films with high transparency and significant optical absorption (α is $\geq 1 \times 10^5$ cm^{-1}) that is tunable across much of the ultraviolet spectrum. In such a structure, tailored optical absorption can potentially be achieved through changes in bilayer architecture. The onset of optical absorption in the constituent oxides was experimentally determined to be ~ 5eVfor ZrO_2 polymorphs [1, 18, 26, 97, 98] and ~ 3 eV for TiO_2 polymorphs [99-102]. The absorption onset of pseudobinary $(ZrO_2)_x$-$(TiO_2)_{1-x}$ structures with different stoichiometry could therefore conceivably span an ultraviolet wavelength range from 248 to 414 nm. In addition, Zr and Ti atoms have the same electronegativity, 1.6 [103], and also form d^o oxides with isovalent cations, Zr^{+4} and Ti^{+4}. This electronic similarity theoretically permits the substitution of one cation for the other in dilute solution [89, 90] with minimal electronic disturbances at an interface, in comparison to TiO_2-Al_2O_3 (Sec. V.A). However, although ZrO_2 and TiO_2 are electronically compatible, a potential complication when combining them in an engineered structure is the large difference in the cations' ionic radii.

Figure 22a shows α versus E for the nanolaminates whose architecture is given in Table 2. The low absorption cutoff value for E (E at which α is = 1×10^5 cm^{-1}) is graphed versus the ZrO_2 volume fraction in Figure 22b. This figure shows that the onset of significant absorption is blueshifted with increasing ZrO_2 VF. Interestingly, films E and A have the same overall stoichiometry but very different architecture, yet they have almost identical FOAE characteristics. This important result demonstrates (again) the primacy of nearest-neighbor bonds in determining initial interband transitions in d^o oxides.

α versus E curves for films C and D have distinctly different shapes than curves for films A, E, F, and G. The α versus E curves for films A, E, F, and G are characterized by a rapid rise at their onset followed by a region in which α slowly varies with E. In contrast, the α versus E curves for films C and D have a long low absorption at their onset, followed by a rapid rise in α at E >~ 5.5 eV. This bimodal shape distribution is confounding at first glance. But notice that the curve for film B has features of both shapes. Remembering that film B is composed entirely of interfaces, therein lies the explanation of the bimodal shape distribution mystery presented in Figure 22a. A general formula for the decomposition of α in terms of all of its constituents is,

$$\alpha = VF(IF)\ \alpha(IF) + VF(XS)\ \alpha(XS) \quad (10)$$

where IF stands for "interfacial material," the fraction of each bilayer that is involved in interfacial electronic interactions such that its FOAE is modified, and XS stands for "excess material," VF(IF) is the volume fraction of each bilayer not whose FOAE is not affected by its partner oxide.

Film B consists entirely of interfacial material and exists in the other nanolaminates as VF(IF). Hence, VF(XS) = 1-VF(IF) for the nanolaminates becomes VF(XS)= 1-VF(film B). The excess material is nanomosaic rutile in films E, F, and G and cubic ZrO_2in films C and D, as recorded in Table 3. Further details and graphs of this decomposition are given in Reference (82). Let it suffice here to say that the calculation shows that the FOAE onset in films E, F, and G, is determined by excess nanomosaic rutile, whereas in films C and D, it is determined by the interfacial structure shown in Figure 21, that is, it is determined by "embedded" film B.

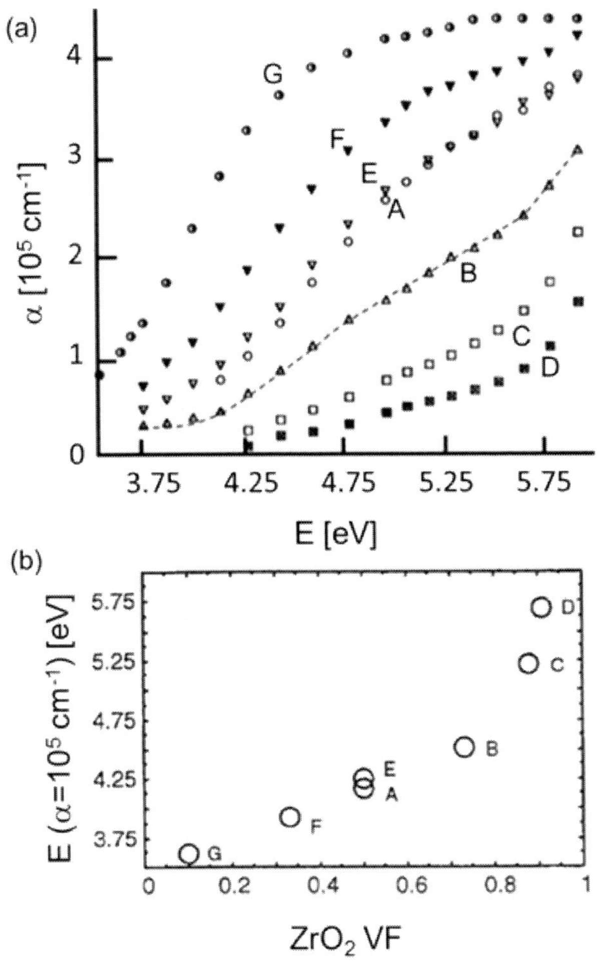

Figure 22. (a) α vs. E for ZrO_2-TiO_2 nanolaminates (Table 2) [82]. E at which $\alpha \leq 1 \times 10^5\ cm^{-1}$ vs. ZrO_2 volume fraction for ZrO_2-TiO_2 nanolaminates (Table 2) [82].

Table 3. Volume fraction of the interfacial and excess material in ZrO_2-TiO_2 nanolaminates

Film	ZrO_2/TiO_2 layer thickness (nm)	VF(film B)	VF(XS)
B	4.0/1.5	1	0
Films with excess cubic ZrO_2			
C	10.8/1.5	0.45	0.55
D	16.2/1.5	0.31	0.69
Films with excess nanomosaic rutile TiO_2			
E	4.0/4.0	0.69	0.31
F	4.0/8.0	0.46	0.54
G	4.0/36.0	0.14	0.86

B. HfO_2-TiO_2

The bulk pseudobinary HfO_2-TiO_2 system [104, 105] shows low miscibility between the end-member oxides, monoclinic baddeleyite-type HfO_2 and rutile TiO_2. Similar to ZrO_2-TiO_2, there is a large difference in ionic radius between Ti^{+4} (0.068 nm) and Hf^{+4} (0.080 nm) [104]. However, only a single mixed cation phase, orthorhombic α-PbO_2-type $HfTiO_4$, exists at STP. An extensive crystallite substitutional solid solution based on a prototypical o-$HfTiO_4$ lattice does not form.

As mentioned in Sec. IV.B, thin film HfO_2 is a candidate for a high dielectric constant replacement material for SiO_2 in integrated circuits [40]. The addition of TiO_2 to HfO_2 has shown promise for producing a $Hf_{1-x}Ti_xO_2$ ternary with an even higher dielectric constant than pure HfO_2 while maintaining thermal stability with Si [106-110]. In addition, by analogy with ZrO_2-TiO_2, significant optical absorption in HfO_2-TiO_2 might be tunable across the ultraviolet spectrum through changes in bilayer architecture.

HfO_2-TiO_2 nanolaminates were studied by XRD, Raman spectroscopy, x-ray photoelectron spectroscopy, and spectrophotometry. The results were reported in a series of papers by Cisneros-Morales et al. [111-117]. Figure 23 shows high resolution XRD patterns from HfO_2–TiO_2 nanolaminates whose architecture is recorded in Table 4. Broad, low resolution XRD patterns of these films yield major peaks solely in the $27° < 2\theta < 34°$ range. Table 5 records the standard 2θ position and interplanar spacing of all relevant peaks [35, 73, 118-120]. The numbers at the top of Figure 23 correspond to the assignments in Table 5. Note the inclusion in Table 5 of phases that are not bulk equilibrium structures at STP. t-HfO_2 is a high temperature polymorph [104, 105]. o-HfO_2 is a high pressure polymorph stable between 4 and 14.5 GPa in bulk material [35, 36]. It was identified along with t-HfO_2 as an initially nucleated phase in thin HfO_2 layers grown by reactive sputter deposition at room temperature [38] (see Section IV.B). m-$HfTiO_4$ is a high pressure polymorph stable between 1 and 9 GPa in bulk material [73] and recently, formed at ambient pressure and elevated temperature in thin films fabricated by the sol-gel method [121]. Nanomosaic rutile is an intergrowth structure formed in thin TiO_2 layers sputter deposited at room temperature [70-72] (see Section V.A). Its structure, shown in Figures 16 and 17, is based on a rutile lattice containing α-PbO_2-type TiO_2 growth defects.

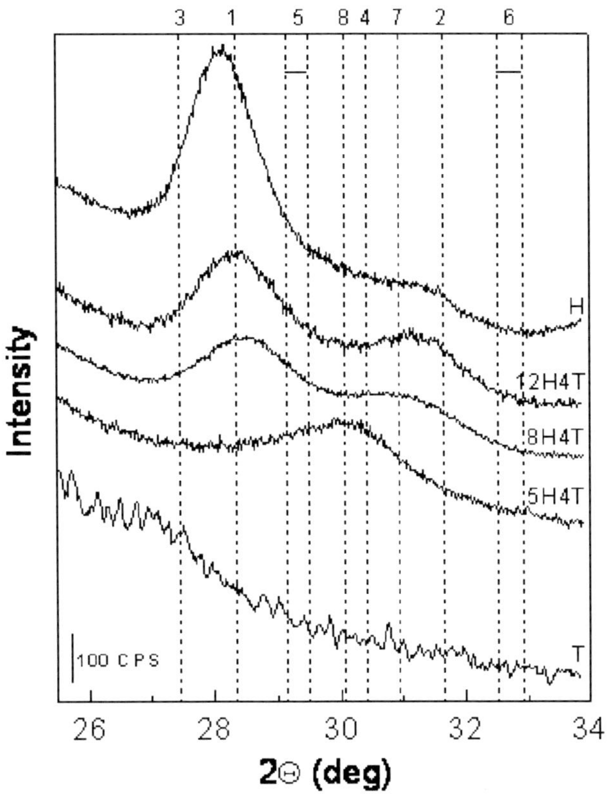

Figure 23. XRD patterns from HfO_2–TiO_2 nanolaminates grown on fused SiO_2 (Table 4) [113].

Table 4. HfO_2-TiO_2 nanolaminate architecture

Film	Thickness (nm)	HfO_2/TiO_2 layer thickness (nm)	No. of bilayers[(a)]	HfO_2 VF
5H/4T	293	5/4	32	0.54
8H/4T	2236	8/4	19	0.67
12H/4T	220	12/4	13	0.74

The XRD pattern of film 5H4T has contributions from (111) o-$HfTiO_4$ (no. 4) due to interfacial mixing by particle bombardment intrinsic to the sputter deposition process [2] or surfaction [122], and possibly contributions from (111) t-HfO_2 (no. 8) and (211) o-HfO_2 (no. 7). The XRD patterns of films 8H4T and 12H4T consist of a primary peak, attributed to (-111) m-HfO_2 planes (no. 1), and a secondary peak at higher angle. Contributions to the secondary peak come from (111) m-HfO_2 (no. 2), and also from (111) t-HfO_2 (no. 8) and (211) o-HfO_2 (no. 7) associated with small HfO_2 crystallite size, and (111) o-$HfTiO_4$ (no. 4). Notice that the primary and secondary peaks shift in different directions with increasing HfO_2 layer thickness in the 8H4T→12H4T→HfO_2 film sequence. This phenomenon occurs because contributions from non-monoclinic HfO_2 and mixed cation phases are eliminated from both primary and secondary peaks as the HfO_2 volume fraction increases and ultimately reaches unity in the single layer HfO_2 film.

From the XRD data presented above, we propose that o-$HfTiO_4$ forms closest to a metallurgical interface, likely at a HfO_2-on-TiO_2 interface [115]. Non-monoclinic HfO_2 and m-HfO_2 are formed when the Ti/Hf ratio available for reaction at that growth interface falls below that required to form o-$HfTiO_4$.

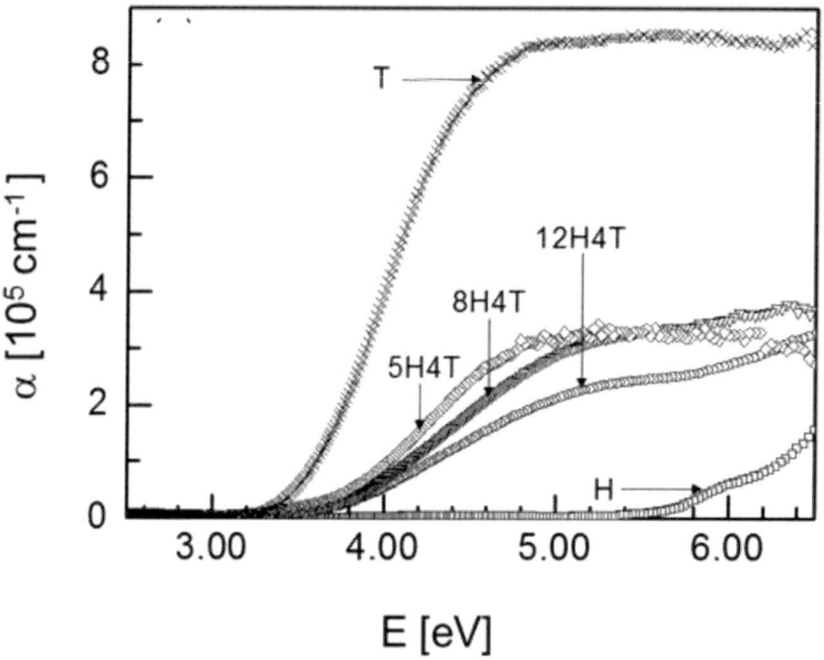

Figure 24. α vs. E for HfO_2–TiO_2 nanolaminates (Table 4) [113].

Figure 24 shows α versus E at the FOAE for the HfO_2-on-TiO_2 nanolaminates whose architecture is recorded in Table 4. Data for 87 nm-thick single layer TiO_2 and 167 nm-thick single layer HfO_2 films are included. Figure 24 shows that α for the nanolaminates as a group lies between the curves for the TiO_2 and HfO_2 films. Band-tailing, that is, additional electronic states that extend the FOAE of the nanolaminates to lower energy than the FOAE of TiO_2, is not observed.

The α versus E curves in Figure 24 are now examined in light of the role of specific functional metal-oxygen units [113]. The primary question is whether α, the absorption coefficient for the mixture can be decomposed into the pure constituents according to Vegard's rule, as was the case for ZrO_2-Al_2O_3 and HfO_2-Al_2O_3 nanolaminates, or whether an interfacial component must be considered, as was the case for TiO_2-Al_2O_3 and ZrO_2-TiO_2 nanolaminates. Decomposing α(5H4T) using a persistence model yields,

$$\alpha(5H4T)=VF(HfO_2)\alpha(HfO_2) + VF(TiO_2)\alpha(TiO_2). \qquad (11)$$

Figure 25 graphs four quantities: 1) experimentally determined α, 2) α(5H4T) calculated from Eq. (11), 3) α(HfO_2), and 4) α(TiO_2). It can be seen that the experimentally determined and calculated curves for the FOAE do not coincide. The same exercise carried out for films 8H4T and 12H4T yields similar results. A persistence model does not work here.

We next try a modified persistence model that decomposes α into contributions from an interfacial region rich in o-$HfTiO_4$ (and also containing some non-monoclinic HfO_2 material), and a region remote from an interface that basically contains m-HfO_2 (see Figure 23). Equating film 5H4T with the interfacial region in the other nanolaminates, Eq. (10) is written as,

$$\alpha = VF(5H4T)\,\alpha(5H4T) + VF(XS)\,\alpha(XS). \quad (12)$$

Table 5. Crystallographic parameters of standards for comparison with XRD peaks in HfO_2-TiO_2 nanolaminates

Phase	Lattice type	Space group (No.)	No.	hkl	2θ (deg)	$d_{(hkl)}$ (nm)	Occur -rence
HfO_2	Monoclinic[l]	*P2₁/c* (14)	1	-111	28.35	0.3148	STP[a]
			2	111	31.65	0.2827	
TiO_2	Tetragonal (nanomosaic)	*P4₂/mnm* (136)-*Pbcn* (60) (intergrowth)	3	110	27.46	0.3248	STP
$HfTiO_4$	Orthorhombic	*Pnab* (60)	4	111	30.43	0.2937	STP
$HfTiO_4$	Monoclinic	*P2₁/c* (14)	5	-111	29.15-29.49[c]	0.3063-0.3929[c]	HP[b]
			6	111	32.53-32.93[c]	0.2752-0.2719[c]	
HfO_2	Orthorhombic[f]	*Pbca* (61)	7	211	30.94	0.2890	HP
HfO_2	Tetragonal	*P4₂/nmc* (137)	8	111	30.06	0.2988	HT[d]

(a) STP, Standard temperature and pressure. (b) HP, high pressure. (c) 1 - 9 GPa. (d)HT, high temperature.

Figure 26 shows the decomposition of α using Eq. (12) for (a) film 8H4T and (b) film 12H4T. In the case of both films, α(EXP) and α(IF) overlap for $E \leq 4.45$eV. This result is important. Keep in mind that all of the TiO_2 in the films is used to form o-$HfTiO_4$ and cannot contribute as rutile to the FOAE. It is therefore concluded that the FOAE onset is associated with o-$HfTiO_4$. We emphasize: electron states associated with o-$HfTiO_4$ solely determine the FOAE onset in the Hf-rich nanolaminates.

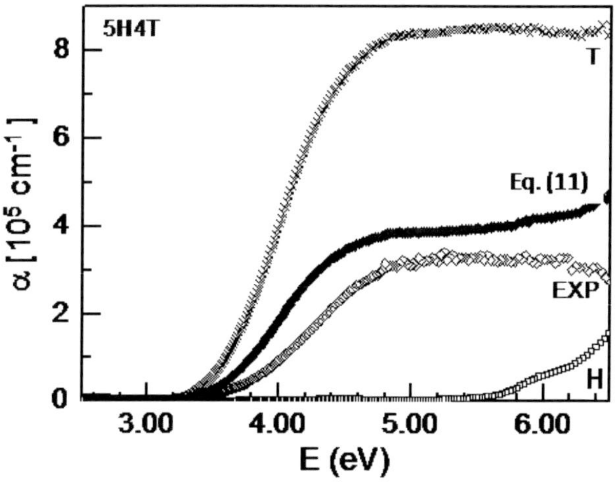

Figure 25. Four quantities, α, experimentally determined, α(5H4T) calculated using Eq.(11), $\alpha(HfO_2)$, and $\alpha(TiO_2)$ vs. E for HfO_2–TiO_2 nanolaminate 5H4T (Table 4) [113].

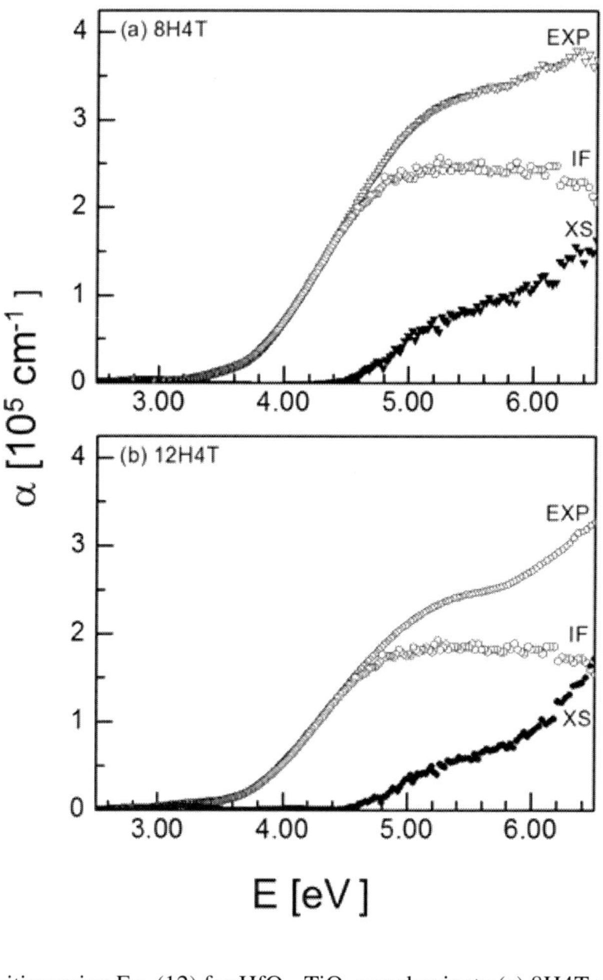

Figure 26. Decomposition using Eq. (12) for HfO_2–TiO_2 nanolaminate (a) 8H4T and (b) 12H4T, where XS = excess HfO_2 and IF = volume fraction of interfacial material [113].

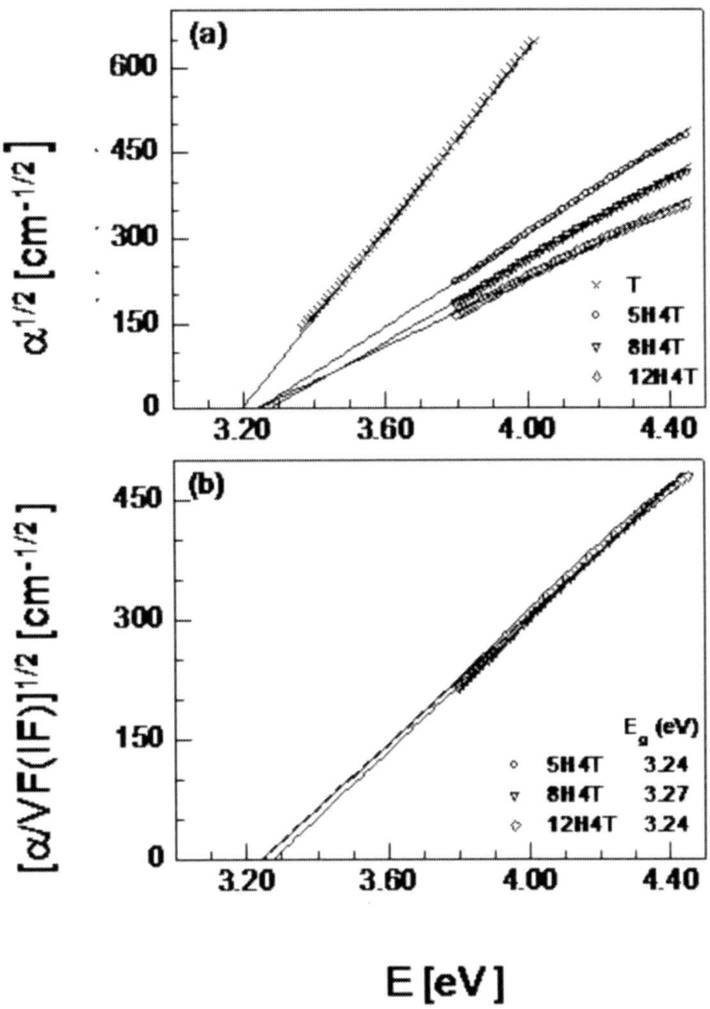

Figure 27. (a) $\alpha^{1/2}$ versus E and (b) [$\alpha^{1/2}$/VF(IF)] versus E for (a) HfO_2–TiO_2 nanolaminates (Table 4).

$\alpha^{1/2}$ versus E is graphed in Figure 27a for the nanolaminates and for single layer TiO_2. Normalization of these curves by the volume fraction of film 5H4T in the nanolaminates brings them into coincidence, as shown in Figure 27b. The values of E_G obtained by regression analysis of the curves in Figure 27a are recorded in Figure 27b, indicating that the interfaces in Hf-rich nanolaminites have a "5H4T film nature." The average value of E_G is 3.25 ± 0.02 eV, very close to but slightly higher than E_G=3.20 eV for TiO_2, in agreement with Lucovsky et al. [123]. (As points of comparison, E_G was found to vary between 3.21 eV and 3.33 eV by Studenyak et al. [124] and 3.42 eV by Domaradzki et al. [125] for nanocomposites, not nanolaminates, of comparable Hf atomic fraction to films in this study. Ye et al.[126] obtained a band gap for nanocomposites that is higher than that reported by the other investigators [124, 125] because Ye et al. neglected the contribution from states at the onset of optical absorption.)

Lucovsky et al. [123] and Fulton et al. [127] demonstrated the localized nature of interband transitions at the FOAE onset in $HfTiO_4$. Applying Tauc's rule [29] (yet another time), we suggest that $E_G = 3.25 \pm 0.02$ eV, given above is characteristic of bulk o-$HfTiO_4$.

CONCLUSION

A nanolaminate architecture was used to produce Group IVB transition metal oxide films with unique nanocrystal phase structures and tailored optical behavior at the fundamental optical absorption edge (FOAE). Five nanolaminate systems, ZrO_2-Al_2O_3, HfO_2-Al_2O_3, TiO_2-Al_2O_3, ZrO_2-TiO_2, and HfO_2-TiO_2, were grown by reactive sputter deposition on unheated substrates with a dissimilar oxide surface, SiO_2. Despite the far-from-equilibrium growth environment, bulk phase diagrams of the corresponding pseudobinary systems served as guides for predicting cation mixing at bilayer interfaces. However, phase structure could not be predicted from bulk thermodynamics because finite crystal size effects were prevalent in the transition metal oxide nanocrystals. Elevated temperature and pressure phases, metastable at ambient conditions, were the rule rather than the exception.

ZrO_2-Al_2O_3 and HfO_2-Al_2O_3 nanolaminates, in which the corresponding bulk oxides are immiscible, had atomically sharp interfaces between constituents, and their absorption coefficient at the FOAE and optical band gap were well described by a persistence model. Useful metastable transition metal oxide phases were captured in these structures by limiting nanocrystal size. For example, ZrO_2-Al_2O_3 nanolaminates solely containing t-ZrO_2 are the essential ingredient in smart, transformation-toughening, transparent coatings for mechanical and anticorrosion protection [129, 130]. HfO_2-Al_2O_3 nanolaminates solely containing non-monoclinic phases [51] quench an undesirable pre-FOAE polaron absorption band that is intrinsic to m-HfO_2.

On the other hand, TiO_2-Al_2O_3, ZrO_2-TiO_2, and HfO_2-TiO_2 nanolaminates in which the corresponding bulk oxides have some miscibility form electronically-complex interfaces. Mixed cation phases and structures may (but not necessarily) form, but in either case, a persistence model for α at the FOAE has to be modified to include the interfacial component. A modified persistence model was successfully used even in these complex cases because of the primacy (and almost exclusivity) of nearest-neighbor bonding in determining interband optical transitions at the FOAE in d^0 transition metal oxides.

In these films, the onset of the FOAE, i.e., the optical band gap, does not move across the energy spectrum between the extrema set by the individual constituents. Rather it is determined by the constituent with the lowest energy band gap. However, the energy at which significant absorption occurs can be tailored by adjusting the amount of each constituent in a bilayer.

Lastly, we emphasize that reactive sputter deposition is ideally suited to produce all kinds of nanolaminate films. Excellent layer thickness control can be easily achieved. Multiple targets can be used to create modular structures of different nanolaminates, where each module has a different purpose [130]. Simple adjustments in deposition process parameters, such as the application of a substrate bias or the type of rare gas used in conjunction with the reactive gas in the discharge produce a plethora of different structures [2].

ACKNOWLEDGEMENTS

The author wholeheartedly thanks the faculty, staff, and students of the Department of Chemistry and Biochemistry at the University of Wisconsin-Milwaukee for graciously welcoming her and making the past four years as one of their faculty the most pleasant of her long career in science.

REFERENCES

[1] Aita, C. R. *J. Phys. Condens. Matter* 2008, *20*, 264006(1)-264006(11).

[2] Aita, C. R. *CRC Rev. Sol. State Mater. Sci.* 1998, *23*, 205-274.

[3] Felner, F.F. *Low Temperature Oxidation;* Wiley-Interscience; New York, NY, 1981; pp 31-49.

[4] Garvie, R. C. *J. Phys. Chem.* 1987, *82,* 218-224

[5] Tu, K.-N., Mayer J. W.; Feldman L. C. *Electronic Thin Film Science;* Macmillan; New York, NY, 1992; pp 100–113.

[6] Birkholz, M. *Z. Phys.* B 1995, *96*, 325-332.

[7] Birkholz, M. *Z. Phys.* B 1995, *96*, 333-340.

[8] Ayyab, P, Palkar, V R, Chattopadhyay, S; Multani, M *Phys. Rev. B* 2006, *73,* 115330(1)-115330(7).

[9] Guangshe, L.; Boerio-Goates, J.; Woodfield, B. F. *Appl. Phys. Lett.* 2004, *85,* 2059-2061.

[10] Bauer, E. In *Single Crystal Films*; Francombe, M. H.; Sato, H. Eds.; MacMillan; New York, NY, 1964; pp 43-67.

[11] Cox, P. A. *Transitions Metal Oxides: An Introduction to Their Electronic Structure and Properties;* Oxford Science Publications; Oxford, UK 1992; pp 29-30, 37-115.

[12] Tromp, R. M.; Hannon, J. B. *Surf. Rev. Lett.* 2002, *9*, 1565-1593.

[13] Equation (1) is an exact expression for the condition of reflection *only* from the film/air interface (and not from the film/substrate interface), which is the case for high α. If refection occurs at both interfaces (at lower α), the antecedent term is $(1\text{-R})^2$. Note that another frequently-used but approximate expression in the limit of high α for a multilayer is given in Pankove, J. I. *Optical Processes in Semiconductors;* Dover; New York, NY 1975; pp 34-36.

[14] Pankove, J. I. *Optical Processes in Semiconductors;* Dover; New York, NY 1975; pp 34-52.

[15] Cody, G. D. In *Hydrogenated Amorphous Silicon: Optical Properties;* Pankove, J. I.; Ed.; Semiconductors and Semimetals Series; Academic, New York, NY, 1984;Vol. 21, pp 11-82.

[16] Bethe H. A.; Jackiw, R. *Intermediate Quantum Mechanics*; 2nded; W. A. Benjamin; New York, NY, 1968; pp 209-215.

[17] Hoppe, E. E.; Aita, C. R.; Sorbello, R. S. unpublished.

[18] Aita, C. R.; Hoppe E. E.; Sorbello, R. S. *Appl. Phys. Lett.* 2003, *82,* 677-679.

[19] Wood, R. W. *Physical Optics;* Dover, New York, NY, 1967; p 101-103.

[20] Green, D. J.; Hannink, R.H.J.; Swain, M. V. *Transformation Toughening of Ceramics*; CRC Press; Boca Raton, FL,1989; pp 1-55.

[21] Scanlan, C.M.; Gajdardziska-Josifovska, M.; Aita, C. R. *Appl. Phys. Lett.* 1994, *64*, 3548-3550.

[22] Aita, C. R.; Wiggins, M. D.; Whig. R.; Scanlan, C. M.; Gajdardziska-Josifovska, M. *J.* Appl. Phys. 1996, *79,* 1176-1178.

[23] Gajdardziska-Josifovska, M; Aita, C. R. *J. Appl. Phys.*1996, *79,* 1315-1319.

[24] Schofield, M.; Gajdardziska-Josifovska, M; Aita, C. R.; Rice, P. M. *Thin Solid Films* 1998, *326,* 106-116.

[25] Azaroff, L. V. *Elements of X-ray Crystallography*; McGraw-Hill; New York, NY, 1968; p. 193.

[26] French, R. H.; Glass, S. J.; Ohuchi, F. S.; Xu, Y.-N.; Ching, W. Y. *Phys. Rev. B* 1994, *49,* 5133-5141.

[27] Morinaga, M.; Adachi, H.; Tsukada, M. *J. Phys. Chem. Solids* 1983, *44,* 301-306.

[28] Zandiehnadem, F.; Murray, Ching, W. Y. *Physica B* 1988, *150,* 19-24.

[29] Tauc, J.; In *Amorphous and Liquid Semiconductors;* J. Tauc; Ed.*;* Plenum; London, UK, 1974, pp 159-220.

[30] Králik, B.; Chang, E. K.; Louie, S. G. *Phys. Rev. B* 1998, *57,* 7027-7036.

[31] Jomard, G.; Petit, T.; Pasturel, A.; Magaud, L.; Kresse, G.;Hafner, J. *Phys. Rev. B* 1999, *59,* 4044-4052.

[32] Orlando, R.; Pisani, C.; Roetti, C.; Stefanovich, E. *Phys. Rev. B* 1992, *45,* 592-601.

[33] Lysenko, V. A.; *Inorg. Mater.*1994, *30,* 930-932.

[34] Kang, J.; Lee, E.-C.; Chang, K. *J. Phys. Rev B* 2003, *68,* 054106(1)-054106(8).

[35] Jaffe, J. E.; Bachorz, R. A.; Gutowski, M. *Phys. Rev. B* 2005, *72,* 144107(1)-144107(9).

[36] Ohtaka, O.; Yamanaka, T.; Kume, S; Hara, N.; Asano, H. *J. Amer. Ceram Soc.* 1995, *78,* 233-237.

[37] Lowther, J. E.; Dewhurst, J. K.; Leger, J. M.; Haines, J. *Phys Rev. B* 1995, *60,* 14485-14488.

[38] Hoppe, E. E.; Aita, C. R.; Gajdardziska-Josifovska, M. *Appl. Phys. Lett.* 2003, *91* 203105(1)-203105(3). Erratum: Hoppe, E. E.; Aita, C. R.; Gajdardziska-Josifovska, M. *Appl. Phys. Lett.*2003, *92,*109903(1).

[39] Zheng, W.; Bowen, Jr., K. H.; Li, J.; Dąbkowska; M. Gutowski *J. Phys. Chem. A* 2005, *109* 11521-11525.

[40] See, for example, the Reviews and references within: Wilk, G. D.; Wallace, R. M.; Anthony, J. M. *J. Appl. Phys.*2001, *89,* 5243-5275. Robertson, J.; Falabretti, B. *J. Appl. Phys.* 2006, *100,* 014111(1)-014111(8).

[41] Kirm, M.; Aarik, J.; Jürgens, M.; Sildos, I. *Nuc. Instrum. Meth. Phys. Res.* 2005, *537,* 251-255.

[42] Takeuchi, H.; Ha, D.; King, T.-J. *J. Vac. Sci. Technol. A*2004, *22,* 1337-1341.

[43] Aarik, J.; Mändar, H.; Kirm, M.; Pung, L. *Thin Solid Films* 2004, *466,*41-47.

[44] Hoppe, E. E.; Sorbello, R. S.; Aita, C. R. *J. Appl. Phys.* 2007, *101,* 123534(1)-123534(6).

[45] Cho, Y.J.; Nguyen, N.V.; Richter, C.A.; Ehrstein, J.R.; Lee, B.H.; Lee, J. C. *Appl. Phys. Lett.* 2002, *80*, 1249-1251.

[46] Nguyen, N.V.; Davydov, A.V.; Chandler-Howowitz, D.; Frank, M. *Appl. Phys. Lett.* 2005, *87,* 192903(1)-192903(3).

[47] Aita, C. R.; Cisneros-Morales, M. C.; Hoppe, E. E. *J. Phys. Chem. C*2012, *116,* 26679-26680.

[48] Shluger, A. L.; McKenna, K. P.; Sushko, P. V.; Muñoz Ramo, D.; Kimmel, A. V. *Modelling Simul. Mater. Sci. Eng.* 2009, *17,* 084004(1)-084004(21).

[49] Muñoz Ramo, D.; Shluger, A.L.; Gavartin, J.L.; Bersuker, G. *Phys. Rev. Lett.* 2007, *99,* 155504(1)-(155504(3).

[50] Muñoz Ramo, D.; Gavartin, J.L.; Shluger, A.L.; Bersuker, G., *Microelectronic Eng.* 2007, *84,* 2362-2366.

[51] Cisneros-Morales, M. C.; Aita, C. R. *Appl. Phys. Lett.* 2010, *96,* 191904(1)-191904(3).

[52] Hoppe, E. E.; Aita, C. R. *Appl. Phys. Lett.* 2008, *92* 141912(1)- 141912(3) (3). Erratum: Hoppe, E. E.; Aita, C. R.; Gajdardziska-Josifovska, M. *Appl. Phys. Lett.* 2010, *97,* 269904(1).

[53] Onodera, Y.; Toyozawa Y. *J. Phys. Soc. Jpn.*, 1968, *24,* 341-355.

[54] Matsuda, H.; Miyata T. *Suppl. Prof. Theo. Phys.*1968, *Extra Edition,* 450-463.

[55] Morosin, B.; Lynch, R. W. *Acta Crystallogr. Sect. B: Struct. Crystallogr. Cryst. Chem.* 1972, *28,* 1040-1046.

[56] Baudin, C.; Sayir, A.; Berger, M. H. *Key Eng. Mater.* 2005, *290,* 199-202.

[57] Thomas, H. A. J.; Stevens, R. *Br. Ceram. Trans. J.*1989, *88,* 144-151.

[58] Zaitsu, S.-I.; Jitsuno, T.; Nakatsuka, M.; Yamanaka, T.; Motokoshi, S. *Appl. Phys. Lett.*2002, *80,* 2442-2444.

[59] Pacheco-Malagon, G; Garcia-Borquez, A.; Coster, D.; Sklyarov, A.; Petit, S.; Fripiat, J. J. *J. Mater. Res.* 1995, *10*, 1264-1269.

[60] Kuo, D. H.; Tzeng,K. H. *Thin Solid Films* 2002, *420–421,* 497-502.

[61] Nishiguchi, H., Zhang, J.-L.; Anpo, M.; Masuhara, M. *J. Phys. Chem. B* 2001, *105,* 3218-3222.

[62] Anpo, M.; Kawamura, T.; Kodama, S.; Msruya, K.; Onishi, T. *J. Phys. Chem.* 1988, *88,* 438-440.

[63] Navrotsky, A. *Am. Mineral.* 1975, *60,* 249-256.

[64] Sanchez-Agudo, M.; Soriano, L.; Quiros, C.; Avila, J.; Sanz, J.M. *Surf. Sci.* 2001, *482–485,* 470-475.

[65] Sanchez-Agudo, M.; Soriano, L.; Quiros, C.; Abbate, M.; Roca, L.; Avila, J.; Sanz, J. M. *Langmuir* 2001, *17,* 7339-7343.

[66] Steveson, M.; Bredow, T.; Gerson, A. R. *Phys. Chem. Chem. Phys.* 2002, *4*, 358-365.

[67] Gesenhues, U.; Rentschler, *T. J. Solid State Chem.* 1999, *143,* 210-218.

[68] Omari, M. A.; Sorbello, R. S.; Aita, C. R. *J. Vac. Sci. Technol. A* 2006, *24,* 317-323.

[69] Omari, M. A.; Sorbello, R. S.; Aita, C. R. *J. Appl. Phys.* 2006, *99,* 123508(1)-123508(6).

[70] Aita, C. R. *J. Vac. Sci. Technol.* 2006, *24,* 2054-2060.

[71] Aita, C. R. *Appl. Phys. Lett.* 2007, *90,* 213112(1)-213112(3).

[72] Aita, C. R. *J. Vac. Sci. Technol.* 2009, *27,* 648-652.

[73] Lynch, R. W.; Morosin, B. *J. Am. Ceram. Soc.* 1972, *55,* 410-413.

[74] McHale A. E.; Roth, R. S. *J. Am. Ceram. Soc.* 1983, *66,* C18-C20.

[75] McHale A. E.; Roth, R. S. *J. Am. Ceram. Soc.* 1986, *69,* 827-832.

[76] Dubrovinskaia, N.; Dubrovinsky, L.S.; Ahuja, R.; Prokopenko, V. B.; Dmitriev, V.; Weber, H.-P.; Osorio-Guillen, J.-M.; Johansson, B. *Phys. Rev. Lett.* 2001, *87*, 275501(1)- 275501(4).
[77] El Goresy, A.; Chen, M.; Dubrovinsky, L.; Gillet, P.; Graup, G. *Science* 2001, *293,* 1467-1470.
[78] DeLoach, J. D.; Aita, C. R. *J. Mater. Sci. Lett.* 2000, *19,* 1123-1125.
[79] DeLoach, J. D.; Shibilski, J. J.; Crape, C. R.; Aita, C. R. *J. Vac. Sci. Technol. A* 2000, *18,* 2922-2927.
[80] DeLoach, J. D.; Aita, C. R.; Loong, C.-K. *J. Vac. Sci. Technol. A* 2002, *20,* 1517-1524.
[81] Aita, C. R.; DeLoach, J. D.; Yakovlev, V. V. *Appl. Phys. Lett.* 2002, *81,* 238-240.
[82] Aita, C. R.; DeLoach, J. D.; Sorbello, R. S. *J. Appl. Phys.* 2003, *94,* 654-663.
[83] Arashi, H.; Yagi, T.; Akimoto, S.; Kudoh, Y. *Phys. Rev. B.* 1990, *41*, 4309-4313.
[84] Krebs, M. A.; Condrate, Sr., R. A. *J. Mat. Sci. Lett.* 1988, *7,* 1327-1330.
[85] Zhilin, A. A.; Petrov, V. I.; YaTsenter, M.;Chuvaeva, T. I. *Opt. Spectrosc.* 1992, *73,* 1151-1157.
[86] Azough, F.; Freer, R.; Petzelt, J.*J. Mat. Sci.* 1993, *28,* 2273-2276.
[87] Kim, Y. K. Jang, H. M. *J. Appl. Phys.* 2001, *89,* 6349-6355.
[88] Chen, H.-R.; Shi, J.-L.; Zhang, W,-H.; Ruan, M.-L.; Yan, D.-S. *Chem. Mater.* 2001, *13,* 1035-1040.
[89] Tanabe, K; Sumiyoshi, T.; Shibata, K.; Kiyoura, T. J. Kitagawa, *Bull. Chem. Soc. Jpn.*, 1974, *47,* 1064-1065.
[90] Kung, H. H. *J. Solid State Chem.*, 1984, *52,* 191-196.
[91] Shin, H.; Agarwal, M.; De Guire, M. R.; Heuer, A. H. *J. Am. Ceram. Soc.* 1996, *79,* 1975-1978.
[92] Gosele, U.; Tu, K. N. *J. Appl. Phys.* 1989, *66,* 2619-2626.
[93] d'Heurle, F. M. *J. Mater. Res.* 1988, *3,* 167-195.
[94] Sonnenberg, N.; Longo, A. S.; Cima, M. J.; Chang, B. P.; Ressler, K. G.; McIntyre, P. C.; Liu, Y. P. *J. Appl. Phys.* 1993, *74,* 1027-1034.
[95] McIntyre, P. C.; Ressler, K. G.; Sonnenberg, N.; Cima, M. *J. Vac. Sci. Technol. A* 1996, *14,* 210-215.
[96] Kingery, W. D.; Bowen, H. K.; Uhlmann, D. R. *Introduction to Ceramics;* Wiley-Interscience; New York, NY, 1976; pp. 714, 727.
[97] Kwok, C.-K.; Aita, C. R.; *J. Vac. Sci. Technol. A* 1989, *7,* 1235-1239.
[98] Bendoraitis, J. G.; Salomon, R. E. *J. Phys. Chem.* 1965, *69,* 3666-3667.
[99] Cronemeyer, D. C. *Phys. Rev.* 1952, *87,* 876-886.
[100] Grant, F. A. *Rev. Mod. Phys.* 1959, *31*, 646-673.
[101] Pascaul, J; Camassel, J.; Mathieu, H. *Phys. Rev. B* 1978, *18*, 5606-5614.
[102] Tang, H.; Levy, F.; Berger, H.; Schmid, P. E. *Phys. Rev. B* 1995, *52,* 7771-7775.
[103] Wells, A. F. *Structural Inorganic Chemistry;* Clarendon; Oxford, UK, 1950; p 35.
[104] Ruh, R.; Hollenberg, G. W.; Charles, E.G.; Patel, V. A. *J. Am. Ceram. Soc.* 1976, *59,* 495-499.
[105] Coutures J. P.; Coutures, J. *J. Am. Ceram. Soc.* 1987, *70,* 383-387.
[106] Li, M.; Zhang, Z.; Campbell, S. A.; Gladfelter, W. L; Agustin, M. P.; Klenov, D. O.; Stemmer, S. *J. Appl. Phys.* 2005, *98,* 054506(1)-054506(8).
[107] Li, M.; Zhang, Z.; Campbell, S. A.; Li, H.-J. Peterson, J. J. *J. Appl. Phys.* 2007, *101*, 044509(1)-044509(9).

[108] Honda, K.; Sakai, A.; Sakashita, M.; Ikeda, H.; Zaima, S.; Yasuda, Y. *Jpn. J. Appl. Phys.* 2004, *43,* 1571-1576.

[109] Chen, F.; Bin, X.; Hella, C.; Shi, X.; Gladfelter, W. L.; Campbell, S. A. *Microelect. Eng.* 2004, *72,* 263-266.

[110] Triyoso, D. H.; Hedge, R. I.; Zollner, S.; Ramon, M. E.; Kalpat, S.; Gregory, R.; Wang, X.-D.; Jiang,J.; Raymond, M.; Rai, R.; Werho, D.; Roan, D.; White, Jr., B.E.; Tobin, P. J. *J. Appl. Phys.* 2005, *98,* 054104(1)-054104(8).

[111] Cisneros-Morales, M. C.; Aita, C. R. *Appl. Phys. Lett.* 2008, *93,* 021915(1)-021915(3).

[112] Cisneros-Morales, M. C.; Aita, C. R. *J. Vac. Sci. Technol. A* 2010, *28,* 1161-1168.

[113] Cisneros-Morales, M. C.; Aita, C. R. *J. Appl. Phys.* 2010, *108*, 123506(1)-123506(8).

[114] Cisneros-Morales, M. C.; Aita, C. R. *Appl. Phys. Lett.* 2011, *98,* 051909(1)-051909(3).

[115] Cisneros-Morales, M. C.; Aita, C. R. *J. Appl. Phys.*2011, *109*, 123523(1)-123523(8).

[116] Cisneros-Morales, M. C.; Aita, C. R. *J. Appl. Phys.*2012, *111*, 109904(1)-109904(3).

[117] Hoppe, E. E.; Cisneros-Morales, M. C.; Aita, C. R. *APL Mater.* 2013, *1,* 022108(1)-022108(6).

[118] Joint Committee on Powder Diffraction Standards Card No. 78-0050.

[119] Joint Committee on Powder Diffraction Standards Card No. 8-0342.

[120] Joint Committee on Powder Diffraction Standards Card No. 40-0794.

[121] Kidchob,T.; Falcaro, P.; Schiavuta, P.; Enzo, S.; Innocenzi, P. *J. Amer. Ceram. Soc.* 2008, *91,* 2112-2116.

[122] Copel, M.; Reuter, M. C.; Kaxiras, E.; Tromp, R. M. *Phys. Rev. Lett.* 1989, *63,* 632-635.

[123] Lucovsky, G.; Fulton, C. C.; Zhang, Y.; Zou, Y.; Luning, J.; Edge, L. F.; Whitten, J. L.; Nemanich, R. J.; Ade, H.; Schlom, D. G.; Afanase'v, V. V.; Stesmans, A.; Zollner, S.; Triyoso, D.; Rogers, B. R. *IEEE Trans. Device Mater. Reliab.* 2005, *5,* 65-83.

[124] Studenyak, I. P.; Nahusko, O. T.; Kranjčec, M. *Vacuum* 2007, *82,* 35-38.

[125] Domaradzki, J.; Kaczmarek, D.; Prociow, E.L.; Borkowska, A.; Kudrawiec, R.; Misiewicz, J.; Schmeisser, D.; Beukert, G. *Surf. Coat. Technol.* 2006, *200,* 6283-6287.

[126] Ye, C.; Wang, H.; Zhang, J.; Ye, J.; Wang, Y.; Wang, B.; Jin, Y. *J. Appl. Phys.* 2010, *107,* 104103(1)-104103(3).

[127] Fulton, C. C.; Lucovsky, G.; Zhang, Y.; Zou, Y.; Nemanich, R. J.; Ade, H.; Whitten, J. L. *J. Electron Spectrosc. Relat. Phenom.* 2005, *144–147,* 913-916.

[128] Christensen, A; Carter, E. A. *Phys. Rev. B* 2000, *62,* 16968-16983.

[129] Aita, C. R.; *U. S. Patent Specification 5472795;* 1995.

[130] Aita, C. R.; Yakovlev, V.; Cayton, M.; Mirhoseini, M.; Aita, M.; *U. S. Patent Specification 6869701;* 2005.

In: Oxide Electronics and Functional Properties …
Editor: Alexander Pergament
ISBN: 978-1-63321-499-6

Chapter 4

OPTICAL AND ELECTRICAL SWITCHING OF THERMOCHROMIC VO_2 SMART COATINGS

Mohammed Soltani*
RSL-Tech, Montreal, Quebec, Canada

ABSTRACT

Thermochromic vanadium dioxide (VO_2) exhibits a reversible semiconductor-to-metallic phase transition (SMT) at a relatively low transition temperature ($T_t \approx 68°C$). The T_t can be easily controlled by doping the VO_2 with impurities such as Al, Cr, W, Ti, F, and Mg, etc. The SMT is accompanied by a strong modification of the electrical and optical properties in the infrared region. The SMT can be controlled by various external stimuli, including temperature, pressure, electric field, photo-excitation, and carrier injection in VO_2 heterostructure. In addition, the time switching of the SMT is ultrafast: the VO_2 film switches on timescales of ~500 femtoseconds. These characteristics make VO_2 an ideal smart material for use in various applications. This chapter gives a brief overview of the electrical and optical properties and some applications of undoped as well as W&Ti doped VO_2.

Keywords: Thermochromic, Vanadium Dioxide (VO_2), semiconductor-to-metallic phase transition (SMT), transition temperature, switching, W-Ti codoped VO_2

1. INTRODUCTION AND PROPERTIES OF VO_2

Thermochromic vanadium dioxide (VO_2) smart coatings present semiconductor-to-metallic phase transition (SMT) at relatively low transition temperature ($T_t \sim 68$ °C) [1]. This phase transition is accompanied by a structural change from a low-temperature monoclinic phase (semiconducting state) to a high-temperature tetragonal phase (metallic state). Figure 1 shows the monoclinic and tetragonal structures of VO_2. The SMT is also accompanied by a

* Email: mosoltani@hotmail.com.

strong modification of the electrical and optical properties in the infrared region: the electrical resistivity decreases by several magnitudes with increasing the temperature (see Figure 8), while the VO_2 is transmitting in the semiconducting state and become more reflective and opaque in the metallic state (Figure 2).

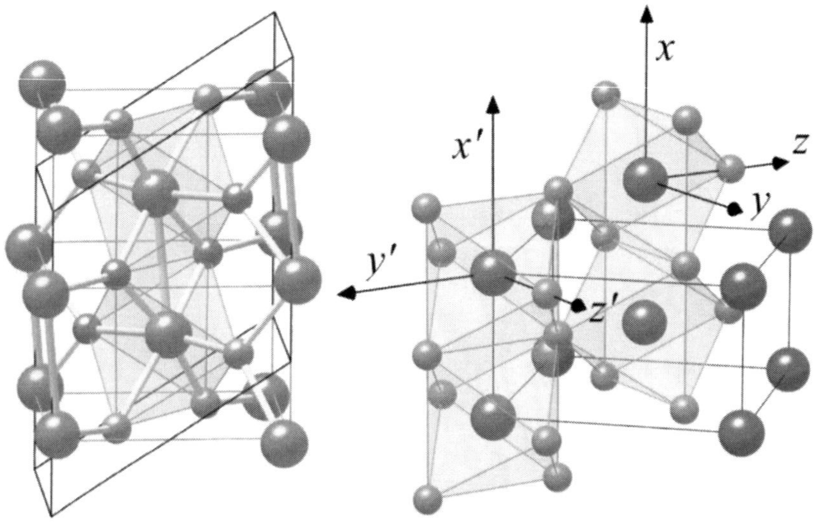

Figure 1. Crystalline structure of VO_2. Low-temperature monoclinic (semiconducting state; left) and high-temperature tetragonal rutile (metallic state; right). The structures are shown with orange/red spheres representing vanadium atoms and blue/purple spheres representing oxygen atoms. Taken from [2] with permission.

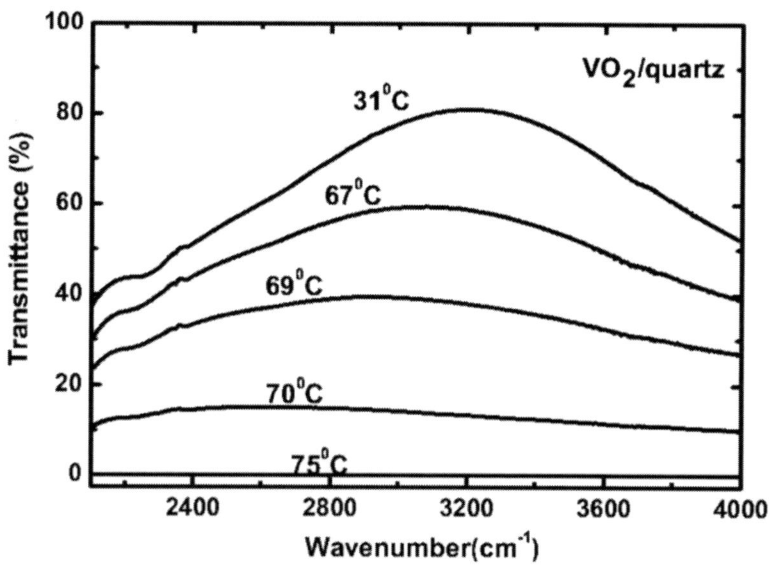

Figure 2. Temperature dependence of the Infrared transmittance during the heating cycle for a VO2-coated quartz substrate. Taken from [3] with permission.

Figure 2 shows the temperature dependence of the infrared transmittance during the heating cycle for VO_2-coated quartz. It is observed that the VO_2 is transmitting in the semiconducting state at room temperature and is completely opaque in the metallic state at 75 °C.

The T_t can be controlled by doping the VO_2 with various dopants such as Al, Ti, W, F, etc. In addition, the SMT of VO_2 can be easily controlled by external parameters including temperature, pressure, photo-carrier injection into VO_2 heterostructure, photo-excitation, and an electric field.

The time switching of the SMT is ultrafast: the VO_2 thin film switches on timescales of ~500 femtoseconds [4, 5]. However, the physical mechanism behind the ultrafast switching of VO_2 is still under debate in the literature. Two models are used to describe the SMT: (i) the Peierls model [6-12], in which the SMT is described in terms of interactions between electrons and phonons and is structurally driven; and (ii) the Mott-Hubbard model [7, 11-13], which describes the SMT in terms of an electron-electron correlation, and is therefore charge driven.

These characteristics make that undoped and doped-VO_2 coatings are excellent materials that can be exploited in various applications such as uncooled IR microbolometers [14], holographic storage systems [15], optical fiber switches [16], ultrafast switching, smart windows [17-20], sunshields for spacecraft [21], optical limiting devices [22], all-optical switches [23], thermo-optical modulator [24], RF-microwave switches [25], metamaterials [26, 27], plasmonic [28-30], electro-optical switches [31], THz devices [32], negative capacitors [33], field effect transistors [34, 35], active shutters [36], photonic resonators [37], smart radiator devices for spacecraft [38, 39], etc.

Figure 3 shows the scheme of the valence band diagram for the semiconducting (monoclinic; right) and metallic (tetragonal rutile; left) phases of VO_2. In the metallic state, VO_2 has one outer d electron per molecule and the two $d_{//}$ and $d_{//}^*$ overlap on one band. In the semiconducting state, the vanadium π^* band is above the Fermi energy level (E_F) and the 3d band is split on one filed $d_{//}$ band and one empty $d_{//}^*$ band. The band gap energy between these bands is about 0.67 eV [40].

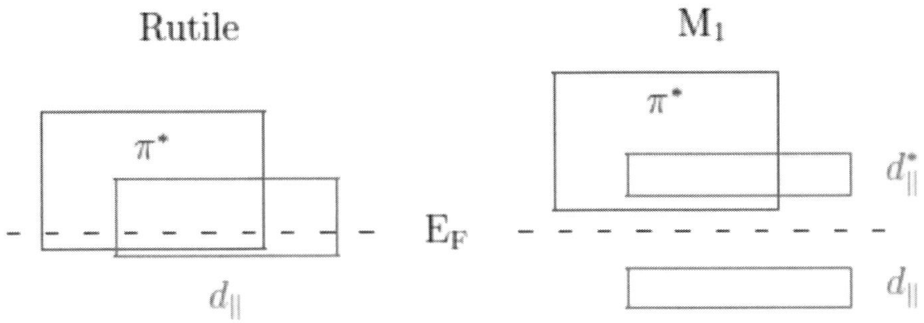

Figure 3. Valence band diagrams for metallic tetragonal (rutile) and semiconducting monoclinic (M_1) states of VO_2. Taken from [2] with permission.

2. Synthesis of VO_2

Chemical vapor deposition [41], reactive electron-beam evaporation [42], reactive magnetron sputtering [43, 44], sol-gel method [45-49], hydrothermal processes [50], physical vapor transport [51], ion-activated reactive evaporation [52], and reactive pulsed laser deposition (RPLD)[53] are currently used for synthesis of VO_2 coatings. High quality of VO_2 can be performed by means of RPLD, which allows a judicious control of the different deposition parameters such as substrate temperature, oxygen concentration and the deposition pressure. Also, the RPLD process of VO_2 coatings at high temperature requires no additional post annealing. In addition, the doping of VO_2 can be achieved easily by using either metal-doped vanadium target or dual-metal-vanadium target [23].

In the present study, the undoped and metal-doped VO_2 coatings were synthetized by means of RPLD. The optimization of the RPLD parameters shows clearly that the deposition pressure and the oxygen-to-argon ratio are the principal parameters that control the formation of the VO_2 single phase. Figure 4 compares the XRD patterns of scans from two vanadium oxide films deposited onto Si(100) at substrate temperature of 520 °C and at different total pressure and different O_2/Ar. It is observed that V_2O_5 phase with (001) preferable orientation is formed at pressure of 200 mtorr and O_2/Ar of 10%, while the VO_2 single phase with (011) preferable orientation is achieved with pressure of 100 mtorr and 5% of O_2/Ar.

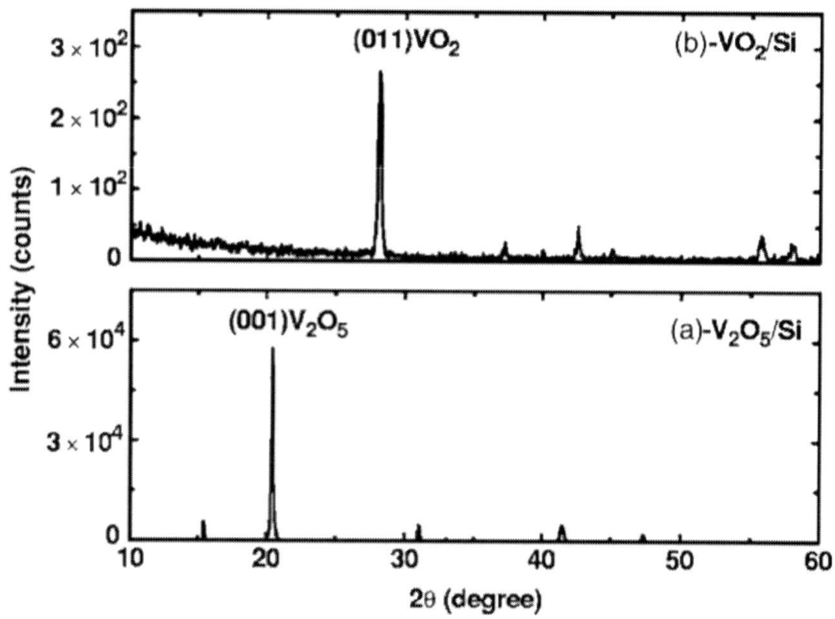

Figure 4. XRD patterns of VO_2 and V_2O_5 coated Si(100) at substrate temperature of 520 °C. (a) V_2O_5 phase achieved at pressure of 200 mtorr with O_2/Ar of 10% ; (b) VO_2 single phase achieved at pressure of 100 mtorr and O_2/Ar of 5%. Taken from [21] with permission.

Figure 5 compares the XRD patterns of VO_2 single phase achieved onto Si(100) with the optimized parameters (100 mtorr and 5% of O_2/Ar) and at substrate temperature of 300, 420 and 520 °C. It is observed that all films are VO_2 single phase with (011) preferable orientation. However, the deposition temperature affects considerably the switching contrast between the semiconducting and metallic states of VO_2: the switching contrast decreases with decreasing the deposition temperature. Figure 6 shows the IR transmittance switching at 2.5 μm for VO_2 deposited onto Si at low substrate temperature of 300 °C. It is observed that the transition temperature is similar to that of single crystal VO_2 ($T_t \approx 68$ °C) and the contrast switching (about 7%) is less than that of VO_2 deposited at higher temperature (see Figure 7).

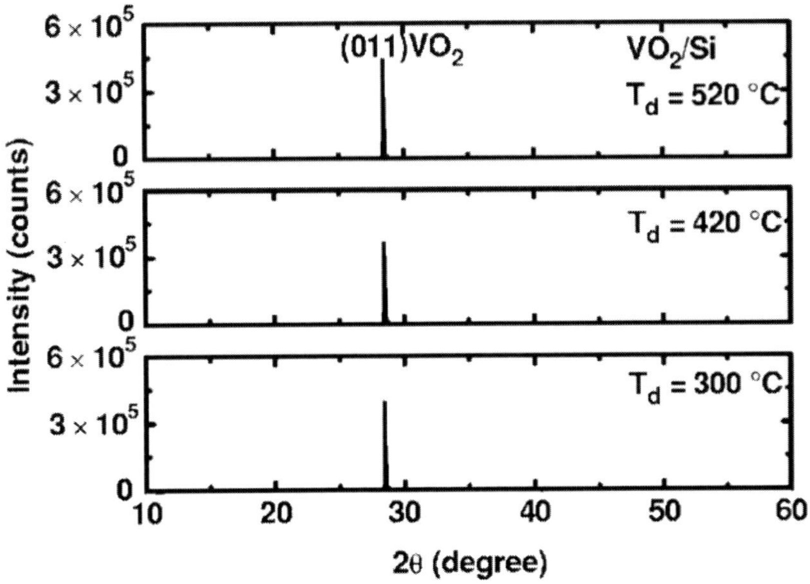

Figure 5. Substrate temperature (300, 420, and 520 °C) effect on the XRD patterns of single phase VO_2 deposited onto (100) Si at 100 mtorr and 5% of O_2/Ar. Taken from [21] with permission.

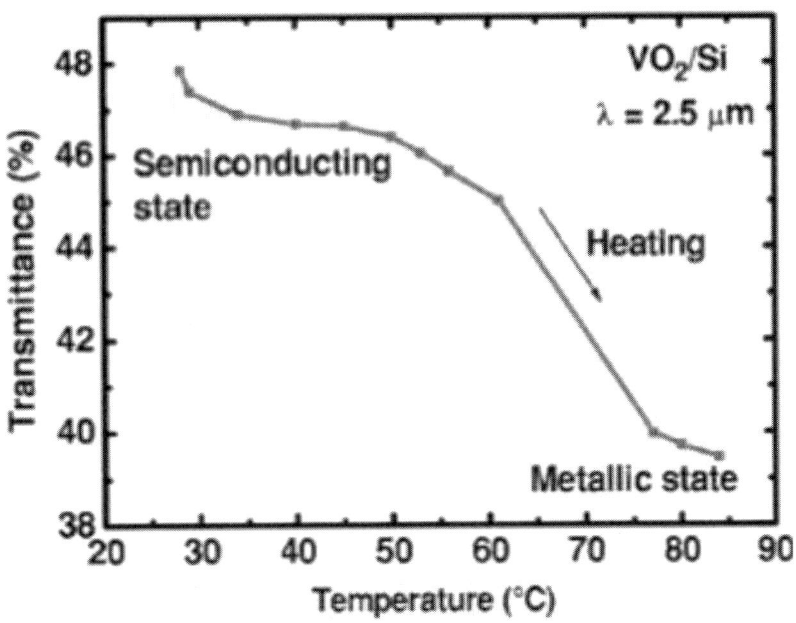

Figure 6. Temperature dependence of the IR transmittance at a wavelength of 2.5 μm in the heating cycle for VO_2 onto Si deposited at 300 °C. Taken from [21] with permission.

3. W-Doping and W-Ti-Co-Doping Effects on Phase Transition of VO_2

As mentioned before, doping can control the transition temperature of VO_2. Figure 7 compares the optical hysteresis at a wavelength of 2.5 μm obtained from the IR transmittance as a function of temperature in both heating and cooling cycles of undoped, W(1.4 at. %)-doped, and Ti(12 at.%)-W(1.4 at. %) co-doped VO_2 deposited onto quartz substrates at 520 °C [3]. It is observed that the transmittance decreases with increasing the temperature for undoped as well as for W-doped and Ti-W-co-doped VO_2. All these coatings are completely opaque in the metallic state ($IR_{trans} = 0$). However, W and Ti dopants affect considerably the transmittance of the coatings in the semiconducting state: the transmittance drops from 50% for undoped VO_2 to about 30 % for W-doped VO_2, while the transmittance is about 40% for the W-Ti co-doped VO_2. We observe also that the W and Ti doping affect strongly the transition temperature: the T_t decreases from 68 °C for undoped VO_2 to 36 °C for W(1.4 at. %)-doped VO_2. Thus, the T_t decreases by about 22.85 °C per 1 at. % of W doping ions added. The optical hysteresis is about 5 °C for either undoped as well as W-doped VO_2 (see Figure 7).

As shown in Figure 7, the Ti-W co-doping affects considerably both the transition temperature and the optical hysteresis: the T_t is about 60 °C, while the hysteresis is completely suppressed. This characteristic can be exploited in the fabrication of optical modulators.

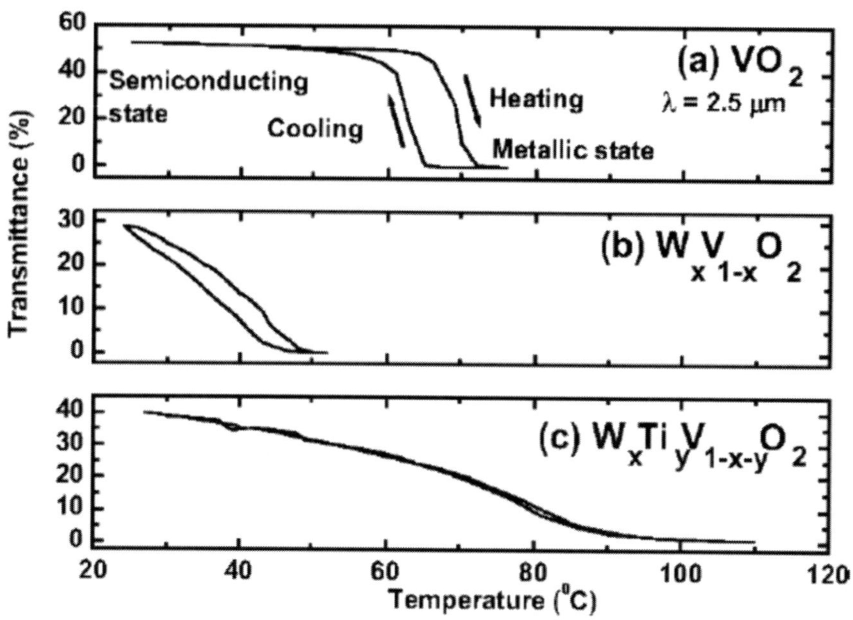

Figure 7. Comparison of the optical hysteresis lops at 2.5 μm of (a) undoped, (b) W(1.4 at. %)-doped, and (c) Ti(12 at.%)-W(1.4 at. %) codoped VO_2 deposited onto quartz substrates at 520 °C. Taken from [3] with permission.

The W and Ti doping affect also the electrical resistivity of VO_2 as shows in Figure 8. The resistivity decreases as the temperature increases for either undoped as well as for W-doped and T-W-codoped VO_2. This decreasing of the resistivity is due to the temperature-induced increasing of the conductivity when the coatings switch from their semiconducting state to their metallic state. The hysteresis width is about 5 °C for both undoped and W-doped VO_2, while the electrical hysteresis is completely suppressed for Ti-W-codoped VO_2 coating.

This temperature dependence of the electrical resistivity is exploited in IR bolometer sensors in which the electrical resistivity is recorded as a function of the IR heating radiation. The thermal coefficient of resistance (*TCR*), defined as the slop of the log of the electrical resistivity *R*, is the principal parameter that must be maximized in order to achieve high sensitivity of IR bolometers:

$$TCR = \frac{dR}{R \cdot dT} \tag{1}$$

In the semiconducting state, both undoped and W-doped VO_2 present relatively comparable TCR: about –1.76%/°C for VO_2 and about –1.90%/°C for W-doped VO_2. However, Ti-W-codoped VO_2 exhibits much higher TCR (about –5.12%/°C) [3]. This TCR value is comparable to that of single crystal VO_2 (–6%/°C) [54] and much higher than that of VOx (≈ –2%/°C), which is currently used in the uncooled IR micro-bolometers.

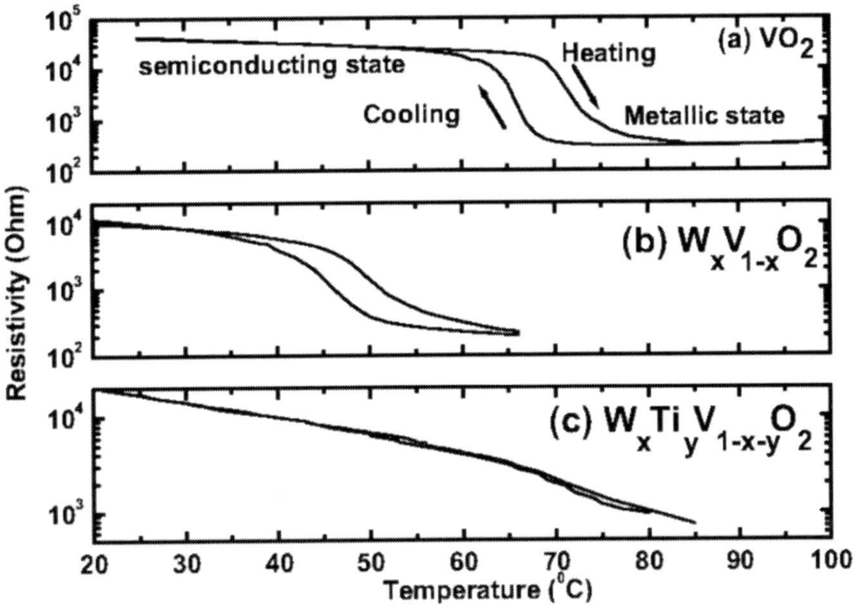

Figure 8. Comparison of the electrical hysteresis of (a) undoped, (b) W(1.4 at. %)-doped, and (c) Ti(12 at.%)-W(1.4 at. %) codoped VO_2 deposited onto quartz substrates at 520 °C. Taken from [3] with permission.

4. ALL-OPTICAL SWITCHING IN VO_2

VO_2 active coatings can be exploited in the fabrication of all-optical switches, in which the optical switching (on/off) of VO_2 is controlled by photo-excitation. Soltani et al. [23]

investigated the all-optical switching in undoped and W(1.4 at.%)-doped VO_2 by means of fibred pump-probe technique. Figure 9 shows the scheme of optical fibred pump-probe experimental setup used to investigate the transmittance switching of undoped and W-doped VO_2 active layers.

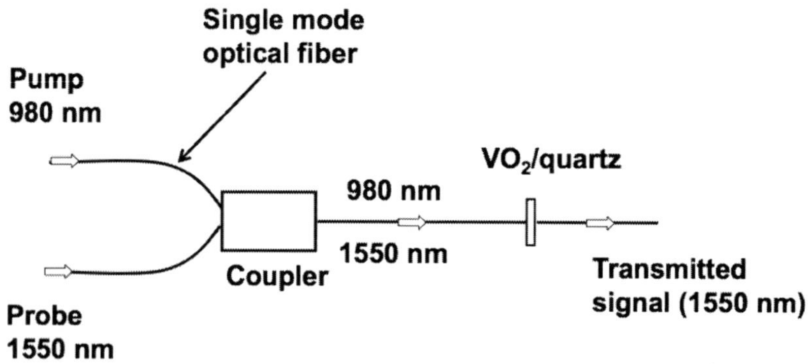

Figure 9. Scheme of optical fibred pump-probe switching setup used to investigate the all-optical switching in undoped and W(1.4 at.%)-)-doped VO_2-coated quartz.

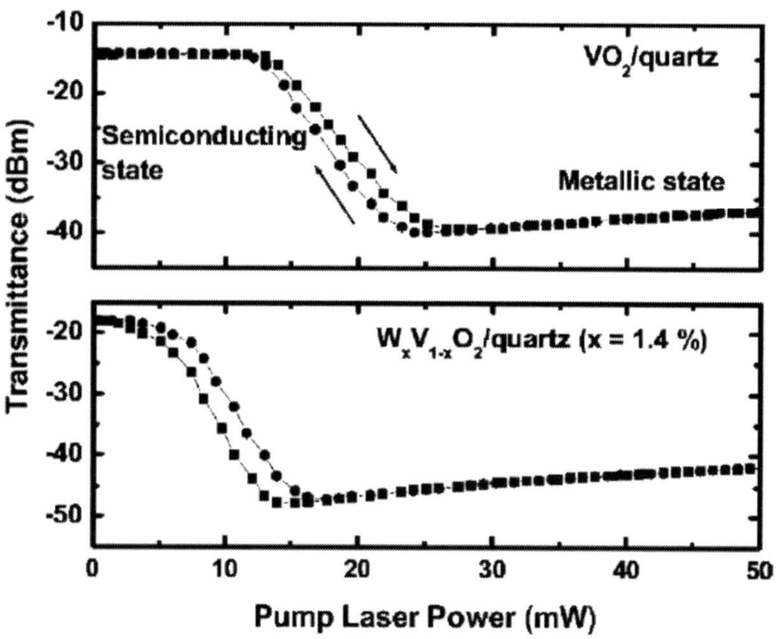

Figure 10. Transmittance switching at $\lambda = 1550$ nm of undoped and W(1.4 at.%)-doped VO_2 coated quartz. Taken from [23] with permission.

In this experiment, a continuous wave beam pump laser ($\lambda = 980$ nm) provided by diode laser with controllable power (up to 60 mW) was used for inducing the SMT by photo-excitation and a probe beam laser ($\lambda = 1550$ nm) provided by tunable laser source was used to inform on the transmitting switching (on/off) state of the VO_2 active coatings. Both beams were coupled in an input single mode optical fiber and excited the coatings at normal incidence. The transmitted light at 1550 nm was recorded as a function of the pump laser power by means of photo-detector.

Figure 10 compares the all-optical switching (on/off) on transmittance mode as a function of the pump laser power for undoped and W-doped VO_2. It is observed that the transmittance decreases with increasing the power of the pump laser: at low power, both coatings are transmitting (i.e., semiconducting state) and become opaque in the metallic state. The pump laser power required to induce the optical switching is about 18 mW for undoped VO_2 and about 10 mW for W-doped VO_2. The optical hysteresis width is about 1.3 mW for VO_2 and 1.95 mW for W-doped VO_2. The contrast switching between the semiconductor (on) and the metallic (off) states is about 25 dB for VO_2 and about 28 dB for W-doped VO_2. Note that the VO_2 can also be used in the fabrication of 1×2 all-optical switches [31]. The mechanism of the switching involved here is due to the change of the energy band gap of VO_2 under photo-excitation with the pump beam laser [33]. Since the photon energy ($h\nu$ = 1.265 eV) of the pump beam laser is higher than that of the energy band gap (~ 0.67 eV) of VO_2, the increasing of the pump beam power induces a photo-excitation of electrons from the filled $d_{//}$ band to the empty $d_{//}^{*}$ band. This creation of excitons in VO_2 causes overlapping of these bands on one half filled valence d band (see Figure 3). As a result, the charge density increases and then the VO_2 switches from its semiconducting (on) to its metallic (off) state.

5. ELECTRO-OPTICAL SWITCHING IN VO_2

The application of an electric field is another interesting means to control reversibly the optical switching of VO_2, which allows the fabrication of 1×2 electro-optical switches devices. In this type of the device, the optical switching is controlled by application of an external switching voltage (either dc or ac). Figure 11 shows the scheme of the VO_2/TiO_2/ITO/glass structure used to investigate both reflectance and transmittance of VO_2 under the application of dc switching voltage between Indium Tin Oxide (ITO) transparent electrode and the VO_2 active coating, which also is used as top electrode. In this structure, the TiO_2 buffer layer is used to improve the crystallinity of the VO_2 layer, while Indium wires were used as electrical contacts [31].

The structure was probed at an incidence angle of 45° by laser beam at λ = 1550 nm provided by an optical fibred tunable laser. The reflected and transmitted lights were collected by two outputs single mode optical fibers and recorded by two photo-detectors. Figure 12 compares the transmittance and reflectance switching of the VO_2/TiO_2/ITO/glass structure as a function of the dc applied voltage. Due to the electrically induced the SMT of VO_2, it is observed that the transmittance decreases, while the reflectance increases with the increasing the applied voltage. The switching contrast between the semiconductor and metallic state is about 12 dB in the transmittance mode and about 5 dB in the reflection mode. The switching voltage is about 11.5 V. Here, the switching mechanism is due to the carrier charge from TiO_2 into VO_2 under the application of the DC voltage. The application of the voltage causes the injection of the carrier charge into VO_2 and then increasing its charge density. As a result, the band gap of VO_2 disappears (see Figure 3) and the VO_2 switches to its metallic reflecting state [31].

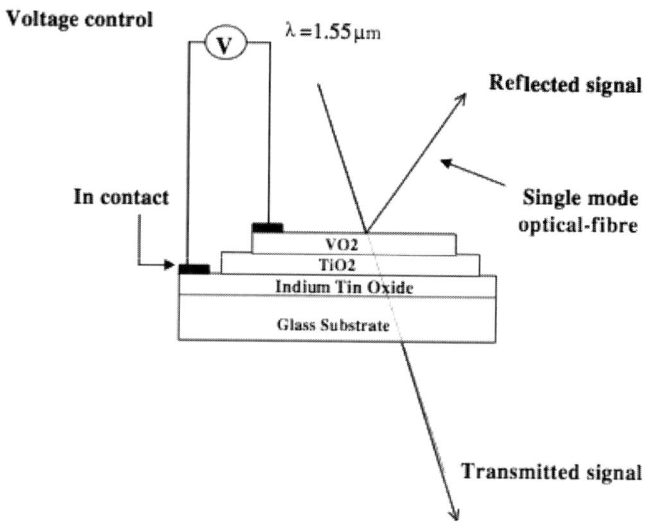

Figure 11. Scheme of the VO_2/TiO_2/ITO/glass structure used to investigate the electro-optical switching of VO2. Taken from [31] with permission.

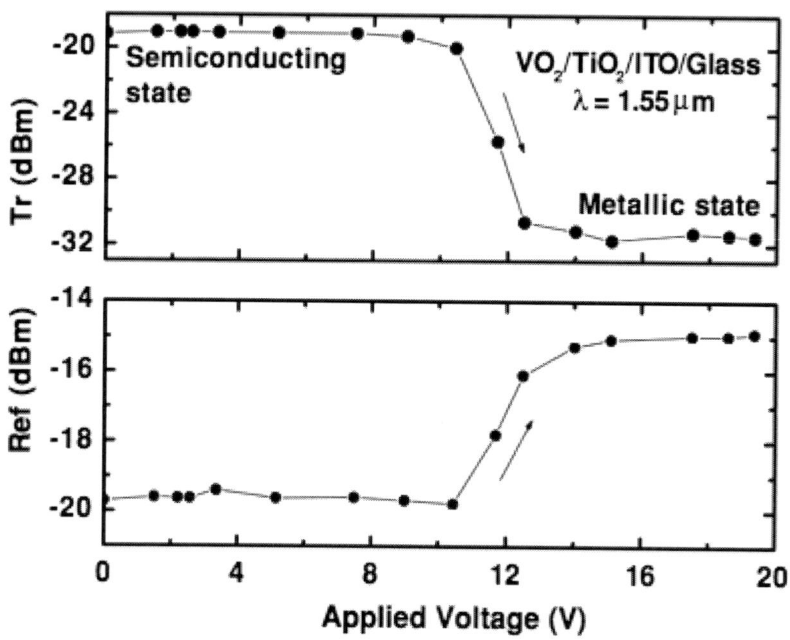

Figure 12. Voltage dependence of both transmittance and reflectance switching at λ = 1550 nm for the VO_2/TiO_2/ITO/glass structure. Taken from [31] with permission.

6. Micro-Optical Switches

W(1.4 at.%)-doped VO_2 was exploited in the fabrication of planar micro-optical switching devices in which the SMT was controlled by an external switching voltage (either dc or ac) [55]. The choice of W-doped VO_2 as active layer was motivated by its low transition temperature (≈ 36 °C), which requiring less voltage to control its optical switching. The starting W-doped VO_2-coated Al_2O_3 presents good transmittance switching as shown in Figure 13: the IR transmittance drops to zero as the W-doped VO_2 switches from its transmitting semiconducting sate to its metallic opaque state.

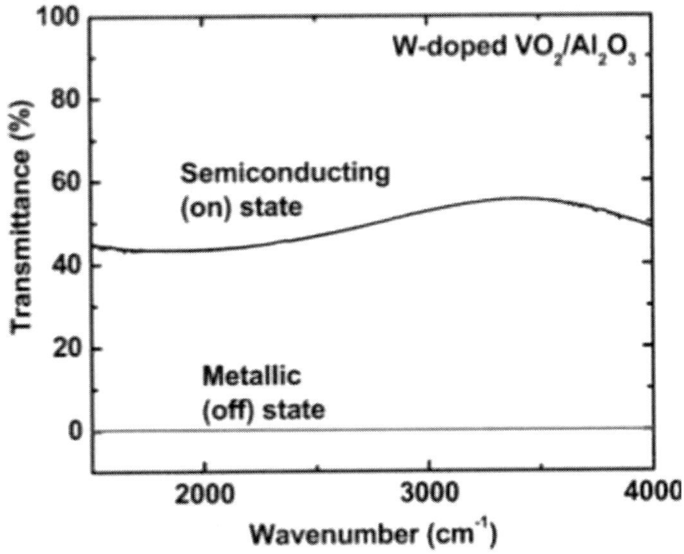

Figure 13. IR transmittance for W-doped VO_2-coated Al_2O_3 in the semiconducting sate at room temperature and in the metallic temperature at 50 °C. Taken from [21] with permission.

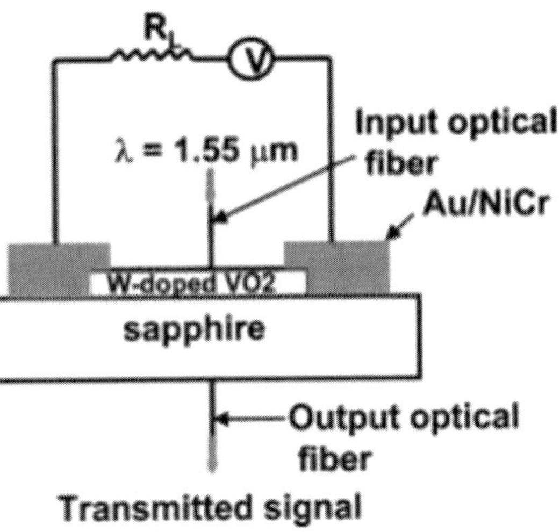

Figure 14. Scheme of the planar micro-optical switch with its electrical circuit used to control the transmittance switching at λ = 1550 nm of W-doped VO_2-coated sapphire. Taken from [21] with permission.

The planar micro-optical slits (100 ×1000 μm^2) were fabricated by the standard photolithography and plasma etching. The fabrication of the micro-optical switches was completed by integrating NiCr electrodes over the micro-slit by means of lift-off process.

Figure 14 shows the scheme of the planar micro-optical switch. The load resistance R_L is used to protect the device from the jump of the current when the W-doped VO_2 switches from its semiconducting state with high electrical resistance to its metallic state with low electrical resistance.

The temperature dependence of the electrical resistance (see Figure 15) of the device showed that the different micro-fabrication steps have no effect on the thermochromic properties of the W-doped VO_2: the resistance decreases with increasing the temperature and the transition temperature (≈ 36 °C) is typically identical to that of the starting W-doped VO_2 layer.

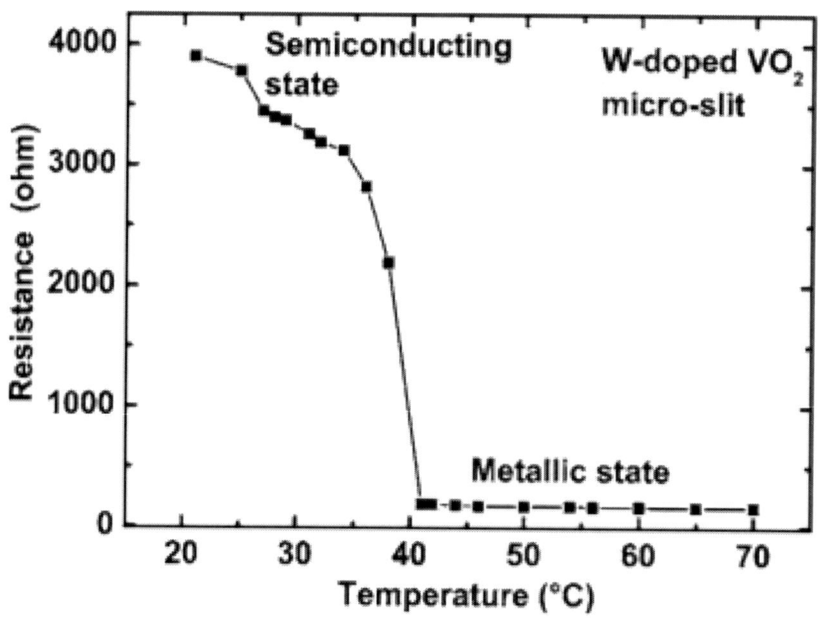

Figure 15. Electrical resistivity as a function of temperature of W-doped VO_2 based planar micro-optical switch. Taken from [21] with permission.

The devise was probed at λ = 1550 nm and the transmitted light was collected by single mode optical fiber and recorded by photodetector. Figure 16 shows the transmittance switching as a function of the applied voltage through the NiCr electrodes. At low voltage, the transmittance remains relatively constant and decreases as the applied voltage increases. The contrast switching between the semiconducting and metallic state is as high as 28 dB. In this case, the switching voltage required to induce the phase transition is about 28 V. The electro-transmittance modulation measurements at λ = 1550 nm were achieved by switching the device with a superposition of dc and ac voltages. The device was switched reversibly about 10 000 cycles without any deterioration of its performance. The mechanism behind of this kind of electrical switching in VO_2 (i.e., electronic or electro-thermal) is still subject of intensive experimental and theoretical investigations in the literature.

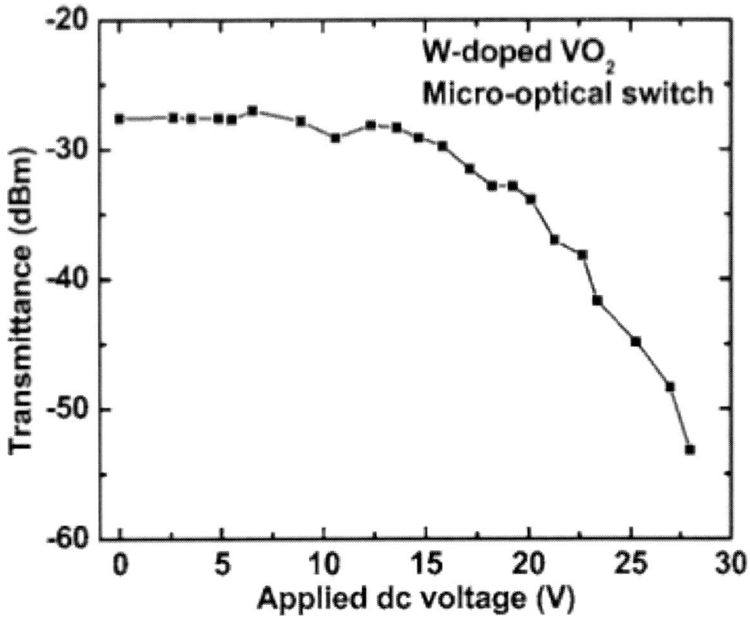

Figure 16. Electro-transmittance switching at for the W-doped VO_2 micro-optical switch device. Taken from [21] with permission.

7. Negative Capacitors Based on Phase Transition of VO_2

Recently, Soltani et al. [33, 56] demonstrated that the SMT of VO_2 can be exploited in the fabrication of planar micro-switch presenting electrically tunable sign of capacitance at room temperature. In this case also, the starting material was high quality of W(1.4 at. %)-doped VO_2 exhibiting good electrical switching. The planar micro-switch (100 μm wide by 1000 μm long) was patterned by photolithography and plasma etching, while the lift-off process was used to achieve the integration of the electrical electrodes over the corners of the W-doped VO_2 planar micro-switch.

The electrical switching of the planar micro-switch was confirmed by measuring its dc current (I)-voltage (V) characteristics at room temperature by using a semiconductor planar analyzer (HP 4145A). Figure 17 (a) shows the measured dc I-V characteristics of the planar micro-switch. It is observed, that the voltage increases monotonously with the current until it reaches the threshold value of 23.5 V for a current threshold of 15 mA. After this threshold, the voltage decreases and the current continue to increase. This is the indication of the negative differential resistance, which occurs when the VO_2 switches from its semiconducting to its metallic state [56]. Figure 17 (b) shows the corresponding variation of the electrical resistance as a function of the applied current. It is observed that the increasing the current induces the switching of the W-doped VO_2 from its semiconducting (high resistance) to its metallic (low resistance) state.

The capacitance and the conductance of the micro-switch were measured as a function of dc voltage and frequency by means of a low frequency analyzer (HP 4192A). Figure 18 shows the frequency dependence (from 1 kHz to 10 MHz) of the conductance and capacitance of the W-doped VO_2 micro switch at 0 V (semiconducting state) and 35 V (metallic state). It is observed that the frequency dependence of either the capacitance as well as the conductance is extremely linked to the semiconducting sate (0 V) and metallic state (35 V). As expected, the conductance of the metallic state is higher than that of the semiconducting state. However, surprisingly, the metallic state presents negative capacitance (NC) values as seeing in Figure 18 (b).

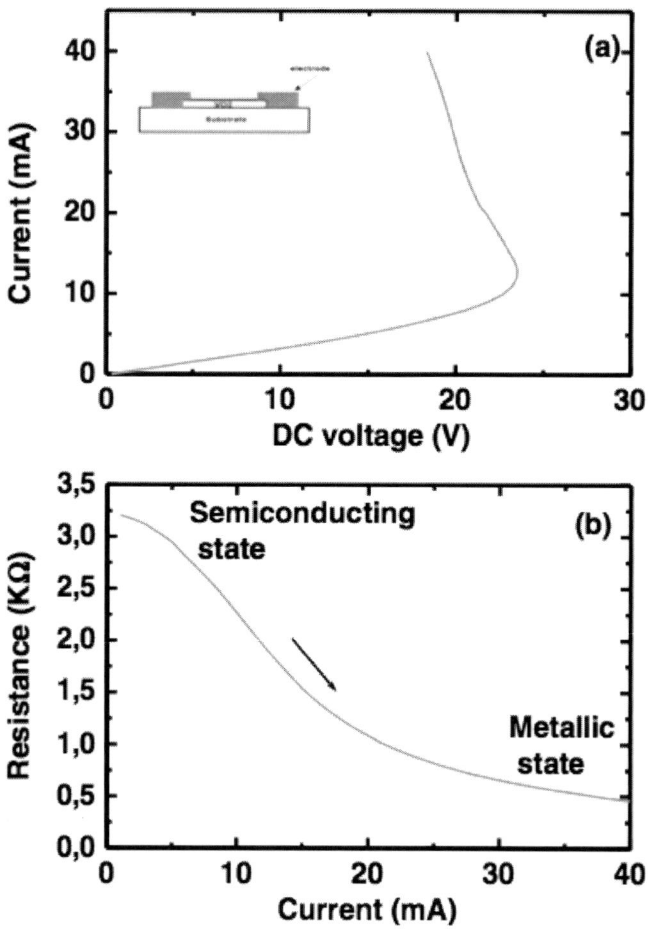

Figure 17. (a) I-V characteristics of the fabricated planar micro-switch based on the semiconductor-to-metallic phase transition of W-doped VO_2. The inset shows the scheme of the planar micro-switch. (b) The corresponding electrical resistance as a function of the applied current to the planar micro-switch. Taken from [56] with permission.

This NC phenomenon was confirmed by the capacitance measurements at different frequencies as a function of the bias voltages (from -35 V up to 35 V). The device was switched from its metallic state at −35 V to its semiconducting sate at 0 V and then to its metallic state at 35 V. Figure 19 shows the measured C-V hysteresis at 1.5 MHz for the W-doped VO_2 micro-switch when the switching voltage was cycled from −35 V to 35V. The capacitance measurements were reproducible as shown by the recorded four C-V curves labeled 1, 2, 3, and 4. Note that first curve was recorder when the W-doped VO_2 was switched directly to its metallic state at –35 V, while the 2, 3, and 4 cures were recorded when the switching of the micro-switch was controlled gradually by the applied voltage. This switching history can explain the observed small difference in the metallic region around –35 V for curve 1. However, this C-V hysteresis is relatively symmetric (see curves 2-4). The hysteresis width is about 6-8 V. This C-V hysteresis memory effect can be used in the fabrication of advanced memcapacitive systems exploiting the SMT of VO_2.

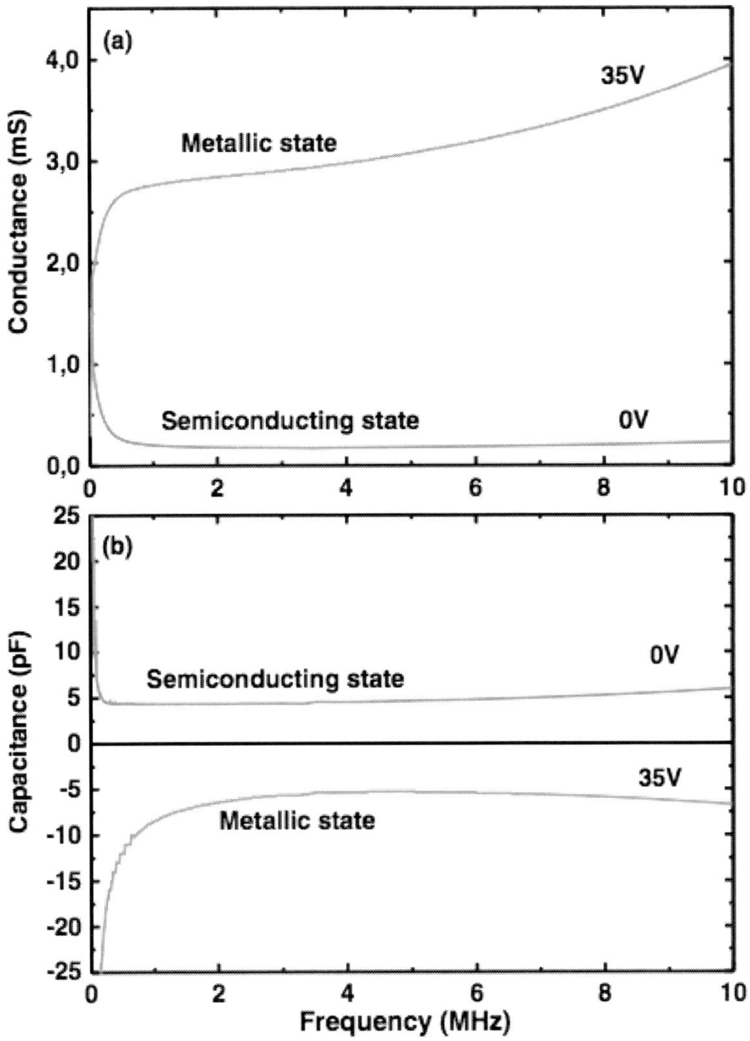

Figure 18. (b) Conductance as a function of frequency for the semiconducting state (0 V) and metallic state (35 V) of the W-doped VO_2 planar micro-switch. (b) The corresponding capacitance as a function of frequency for the two states at 0 V and 35 V. Taken from [56] with permission.

Negative capacitance has been observed in various materials and devices, such as gallium nanoparticles embedded in dielectric matrix [57], PbS nanocrystals embedded in conducting polymers [58], $In_{0.3}Ge_2Sb_2Te_2$ thin films [59], hydrogen-doped amorphous barium titanate device [60], GaN/AlGaN heterojunction dual-band detectors [61], GaAs homojunction far-infrared detectors [62], nanocomposite and polycrystalline solar cells [63], conducting polymer nanowires [64], organic semiconductor devices [65], metal-a-$C_{1-x}N_x$:H-metal devices [65], porous TiO_2 layers [66], $La_{0.8}Sr_{0.2}MnO_3$/Nb-doped $SrTiO_3$ heterojunctions [67]. The origin of the NC was attributed to minority carrier flow, interface states, slow transition time of injected carriers, charge trapping, and space charge [68-72]. Recently, it was proposed that the observed NC in ferroelectric layers can improve considerably the gain of field-effect transistors [72-74].

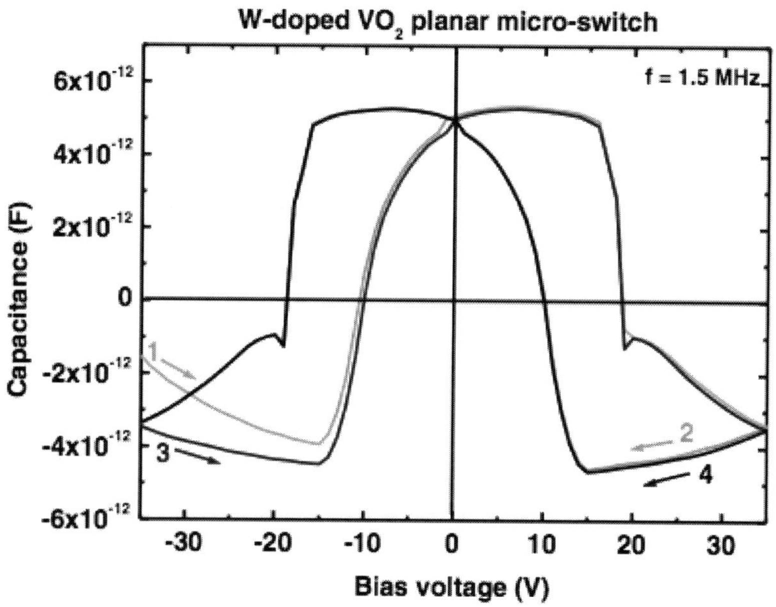

Figure 19. C-V hysteresis memory effect at 1.5 MHz as function of the applied switching voltage (from −35 V up to 35 V). The curves 1, 2, 3, and 4 indicate the four switching measurements sequences. Taken from [56] with permission.

As shown in Figure 18 and Figure 19, the negative capacitance in W-doped VO_2 is clearly dependent on the metallic state for which the conductivity is higher than that of the semiconducting state. In this case, the NC can be explained easily by considering the dependence of the capacitance on both the frequency and the applied voltage [56]:

$$C(\omega,V) = C_0 - \frac{A4\pi\tau}{d(1+\omega^2\tau^2)}\sigma_0 \exp\left[-(V_{th}-V)E_a / V_{th}KT\right] \quad (2)$$

where C_0 is the geometric capacitance, τ is the dielectric relaxation time, A the area of the semiconductor, d its thickness, σ_0 is the conductivity at threshold voltage (V_{th}), K is the Boltzmann constant, T is the temperature, and E_a is the activation energy.

The capacitance could be negative when the second term of equation (2) becomes larger than the geometric capacitance C_0, which occur when V is higher than V_{th}. This is the case for the metallic state, which exhibits with high conductivity and then negative capacitance [see Figure 18].

This type of the VO_2-negative capacitor can be combined with standard capacitors to fabricate tunable capacitors presenting C-V hysteresis memory effect with positive capacitance values (see Figure 20).

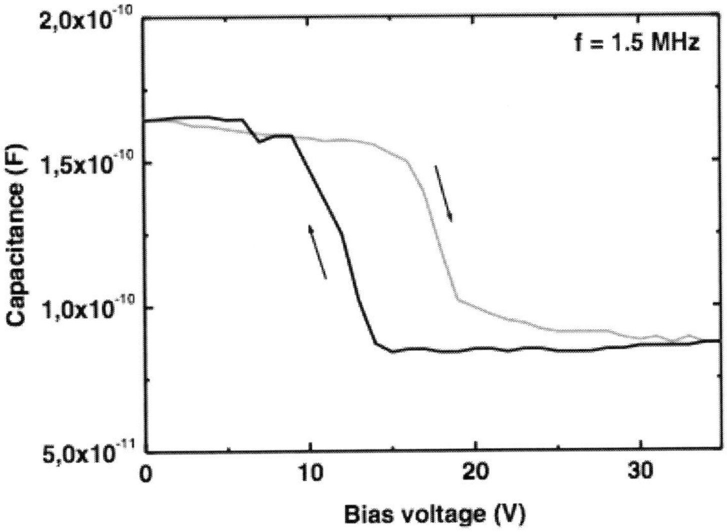

Figure 20. C-V hysteresis memory effect of standard capacitor ($C = 1.59 \times 10^{-10}$ F) in parallel with the W-doped VO_2-negative capacitor. Taken from [56] with permission.

CONCLUSION

This brief review reports on the electrical and optical properties of undoped VO_2, W-doped VO_2, and Ti-W codoped VO_2 smart coatings. The judicious control of the metal dopants (W and as well as Ti-W) allowed the control of the transition temperature, while the SMT was easily controlled by temperature, photo-excitation (all-optical switches), and an external voltage (electro-optical switches). The highlights of the present study are: (i) the suppression of the optical and electrical hysteresis for the Ti-W codoped VO_2, (ii) the thermal coefficient of resistance (TCR) as high as −5.12 %/°C for Ti-W codoped VO_2, (iii) the high optical switching contrast at a wavelength of 1550 nm for large VO_2 layers as well as for VO_2 based planar micro-optical switches, (iv) the fabrication of negative capacitor device exploiting the SMT of W-doped VO_2 planar micro-switch.

Finally, these results will be helpful in the development and fabrication of innovative devices exploiting the interesting thermochromic properties of this fascinating VO_2 smart material.

REFERENCES

[1] Morin, F. J. *Phys. Rev. Lett.* 1959, *3*, 34-36.
[2] Eyert, V. *Ann. Phys. (Leipzig)* 2002, *11*, 650-702.

[3] Soltani, M.; Chaker, M.; Haddad, E.; Kruzelecky, R..V.; Margot, J. *Appl. Phys. Lett.* 2004, *85*, 1958-1960.

[4] Becker, M. F.; Buckman, A. B.; Walser, R. M. *Appl. Phys. Lett.* 1994, *65*, 1507-1509.

[5] Cavalleri, A.; Tóth, Cs.; Siders C. W.; Squier, J. A.; Ráksi, F.; Forget, P.; Kieffer, J. C. *Phys. Rev. Lett.* 2001, *87*, 237401.

[6] Goodenough, J. B. *J. Solid State Chem.* 1971, *3*, 490-500.

[7] *Zylbersztejn,* A.; Mott, N. F. *Phys. Rev. B 1975, 11, 4383-4395.*

[8] Wentzcovitch, R. M.; *Schulz,* W. W.; Allen, P. B. *Phys. Rev. Lett.* 1994, *72*, 3389-3392.

[9] Cavalleri, A.; Dekorsy, Th.; Chong, H. H. W.; Kieffer, J. C.; Schoenlein, R. W. *Phys. Rev. B*, 2004, *70*, 161102(R).

[10] Hubbard, J. *Proc. R. Soc. Lond. A* 1963, *276* (no. 1365), 238-257.

[11] Hubbard, J. *Proc. R. Soc. Lond. A* 1964, *277* (no. 1369), 237-259.

[12] Hubbard, J. *Proc. R. Soc. Lond. A* 1964, *281* (no. 1386), 401-419.

[13] Rice, T.M.; Launois, H.; Pouget, J. P. *Phys. Rev. Lett.* 1994, *73*, 3042.

[14] Flannery, R.; Miller, J. E. *Proc. SPIE* 1992, *1689*, 379-395.

[15] Roach, W. R. *Appl. Phys. Lett.* 1971, *19*, 453.

[16] Lee, C. E.; Atkins, *R. A.;* Giler, W. N.; Taylor, H. F. *Appl. Opt.* 1989, *28*, 4511-4512.

[17] Li, Y.; Ji, S.; Gao, Y.; Luo, H.; Kanehira, M. *Sci. Rep.* 2013, *3*, 1370(1)-1370(13).

[18] Zhou, M,; Bao, J.; Tao, M.; Zhu, R.; Lin, Y.; Zhang, X,; Xie Y. *Chem. Commun.* 2013,*49*, 6021-6023.

[19] Zhou, J.; Gao, Y.; Liu, X.; Chen, Z.; Dai, L.; Cao, C.; Luo, H.; Kanahira, M.; Sun, C.; Yan, L. *Phys. Chem. Chem. Phys.* 2013, *15*, 7505-7511.

[20] Dai, L.; Chen, S.; Liu, J.; Gao, Y.; Zhou, J.; Chen, Z.; Cao, C.; Luo, H.; Kanehira, M. *Phys. Chem. Chem. Phys.* 2013, *15*, 11723-11729.

[21] Soltani, M.; Chaker, M.; Haddad, E.; Kruzelesky, R. V. *Journal of Vacuum Science & Technology A: Vacuum, Surfaces, and Films* 2006, *24*, 612.

[22] Kaye, A. B.; Haglund Jr., R. F. *Phase-change materials and optical limiting devices utilizing phase-change materials* 2012, US patent 8259381 B2.

[23] Soltani, M.; Chaker, M.; Haddad, E.; Kruzelecky, R. V.; Nikanpour, D. *Journal of Vacuum Science & Technology A: Vacuum, Surfaces, and Films* 2004, *22*, 859.

[24] Jiang, L.; Carr, W. N. *J. Micromech. Microeng.* 2004, *14*, 833.

[25] Dumas-Bouchiat, F.; Champeaux, C.; Catherinot, A.; Crunteanu, A.; Blondy, P. *Appl. Phys. Lett.* 2007, *91*, 223505.

[26] Crunteanu, A.; Leroy, J.; Humbert, G.; Ferachou, D.; Orlianges, J.-C.; Champeaux, C.; Blondy, P. *Microwave Symposium Digest (MTT), 2012 IEEE MTT-S International.*

[27] Driscoll, T.; Kim, H.-T.; Chae, B.-G.; Kim, B.-J.; Lee Y.-W.; Jokerst, N. M.; Palit, S.; Smith, D. R.; Di Ventra, M.; Basov, D. N. *Science* 2009, 325, 1518-1521.

[28] Orlianges, J. C.; Leroy, J.; Crunteanu, A.; Mayet, R.; Carles, P.; Champeaux, C. *Appl. Phys. Lett.* 2012, *101*, 133102.

[29] Xu, G.; Huang, C.-M.; Tazawa, M.; Jin, P.; Chen, L.-H. *Optics Commum.* 2009, *282*, 896-902.

[30] Ferrara, D. W.; MacQuarrie, E. R.; Nag, J.; Kaye, A. B.; Haglund, R. F. *Appl. Phys. Lett.* 2011, *98*, 241112.

[31] Soltani, M.; Chaker, M.; Haddad, E.; Kruzelesky, R. *Measurement Science and Technology* 2006, *17*, 1052-1056.

[32] Shi, Q.; Huang, W.; Lu, T.; Zhang, Y.; Yue, F.; Qiao, S.; Xiao, Y. *Appl. Phys. Lett.* 2014, *104*, 071903.
[33] Soltani M.; Chaker M. *System and method for generating a negative capacitance* 2012, Patent application US20120286743.
[34] Yang, Z.; Zhou, Y.; Ramanathan, S. *J. Appl. Phys.* 2012, 111, 014506-1 – 014506-5.
[35] Nakano, M.; Shibuya, K.; Ogawa, N.; Hatano, T.; Kawasaki, M.; Iwasa, Y.; Tokura, Y. *Appl. Phys. Lett.* 2013, *103*, 153503.
[36] Hillman, C. E.; De Natale, J. F.; Hacker, J. B.; Higgins, J. A.; Kobrin, P. H. V*anadium-dioxide front-end advanced shutter technology* 2011, US patent 8067996 B2.
[37] Jayaraman, L. V.; Jackson, B. L.; Li, Z. *Photonic device including at least one electro-magnetic resonator operably coupled to a state-change material* 2008, US patent 7446929 B1.
[38] Jiang, X.; Soltani, M.; Haddad, E.; Kruzelecky, R.; Nikanpour, D.; Chaker, M. *Journal of Spacecraft and Rockets* 2006, *43*, 497-500.
[39] Soltani, M.; Chaker, M.; Haddad, E.; Kruzelecky. R. In *Applied Physics in the 21st Century*; Chen, X.; Ed.; Old City Publishing: Philadelphia, PA, 2008, pp.291-314.
[40] Rini, M.; Hao, Z.; Schoenlein, R. W.; Giannetti, C.; Parmigiani, F.; Fourmaux, S.; Kieffer, J. C.; Fujimori, A.; Onoda, M.; Wall, S.; Cavalleri, A. *Appl. Phys. Lett.* 2008, *92*, 181904.
[41] Ibisate, M.; Golmayo, D.; López, C. *J. Opt. A: Pure Appl. Opt. 2008, 10,* 125202.
[42] Marvel, R. E.; Appavoo, K.; Choi, B. K.; Nag, J.; Haglund Jr., R. F. *Appl. Phys. A* 2013, *111*, 975–981
[43] Guinneton, F.; Sauques, L.; Valmalette, J.-C.; Cros, F.; Gavarri, J.-R. *Thin Solid Films* 2004, *446(2)*, 287-295.
[44] Lim, S. P.; Long, J. D.; Xu, S.; Ostrikov, K.. *Journal of Physics D: Applied Physics* 2007, *40*, 1085-1090.
[45] Chae, B. G.; Youn, D. H.; Kim, H. T.; Maeng, S. L.; Kang, K. Y. *Journal of the Korean Physical Society* 2004, *44*, 884-888.
[46] Béteille, F.; Livage, J. *Journal of Sol-Gel Science and Technology* 1998, *13*, 915-921.
[47] Chen, H.-K.; Hung, H.-C.; Yang, T. C.-K.; Wang, S.-F. *J. Non-Cryst. Solids* 2004, *347*, 138-143.
[48] Appavoo, K.; Lei, D. Y.; Sonnefraud, Y.; Wang, B.; Pantelides, S. T.; Maier, S. A.; Haglund Jr., R. F. *Nano letters* 2012, *12*, 780-786.
[49] Wu, Y. F.; Fan, L. L.; Chen, S. M.; Chen, S.; Zou, C. W.; Wu, Z. Y. *AIP Advances* 2013, *3*, 042132.
[50] Popuri, S. R.; Miclau, M.; Artemenko, A.; Labrugere, C.; Villesuzanne, A.; Pollet, M. *Inorg. Chem.* 2013, *52* 4780–4785.
[51] Tselev, A.; Luk'yanchuk, I. A.; Ivanov, I. N.; Budai, D.; Tischler, J. Z.; Strelcov, E.; Kolmakov, A.; Kalinin, S. V. *Nano Letters* 2010, *10*, 4409-4416.
[52] Case, F.C. *Journal of Vacuum Science and Technology. A, Vacuum, Surfaces and Films* 1987, *5*, 1762-1766.
[53] Kim, D. H.; Kwok, H. S. *Appl. Phys. Lett.* 1994, *65*, 3188-3190.
[54] Adler D. In *Solid State Physics: Advances in Research and Applications*; Seitz, F.; Ed.; Academic: NY, 1968; Vol. 21, pp. 1–113.
[55] Soltani, M.; Chaker, M.; Haddad, E.; Kruzelecky. R.; Margot, J. *Journal of Vacuum Science & Technology A: Vacuum, Surfaces, and Films*. 2007, *25*, 971.

[56] Soltani, M.; Chaker, M.; Margot, J. *Science and Technology of Advanced Materials*. 2011, *12*, 045002.
[57] Parravicini, G. B.; Stella, A.; Ungureanu, M. C.; Kofman, R. *Appl. Phys. Lett.* 2004, *85*, 302-304.
[58] Bakueva, L.; Konstantatos, G.; Musikhin, S.; Ruda, H. E.; Shik, A. *Appl. Phys. Lett.* 2004, *85*, 3567.
[59] Mahmoud, S.T.; Ghamlouche, H.; Qamhieh, N.; Ahmed, S. *J. Non-Cryst. Solids* 2008, *354*, 1976-1980.
[60] El Kamel, F.; Gonon, P.; Jomni, F.; Yangui, B. *Appl. Phys. Lett.* 2008, *93*, 042904.
[61] Byrum, L. E.; Ariyawansa, G.; Jayasinghe, R. C.; Dietz, N.; Perera, A. G. U.; Matsik, S. G.; Ferguson, I. T.; Bezinger, A.; Liu, H. C. *J. Appl. Phys.* 2009, *106*, 053701.
[62] Perera, A. G. U.; Shen, W. Z.; Shov, M. E.; Liu, H. C.; Buchanan, M.; Schaff, W. J. *Appl. Phys. Lett.* 1999, *74*, 3167.
[63] Mora-Sery, I.; Bisquert, J.; Fabregat-Santiago, F.; Garcia-Belmonte, G. *Nano Letters* 2006, *6*, 640-650.
[64] Rahman, A.; Sanyal, M. K. *Appl. Phys. Lett.* 2009, *94*, 242102.
[65] Ehrenfreund, E.; Lungenschmied, C.; Dennler, G.; Neugebauer, H.; Sariciftci, N. S. *Appl. Phys. Lett.* 2007, *91*, 012112.
[66] Kytin, V.; Dittrich, Th.; Koch, F.; Lebedev, E. *Appl. Phys. Lett.* 2001, *79*, 108.
[67] Wang, C. C.; Liu, G. Z.; He, M.; Lu, H. B. *Appl. Phys. Lett.* 2008, *92*, 052905.
[68] Shulman, J.; Xue, Y. Y.; Tsui, S.; Chen, F.; Chu, C. W. *Phys. Rev. B* 2009, *80*, 134202.
[69] Ershov, M.; Liu, H. C.; Li, L.; Buchanan, M.; Wasilewski, Z. R.; Jonscher, A. K. *IEEE Transactions on electron devices* 1998, *45*, 2196-2206.
[70] Kopp, T.; Mannhart, J. *Journal of Applied Physics* 2009, *106*, 064504.
[71] Salvatore G.A.; Bouvet, D.; Ionescu, A. M. *Technical Digest - International Electron Devices Meeting* 2008, art. no. 4796642.
[72] Theis. T. N.; Solomon, P. M. *Science* 2010, *327*, 1600-1601.
[73] Zhirnov V. V.; Cavin, R. K. *Nature Nanotechnology* 2008, *3*, 77-78.
[74] Salahuddin, S.; Datta S. *Technical Digest - International Electron Devices Meeting* 2008, art. no. 4796789.

EDITOR CONTACT INFORMATION

Dr. Alexander Pergament
Professor
Petrozavodsk State University
185910 Petrozavodsk, Russia
Tel: 7 8142 719680
E-mail: aperg@psu.karelia.ru

INDEX

A

B

C

D

E

F

G

H

I

N

O

P

Q

R

S

T

U

V

W

X

Y

Z